The European Gas Markets

Manfred Hafner · Simone Tagliapietra
Editors

The European Gas Markets

Challenges and Opportunities

palgrave
macmillan

Editors
Manfred Hafner
Fondazione Eni Enrico Mattei
Milan, Italy

Simone Tagliapietra
Fondazione Eni Enrico Mattei
Milan, Italy

ISBN 978-3-319-55800-4 ISBN 978-3-319-55801-1 (eBook)
DOI 10.1007/978-3-319-55801-1

Library of Congress Control Number: 2017937507

Printed on acid-free paper

This Palgrave Macmillan imprint is published by Springer Nature
The registered company is Springer International Publishing AG
The registered company address is: Gewerbestrasse 11, 6330 Cham, Switzerland

Foreword by Maroš Šefčovič, European Commission's Vice President in charge of the Energy Union

The Role of Gas in the Energy Union

As one of the flagship projects of the European Commission, the Energy Union is doing today what the European Coal and Steel Community (ECSC) did after the Second World War: leading the way and providing the framework for a long-term transformation of Europe for the better. Whereas the ECSC in the spirit of Robert Schuman and Jean Monnet made war between Germany and France materially impossible, Energy Union moves us away from centralised fossil-fuel-based systems towards decentralised, clean power production with consumers at the centre stage. Right from the launch of Energy Union in February 2015, we made it very clear that Europe has chosen its path, its objectives, and its future. And that future is clean. We reiterated this commitment in Paris, and now we are swiftly moving into implementing our vision. The latest major step in making Energy Union a reality was the presentation of the holistic Energy Union Package—Clean energy for all Europeans last December. And we are now making sure that this set of interrelated and mutually reinforcing legislation will be binding before the end of the mandate of this Commission.

In this great adventure, gas plays a crucial role. In the context of Energy Union, gas is not just a fossil fuel among others, notably as a transitional fuel given the intermittent nature of renwables. Gas can contribute to the decarbonisation of Europe and to the emergence of future electricity systems. The gas markets of tomorrow will impact on demand and supply for gas. This in turn effects on how we plan and build our interconnectors. And on how much and from where we'll be importing gas.

Within this overall framework, the EU strategy on liquefied natural gas (LNG) and gas storage adopted in February 2015 aims at improving security of supply and competitiveness through increased diversification, focusing in particular on the most vulnerable regions such south-east Europe, the Baltic region and the Iberian Peninsula.

Gas will continue to play a role in our energy mix during the transition phase to a decarbonised energy system. This strategy has been conceived with the implementation of the Paris Agreement in mind. Of course, we have to ensure secure and affordable gas supply to all EU consumers, in parallel to the transition to a more sustainable energy system. This is particularly important as the EU imports already more than half of the gas it consumes and remains vulnerable to external shocks.

In this context, regional cooperation is important to ensure an adequate and effective development of LNG and storage in Europe. It actively contributes to the completion of the internal gas market and facilitates the identification and development of the infrastructures necessary in terms of security of supply. Our high-level groups are leading the way in promoting such regional cooperation, for instance:

- Under the CESEC high-level group, 15 south-east European countries have committed to eliminate all barriers to full integration of their gas markets.
- The BEMIP group consisting of the Baltic States is already implementing an action plan to fully liberalise their markets.

Further improvements also take place through the regional plans proposed in the revision of the security of gas supply regulation. The regulation also focuses on how to make sure storage facilities can be seamlessly accessed and used across borders under normal but also under more strained supply situations. Energy security is linked to competitiveness. They interplay and reinforce each other. By making sure that gas supplies come at fair and competitive prices and are readily available, we make sure that gas will play a key role in tomorrow's Energy Union.

Brussels
February 2017

Maroš Šefčovič

Acknowledgements

The editors gratefully acknowledge the support of the Fondazione Eni Enrico Mattei (FEEM) to the inception and realisation of this editorial project. They particularly acknowledge the precious assistance of Barbara Racah, FEEM Publications Office Manager, throughout the whole editorial process.

Manfred Hafner
Simone Tagliapietra

Contents

Editors and Contributors

About the Editors

Manfred Hafner is Coordinator of the Energy Scenarios and Policy research programme at Fondazione Eni Enrico Mattei (FEEM), Italy. He is also Professor in international energy studies teaching at the Johns Hopkins University School of Advanced International Studies (SAISEurope) and at the Sciences-Po Paris School of International Affairs (PSIA). Furthermore, he is Senior Scientist at the Euro-Mediterranean Center for Climate Change (CMCC) and Associate Researcher at the Centre for Research on Energy and Environmental Economics and Policy (IEFE) at Bocconi University. Moreover, he is Vice-President Consulting at Enerdata and President of International Energy Consultants (IEC).

Simone Tagliapietra is Senior Researcher at the Fondazione Eni Enrico Mattei (FEEM), Italy, and Research Fellow at Bruegel, Belgium. An expert in international energy issues, he has published widely on EU energy markets, the EU energy and climate policy, and Euro-Mediterranean energy relations. He is the author of the book *Energy Relations in the Euro-Mediterranean: A Political Economy Perspective* (Palgrave, 2017). He received his Ph.D. from the Università Cattolica del Sacro Cuore, Italy, where he lectures on energy economics and policy.

Contributors

Thierry Bros SciencesPo, Paris, France

Carlo Carraro FEEM, Venice, Italy

Anne-Sophie Corbeau KAPSARC, Riyadh, Saudi Arabia

Fabio Genoese SciencesPo, Paris, France

Manfred Hafner FEEM, Milan, Italy

Anouk Honoré Oxford Institute for Energy Studies, Oxford, UK

Chris Le Fevre Oxford Institute for Energy Studies, Oxford, UK

Thomas Longden FEEM, Milan, Italy

Daniela De Lorenzo Eni, Rome, Italy

Giacomo Marangoni FEEM, Milan, Italy

Tatiana Mitrova Energy Research Institute of the Russian Academy of Sciences, Moscow, Russia

Ludwig Moehring Wingas, Kassel, Germany

Oeystein Noreng BI Norwegian School of Management, Oslo, Norway

Francis Perrin Stratégies et Politiques Energétiques and Energy Industries Strategies Information, Paris, France

John Roberts Atlantic Council, Washington, DC, USA

Howard Rogers Oxford Institute for Energy Studies, Oxford, UK

Dario Speranza Eni, Rome, Italy

Jonathan Stern Oxford Institute for Energy Studies, Oxford, UK

Simone Tagliapietra FEEM, Milan, Italy

Massimo Tavoni FEEM, Milan, Italy

Jean-Arnold Vinois Notre Europe Jacques Delors Institute, Paris, France

Cyril Widdershoven Verocy, Heerenberg, The Netherlands

List of Figures

List of Tables

Introduction

Natural gas represents a pivotal element of the European energy architecture. Covering about a quarter of the European total primary energy supply, it indeed represents the second largest fuel in the European energy mix after oil.

The European gas markets are nowadays rapidly changing due to political, commercial and geopolitical evolutions. The still-uncertain role of gas in the European energy mix after the COP-21 Paris Agreement, the evolution of the EU Energy Union, the EU-Russia gas relations after the Ukraine crisis, the new dynamics of global LNG market and the new supply potential for Europe emerging in areas such as the Eastern Mediterranean are just few examples of these ongoing dynamics.

In this shifting context, both new challenges and new opportunities are emerging for Europe. A clear understanding of these new realities is a fundamental prerequisite to ensure the adoption of the right decisions in both European private and public policy circles.

The aim of this book is to explore in detail these new challenges and opportunities for European gas markets, not only with the intention of providing a clear snapshot of the current situation and future outlook, but also with the intention of presenting policy recommendations to the sector's key stakeholders.

In order to meet this ambition, this book gathers together some of the most prominent gas experts of Europe, coming not only from academia but also from industry and key public institutions. This reflects the idea that only a mix of different backgrounds and perspectives might provide a comprehensive analysis of the various factors—economics, politics and technology—interacting with the sector.

This book develops on 16 chapters, presented hereafter.

Chapter 1 explores the role that gas might play in making the EU decarbonisation path more balanced and secure up to 2030 and beyond. To do so, the chapter looks at the potential role of gas as a substitute for coal in power generation, and at the complementary role of gas as a balancing tool to manage the variability of wind and solar energy.

Chapter 2 analyses a set of new scenarios for energy markets in Europe to evaluate the role of gas across a range of assumptions on climate policy with the aim to identify whether current trend and policies are leading to an economically efficient and, at the same time, climate friendly, energy mix in Europe.

Chapter 3 provides an insight into the future of gas demand in Europe, arguing that even in the context of slow economic growth and decarbonisation of the energy sector, there is a potential for gas in the European energy mix, especially in the 2020s when lots of firm—coal and nuclear—capacity closes down.

Chapter 4 looks at the case of decarbonisation in Germany—Europe's central and largest gas market—to outline that only with a new proactive communication strategy, postulating effective and affordable climate protection through integration of gas and renewables, the gas industry can sustain the future role of gas in the EU decarbonisation path.

Chapter 5 provides an insight into the drivers of gas-fired power generation in the current electricity system of Europe, to explain its potential future role.

Chapter 6 looks at the financial and environmental case for gas as a transport fuel. It considers these advantages in the main transport sectors of passenger cars, road haulage vehicles and marine shipping and the resulting levels of demand that might emerge.

Chapter 7 analyses the global LNG market outlook and its repercussions for Europe. It explains how Europe plays the role of a shock absorber in global LNG markets, and it argues that there will likely be limits on how much LNG can be absorbed, particularly if large volumes of surplus LNG are stranded at low prices.

Chapter 8 argues that five revolutions are currently reshaping the world's energy balance: US shale gas, US shale oil, renewable energy growth, energy efficiency growth and rapid evolution of energy storage. It then provides an insightful discussion of how these revolutions are also impacting global LNG markets, the way LNG is priced globally and the overall impact of these trends on Europe under both the economic and geopolitical perspectives.

Chapter 9 discusses Russia's gas strategy. It argues that in the recent past, the country has had to weather a perfect storm of economic, market, domestic political and foreign policy-related upheavals. These dramatic changes not only impact the state budget and the country's macroeconomic stability, but also its gas industry. The fundamental shift in all major components of the country's gas balance (stagnant domestic demand and exports, weak production and imports) and their impact on European gas supply are thus analysed in the chapter.

Chapter 10 offers an in-depth analysis of the latest developments of the Groningen gas field production future. The growing concern of north-western European gas importers, combined with the growing pressure of Dutch NGOs andpolitical parties to end the Dutch gas adventure, presents the European gas market with a new security of supply issue in the future. The future of the Dutch gas roundabout is in doubt, while gas producers and society are in full confrontation. The Dutch disease, based on a government budget policy based on hydrocarbon revenues, is now being substituted for green policies on a confrontation course with the economy. On this basis, the chapter argues, a Dutch Disease 2.0 is being born.

Chapter 11 looks at the role of Norway as a gas supplier to Europe. It argues that Norway could well continue to play this role in a reliable manner in the future, but only if the EU will give incentives in the form of market opening, easy and inexpensive access to infrastructure and policy measures that welcome gas in the energy mix.

Chapter 12 analyses the challenges and opportunities related to the development of gas markets in north Africa. From Algeria's increased production and exports after a period of stagnation, to the confirmation of the potential of the Eastern Mediterranean, the chapter reviews the regional gas production and export outlook, also taking into consideration the political and security risks related to it.

Chapter 13 looks at the gas developments in the Levant Basin, outlying the two key challenges for the energy sector in the region: to satisfy increasing internal demand with affordable and reliable supplies, and to create conditions to efficiently export excess resources and to craft a sustainable energy mix. In this context, the chapter argues that strengthening the South–North corridor could contribute to the recovery of the Mediterranean and boost the security and the development of the whole region.

Chapter 14 discusses the evolution of the Southern Gas Corridor (SGC). It starts by detailing the progress achieved to date in implementing the project, and it then continues by discussing the system's ability to attract further potential input from a 'Next Wave' of offshore Azerbaijani gas-fields, Turkmenistan, Iran, northern Iraq, and the Eastern Mediterranean. Finally, the chapter addresses the issue of whether Russia might seek to utilise part of the Corridor to enable it to use its planned Turkish Stream project for at least some deliveries to Italy, and the question of whether unrest within Turkey might pose serious dangers to transit pipelines.

Chapter 15 illustrates the progressive transition in European gas price formation from dominance of oil-related pricing to a situation in 2015 where nearly two-thirds of gas in European wholesale markets was sold at hub prices. It argues that the main gas hubs in Europe are already well integrated, and in general correlation is high and continues to improve.

Chapter 16 examines, from an EU policy perspective, the current position and future prospects of gas in the European energy mix, the evolution of the EU internal gas market and, finally, the key challenges facing gas in the Energy Union—also in the light of the COP-21 Paris Agreement.

The aspiration of this book is to represent a blueprint for both private and public policy makers. However, it also seeks to be a useful guide for everyone interested in better understanding the dynamics of a complex and fascinating world such as one of the European gas markets.

Manfred Hafner
Simone Tagliapietra

1

The Role of Natural Gas in the EU Decarbonization Path

Manfred Hafner and Simone Tagliapietra

1.1 The EU's Quest for Decarbonization: From Kyoto to the 20-20-20 Targets

Over the last decade, the decarbonization of the energy system has progressively become a key priority for the European Union (EU).[1]

The first steps in this direction were taken by the EU in the framework of the international negotiations on climate change. In 2002, the EU (then still called the European Community) adopted a legislation approving the Kyoto Protocol, stating that it would jointly fulfil with its Member States[2] the commitment to reduce the collective greenhouse gas (GHG) emissions in the 2008–2012 period to 8% below 1990 levels.[3]

In this new international context, EU Member States agreed for the first time on the need for a comprehensive common action towards the

M. Hafner (✉) · S. Tagliapietra
FEEM, Milan, Italy
e-mail: manfred.hafner@feem.it

M. Hafner and S. Tagliapietra (eds.), *The European Gas Markets*,
DOI 10.1007/978-3-319-55801-1_1

increasingly challenging energy issues at the Hampton Court informal EU summit held in October 2005.[4]

Following the political momentum emerged at the summit, the European Commission published in early 2006 a Green Paper on developing a common and coherent European energy policy entitled "A European Strategy for Sustainable, Competitive and Secure Energy".[5] As the title suggests, the paper delineated a European energy policy structured on three key pillars, which continue to remain fundamental also today.

The Green Paper received the praise of the European Council of March 2006, which called for «an Energy Policy for Europe, aiming at effective Community policy, coherence between Member States and consistency between actions in different policy areas and fulfilling in a balanced way the three objectives of security of supply, competitiveness and environmental sustainability.»[6] The European Council, therefore, invited the European Commission to prepare further actions.

The Commission reacted to this endorsement by issuing in January 2007 the so-called Energy and Climate Package, a set of measures centred on the Communication "An Energy Policy for Europe"[7] aimed at establishing a new European energy policy in line with the one proposed in the Green Paper (and thus focused on combat climate change, increase the EU' energy competitiveness and boost the EU's energy security of supply).

The European Council of March 2007 endorsed the package,[8] which was then finally adopted by the European Parliament in December 2008 after months of tough negotiations between Member States.

In addition to the definition of the triple paradigm sustainability–competitiveness–security characterizing the European energy policy, an important advancement included in the "Energy and Climate Package" was represented by the EU's commitment to reach specific targets related to GHG emissions reduction, renewable energies and energy efficiency: the well-known 20-20-20 targets. These targets encompassed a 20% reduction in GHG emissions compared to 1990, a 20% decrease of final energy demand compared to a baseline scenario and the obtainment of a level of 20% of renewable energy in total energy consumption, by 2020.

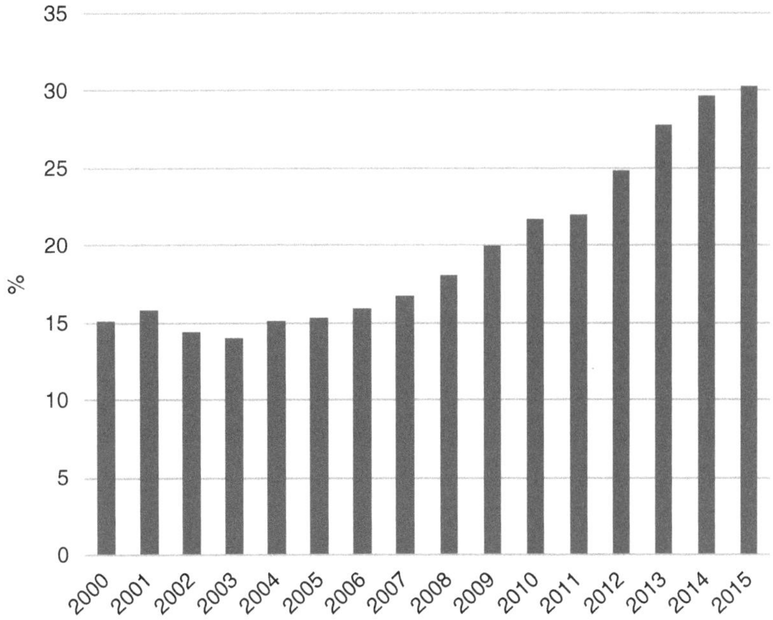

Fig. 1.1 Share of renewable energy in the EU electricity production—including hydro—(2000–2015). *Source* Own elaboration on Enerdata Global Energy and CO_2 Database (accessed September 2016)

These targets had a substantial impact on the EU energy system, particularly as far as the penetration of renewable energy in the system is concerned. As Fig. 1.1 illustrates, the share of renewable energy in the EU electricity production grew substantially over the last years, doubling from 15% in 2000 to 30% 2015.

1.2 After the 20-20-20 Targets: The 2050 Roadmaps and the 2030 Framework

In 2011a, the European Commission adopted the Communication "A Roadmap for Moving to a Competitive Low Carbon Economy in 2050"[9] with the aim to outline its new long-term decarbonization

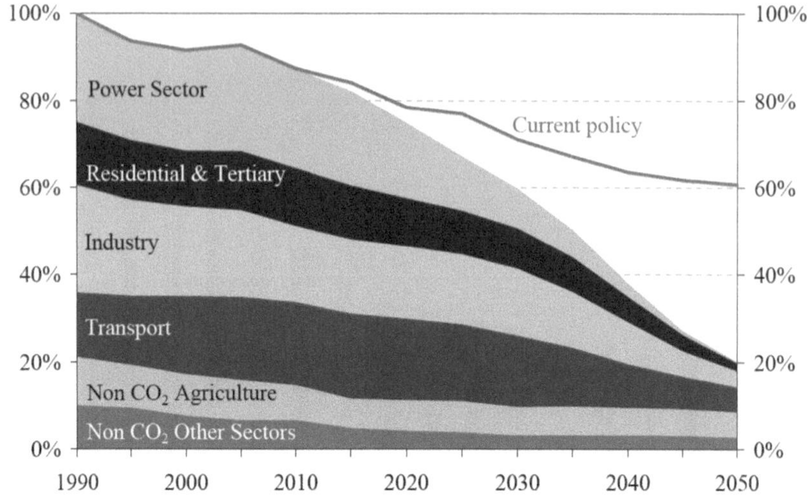

Fig. 1.2 EU GHG emissions towards an 80% domestic reduction (100% = 1990). *Source* European Commission (2011a, p. 5)

targets. The document strengthened the previous targets to the level of envisaging a domestic GHG emissions' cut-off for about 80% by 2050 compared to 1990 (Fig. 1.2). Furthermore, the document reiterated the concept that the decarbonization of the energy system is possible and could even be less costly in the long run than business-as-usual policies.

In order to explore the challenges posed by delivering this decarbonization objective while at the same time ensuring security of energy supply and competitiveness, the Commission adopted in the same year the Communication "Energy Roadmap 2050".[10] As far as natural gas is concerned, the Roadmap underlined that the fuel will be critical for the transformation of the EU energy system. According to the decarbonization scenarios underpinning the document, natural gas will perform better than other fossil fuels and will basically maintain its 2005 share in the EU primary energy consumption up to 2050 (Fig. 1.3).

In line with the two 2050 road maps, the European Commission further detailed its long-term decarbonization strategy in 2014, with the adoption of the Communication "A Policy Framework for Climate

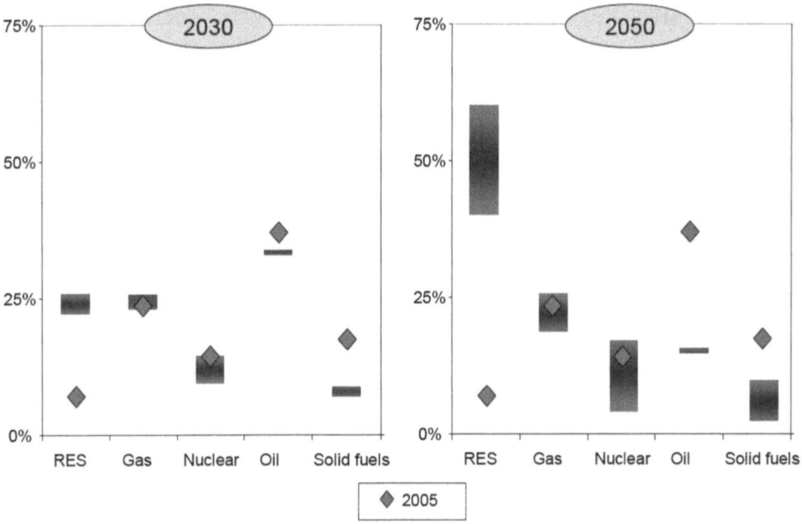

Fig. 1.3 EU decarbonization scenarios—2030 and 2050 range of fuel shares in primary energy consumption compared with 2005 outcome (in %). *Source* European Commission (2011b, p. 5)

and Energy in the Period from 2020 to 2030".[11] This new document focuses on the reduction in GHG emissions (by 40% below the 1990 levels by 2030), on the increase of renewable energy use (at least 27% of the EU's energy consumption by 2030), on the increase of energy efficiency (27% energy savings target for 2030) and on the reform of the EU Emissions Trading System (ETS). This set of provisions was endorsed by the European Council of October 2014.[12] Following this approval, the Commission made its initial legislative proposals to implement the 2030 climate and energy framework at the end of February 2015. The proposals, set out in the "Energy Union Package",[13] aim to provide a coherent approach to climate change, energy security and competitiveness, also to achieve the goals agreed under the 2030 framework.

It is important to underline that, unlike in the previous 2020 framework, the new EU targets will not be translated into national targets via EU legislation. Officially, this is due to the willingness of leaving «greater flexibility for Member States to meet their GHG reduction targets in the most cost-effective manner in accordance with their specific circumstances, energy mixes and capacities to produce renewable energy.»[14] In reality, this seems to be mainly due to the lack of a common vision among the Member States on the future trajectory of the decarbonization path, with certain countries (from the UK to Poland) being reluctant to afford its high costs and being more sensitive to the competitiveness and security pillars of EU energy policy. This situation clearly raises questions on how the new 2030 framework will concretely be implemented.

In this uncertain situation, the role of natural gas in the EU decarbonization path basically remains undefined, like the one of all the other components of the energy system with the notable exception of renewable energy and energy efficiency. This uncertainty opens a wide debate on the future role of natural gas in the EU energy system, particularly *vis-à-vis* the progressively stronger role of renewable energy in the EU energy mix. Considering that following the decarbonization path renewable energy will further consolidate its position of key independent variable in the EU energy equation, the next section will provide a critical assessment of the challenges and opportunities related to a major scale-up of variable renewable energy sources in the EU energy system by 2030.

1.3 Towards the Achievement of the 2030 Renewables Target: The Way Ahead

According to the European Commission, the increase of renewable energy use to 27% of overall EU energy consumption by 2030 will imply that in the same year about 45% of electricity in the EU will have to be generated by renewable energy sources.[15]

This clearly represents a substantial expansion of the current contribution level of renewable energy to the EU electricity generation, estimated by Eurostat at about 25%.[16]

The EU renewable electricity generation mix still continues to be largely composed by hydro. Considering that the hydro potential in the EU is already well exploited, the new 2030 target will thus require an extensive development of variable renewable energy sources such as wind energy and solar energy (namely photovoltaic PV).

As outlined in a recent study by EDF, up-to-date wind and PV have been developed with "fit and forget" logic, being not integrated into the electricity market and having priority dispatch and access to network.[17] However, a massive integration of such variable renewable energy sources into the system will require profound changes in terms of power system operation, market design, infrastructure development and transformation of conventional generation mix.

Being dependent on uncertain weather conditions, wind and PV are variables by definition and their output is both intermittent and non-dispatchable. For this reason, more flexibility will be required in the system, in order to reduce this intermittency and ensure the overall stability of the system. Flexible resources include dispatchable backup power plants, demand-side management and response, energy storage facilities and interconnections with adjacent markets.

The main tool to reduce the intermittency of wind and PV electricity generation is to aggregate their outputs over a wider geographical area. In fact, intermittency at site level is progressively smoothed at regional, national and continental levels as a result of the diversity of outputs.

In other words, the integration of EU electricity systems can mitigate flexibility needs arising from wind and PV, due to different weather patterns across Europe that decorrelate single electricity generation peaks, yielding geographical smoothing effects that ultimately transform intermittency at local level in variability at EU level.

In addition to this, a strong integration of EU electricity systems can allow the cross-border exchanges necessary to minimize surplus renewable electricity generation. As outlined by Fraunhofer IWES, « when no trading options exist, hours with high domestic wind and PV generation require that generation from renewables be stored or curtailed in part. With market integration, decorrelated production peaks across countries enable exports to regions where the load is not covered. By contrast, a hypothetical national autarchy case has storage or curtailment requirements that are ten times as high.»[18]

The process of integration of EU electricity systems will require the development of an appropriate network infrastructure, and particularly of interconnections not only able to transport wind and PV electricity production to consuming centres but also able to share thermal generation capacity between EU countries.

The development of an appropriate infrastructure is thus not only crucial to reduce variability of wind and PV at system level but also crucial to reduce the overall need for backup electricity generation. This represents a vital element, particularly if considering that by displacing baseload generation (i.e. from conventional sources) wind and PV do increase the need for backup capacity.[19]

With an increased role of wind and PV in EU electricity systems, conventional plants are thus progressively switching from their traditional roles to a new backup role, essential to guarantee the stability of the overall system *vis-à-vis* the variability of wind and PV.

In addition to interconnections, flexibility in the system could theoretically be enhanced with demand-side management and demand response mechanisms as well as energy storage. However, these solutions face major challenges. Demand mechanisms are partially challenged by socio-economic issues such as consumer behavioural changes, albeit can well be implemented in the industry and services sectors first. Energy storage is challenged by a persistent technological gap; in fact, to date the only operative option is represented by pumped storage hydropower, as other technologies such as battery systems, compressed air energy storage, flywheels and hydrogen storage continue to be highly expensive. In sum, in the medium term, these solutions will unlikely provide a substantial contribution for backup in the system.

In this framework, exploiting the complementary roles of renewable and conventional electricity generation sources will be even more important in the future EU electricity systems. In particular, conventional sources will continue to play a key role in guaranteeing system stability and security of supply by being able to provide larger and more rapid increases and decreases in output in order to accommodate increasing amounts of variable renewables-based generation.

With regard to this specific aspect, the International Energy Agency (IEA) points out that «the integration of high levels of wind and PV into electricity systems may require market framework reforms to guarantee a sufficient level of investment in the conventional power plants needed to keep the system in balance, together with other measures to shift demand when sun is not shining or the wind is not blowing. Failing to address these needs in advance will negatively impact the reliability of the electricity system.»[20]

On the basis of the situation illustrated in this section, this seems to be particularly urgent in the case of the EU, where variable renewables are set to become the cornerstone of the electricity system, increasing the variability that the rest of the system has to manage. Of course, a new EU electricity market design should also be able to provide adequate economic incentive for investments in the previously mentioned flexibility options (i.e. network infrastructure expansion, development of smart grids, adoption of demand-side measures and development of energy storage technologies), crucial to ensure the sustainability of the EU decarbonization path also beyond the 2030 horizon.

To make a long story short, in order to achieve its 2030 renewable energy target, the EU will need to rethink its electricity system beyond renewable energy itself. The role of conventional fuels in the future system should be better investigated, also to provide investors with the minimum grade of certainty needed to make today investments that will define the EU electricity system of 2030 and beyond. This is particularly true considering the recent, controversial, evolution of the EU electricity generation mix, which will be described in the next section.

1.4 The EU Decarbonization Path and the (Unwelcome) Renaissance of Coal

As the previous sections illustrated, over the last decade the EU has successfully promoted the expansion of renewable energy sources in the European electricity generation mix.

However, part of the environmental benefits generated by this complex and costly expansion has been nullified by the parallel growth of coal in the mix, a trend particularly emerged after 2010 (Fig. 1.4).

The key driver underpinning this trend was the US shale gas revolution. In fact, as US utilities progressively shifted into natural gas, American coal miners had to look for new markets abroad.

In the meantime, many new large mining capacities that were committed in Indonesia and Australia during the boom period of Asian demand (2008–2011) progressively came online between 2012 and 2014, adding even more low-cost supply to the international coal market. Considering that in the meantime coal demand growth in Asia resulted to be lower than expected, the global coal market entered a situation of oversupply.

As a result of this trend, overall EU coal imports increased from 104 million tons of oil equivalent (Mtoe) in 2010 to 119 Mtoe in 2014 and coal import prices plunged from EUR 130 per tonne (t) in March 2008 to below EUR 60 per t in May 2014 at the EU import reference price.[21] Due to a progressive transition from oil indexation to spot pricing, natural gas prices in the EU also decreased over the last years, but at a far slower pace than coal. In this framework, coal became more

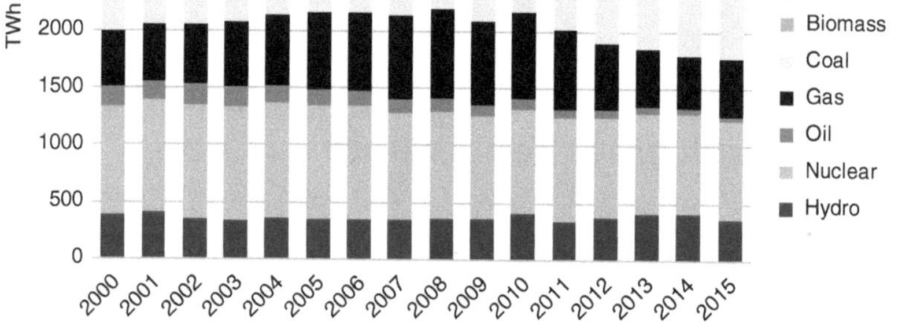

Fig. 1.4 Electricity generation in the EU by source (2000–2015). *Source* Own elaboration on Enerdata Global Energy and CO_2 Database (accessed September 2016)

competitive against natural gas in the EU electricity generation sector. This led to a significant gas-to-coal switch in the EU, not only in the UK, Spain and Germany, but also in the Netherlands.

Considering that coal-fired electricity generation emits more CO_2 per kWh than other power plants, this situation represents a substantial challenge to the EU decarbonization path. This is particularly true if considering that the efficiency of the EU's coal-fired power plants fleet is on average low, with a level of 36% compared to the one of 45% characterizing the most efficient plants, such as the ultra-super critical power plants in Germany.[22]

The existing EU environmental regulation has not had a relevant impact on the cost advantage that coal-fired generators have enjoyed over their competitors. In fact, according to the IEA, the EU Large Combustion Plant Directive (LCPD)[23] «is expected to lead to retiring 8 GW of coal-fired power capacity in the United Kingdom. In other EU countries, reductions are expected to be much lower, totaling around 10 GW. All in all, the shutdowns would affect around 2% of EU total generating capacity.»[24]

In short, a global market oversupply combined with a lack of proper environmental regulation at the EU level, allowed coal to stage a renaissance in EU electricity market over the last few years.

In this context, natural gas found itself in the uncomfortable position to be squeezed by subsidized renewable energy sources on the one side and cheap coal on the other side. This added additional pressure to the already weak natural gas demand conditions in the EU (due to the economic crisis and mild winters), resulting in a dramatic plunge of EU natural gas demand from a peak level of 540 Bcm in 2010 to a level of 438 Bcm in 2015: the level recorded in the mid-1990s (Fig. 1.5).

At a first view, this graph might suggest that the role of natural gas in the EU energy system is irreversibly in decline, particularly if taking into consideration the EU's quest to further advance renewable energy in electricity generation. However, considering that the previous analysis clearly illustrated that a strong expansion of renewable energy by 2030 and beyond will not exclude a key role of conventional electricity generation in the system, the future role of natural gas in the EU decarbonization path does deserve to be better explored.

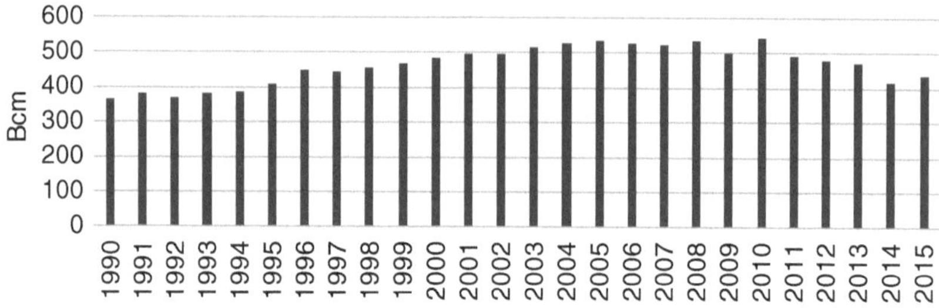

Fig. 1.5 EU natural gas demand (1990–2015). *Source* Own elaboration on Enerdata Global Energy and CO_2 Database (accessed September 2016)

1.5 Exploring the Future Role of Natural Gas in the EU Decarbonization Path

Albeit technically feasible, a further large-scale development of wind and PV in the EU electricity system might potentially encounter economic barriers due to increasing system integration costs. This issue is particularly relevant if considering that the EU itself acknowledges «that in the future, the benefits of renewable energy must be exploited in a way which is to the greatest extent possible market driven»[25] and thus not based on support schemes that ultimately hinder market integration and reduce cost efficiency.

In this context, assessing the future role of conventional electricity generation is of vital importance for the stability and security of the EU electricity system. As an overall trend, considering the previously illustrated characteristics of an electricity system centred on variable renewable energy sources, what will be needed is primarily a park of flexible power plants, where flexibility of a power plant is defined as its ability to run in partial load as well as by parameters such as ramping rates, start-up time and minimum downtime.[26]

Among the various possible options of conventional electricity generation (natural gas, coal, nuclear and oil), natural gas seems to be the fuel better placed to play a key complementary role to wind and PV in the decarbonization path for the following four reasons:

(1) First of all, natural gas-fired power plants can provide the flexible backup capacity needed in a system with high share of variable renewable energy sources. An analysis carried out by Eurelectric (see Fig. 1.6) shows that among conventional electricity generation technologies pumped storage is the most responsive one, as it can be called upon to generate electricity almost instantaneously and as it can ramp up and down by more than 40% of the nominal output per minute. However, being contingent on specific geographical conditions, pumped storage cannot provide the flexible backup capacity needed at system level. Among other technologies, combined-cycle gas turbines (CCGTs) are particularly suitable for load-following operation as they have both fast load gradients (4%/min) and can be brought online fairly quickly (less than 1.5 h from warm conditions). These performances are far beyond those of coal-fired power plants (which are less responsive than any other technologies) and of nuclear power plants (which cannot be brought online from cold and warm conditions in time frames similar to those of other technologies). For this reason, natural gas-fired power plants can well play an important role in meeting the flexibility challenge arising from variable renewable energy sources.[27]

(2) By displacing coal in the EU electricity generation systems, natural gas has the potential to generate immediate and substantial GHG

	NPP	HC	Lign	CCG	PS
Start-up Time "cold"	~ 40h	~ 6h	~ 10h	< 2h	~ 0,1h
Start-up Time "warm"	~ 40h	~ 3h	~ 6h	< 1,5h	~ 0,1h
Load Gradient ↗ "nominal Output"	~ 5%/m	~ 2%/m	~ 2%/m	~ 4%/m	> 40%/m
Load Gradient ↘ "nominal Output"	~ 5%/m	~ 2%/m	~ 2%/m	~ 4%/m	> 40%/m
Minimal Shutdown Time	NO	NO	NO	NO	~ 10h
Minimal possible Load	50%	40%	40%	<50%	~ 15%

Fig. 1.6 Flexibility of conventional electricity generation technologies *Note*: *NPP* nuclear power plants; *HC* hard coal-fired power plants; *Lign* lignite-fired power plants; *CCG* combined-cycle gas-fired power plants; *PS* pumped storage power plants. *Source* Eurelectric (2011, p. 19)

emissions' reductions. In fact, modern CCGTs produce about half the CO_2 emissions per unit of electricity generated compared with coal-fired plants.[28] Considering that coal still plays a key role in the EU electricity system, the scale of this switch might provide a consistent contribution to the EU 2030 GHG emissions reduction target.

(3) A switch from coal-fired power plants to natural gas-fired power plants will positively impact the EU environmental effort not only at macro level (i.e. climate change mitigation) but also at micro level. In fact, as the IEA outlines, «compared with coal and oil, natural gas avoids or reduces much of the local environmental damage arising from fossil-fuel use. Gas gives off fewer pollutants when burned, including the nitrogen oxide (NO_x) that contributes to acidification and ground-level ozone formation; the sulphur dioxide (SO_2) that (with NO_x) causes acid rain; and the particulate matter that (again with NO_x) causes smog and poor air quality. Consequently, using natural gas instead of other fossil fuels in electricity generation (and other sectors) offers the opportunity to improve air quality, especially in and around cities, where this problem is most acute.»[29]

(4) Being the second-largest emitter of CO_2 after the electricity generation sector, the transport sector has an important role to play in the EU decarbonization path. GHG emissions from the transport sector decreased since 2007 due to high oil prices, increased efficiency of passengers' cars and slower growth in mobility. The European Commission[30] expects this trend to continue but this will still not be sufficient to meet the goal to reduce GHG emissions from the sector by 60% by 2050 compared to 1990 and by 20% by 2030 compared to emissions in 2008 as set by the Transport White Paper[31] adopted in 2011.

Notwithstanding their current difficulties (e.g. relatively high costs, low energy density of batteries and lack of recharging infrastructure), electric vehicles will most likely play a key role in the future decarbonization of the transport sector. However, natural gas can also play a role in the field, not only in terms of compressed natural gas

(CNG) vehicles, but also in terms of liquefied natural gas (LNG) for trucks and for ships.

For instance, ExxonMobil does not expect a significant growth in natural gas as a transportation fuel for light-duty vehicles (as CNG vehicles cost more than comparable gasoline-powered cars, have a shorter driving range due to CNG's lower energy intensity, and challenging remains the development of a large network of easily accessible refuelling stations) but it does expect a significant development in terms of LNG for trucks, as LNG-fuelled long-haul trucks have the capacity to travel up to 1,200 km between fill-ups while pulling heavy loads and fuel cost savings could recoup the higher investment costs for an LNG truck (US$70,000 to US$90,000 compared to diesel) within about 3 years.[32] The same rationale also applies to LNG-fuelled buses.

Furthermore, LNG is also expected to play an important role as a ship fuel. According to DNV GL, the world's largest classification society, 63 LNG-fuelled ships (excluding LNG carriers) already operate worldwide, while another 76 are on order (as of May 2015).[33] The key driver behind the choice of LNG as ship fuel relates to its environmental advantages. In fact, ships are generally fuelled by highly polluting fuels such as heavy fuel oil, marine gas oil or distillate fuels. The utilization of LNG allows to significantly reduce local pollution, and thus to safeguard the ecosystems on which ships are operating. This is the reason why the use of LNG as a ship fuel is increasingly encouraged by the authorities of major European harbours, from Rotterdam to Hamburg, from Antwerp to Bremerhaven.[34]

1.6 Conclusions: Towards a More Balanced and Secure Decarbonization Path

As illustrated in the paper, over the last decade the EU has made consistent progress towards the decarbonization of its energy system. However, this process has also brought new challenges to the EU energy markets, generating certain paradoxes (such as the parallel growth of renewable energy and coal in the mix) that need to be addressed in order to ensure the sustainability of the EU decarbonization path.

Considering that after the first rump-up phase—occurred over the last decade—the future integration of more variable renewable energy sources into the system will be more complex under both the technical and economic perspectives, the EU decarbonization path should indeed now find a more balanced and secure trajectory. In particular, a new EU electricity market design should be able to provide adequate economic incentive for investments in the flexibility options (i.e. network infrastructure expansion, development of smart grids, adoption of demand-side measures and development of energy storage technologies) that will be crucial to ensure the sustainability of the EU decarbonization path also beyond the 2030 horizon.

As the paper illustrated, in order to achieve its 2030 renewable energy target, the EU will need to rethink its electricity system beyond renewable energy itself, with a particular focus on the role that natural gas might play in the future of the EU energy system.

Considering its previously illustrated characteristics, and particularly taking into consideration the potential to generate immediate and substantial GHG emissions' reductions by displacing coal with it, natural gas might well play an important role in the future EU decarbonization path. Its role does not need to be supported by dedicated public policies but, on the contrary, what is needed is a more general EU action aimed at rebalancing the overall energy system along the lines of a truly sustainable decarbonization path.

Such an action should be carried out by making use of two specific tools: (i) Carbon pricing; and (ii) Environmental regulation.

1.6.1 Speeding up the Reform of the Emissions Trading Scheme (ETS)

The development of a well-functioning (and technology-neutral) carbon-pricing system, able to discourage high carbon options and to promote most cost-efficient ways of reducing GHG emissions, is theoretically the essential component of a sustainable decarbonization path.

In fact, this system would create the basis of an automatic readjustment of EU electricity markets ideally composed by a progressive phase-out of highly polluting coal-fired power plants, a strong development of renewable energy sources (even in absence of incentives) and a larger utilization of natural gas in electricity generation.

In 2005, the EU adopted the Emission Trading Scheme (ETS) as its flagship GHG emissions' reduction initiative. The scheme, based on the "cap and trade" principle, aims at providing appropriate incentives for investments in low-carbon technology via a carbon emissions price.[35] After two initial phases,[36] the ETS entered its third trading phase at the beginning of 2013, with the introduction of a EU-wide cap on emissions (reduced by 1.74% each year) and a progressive shift towards auctioning of emission allowances (EUAs) in place of cost-free allocation. However, low levels of industrial output and power generation due to the economic crisis have resulted in an increasingly large surplus of EUAs in the ETS, leading to a significant downward pressure on the carbon emissions price (Fig. 1.7).

Considering the current inability of the ETS to send sufficient price signals to investors in low-carbon technologies, with the 2030

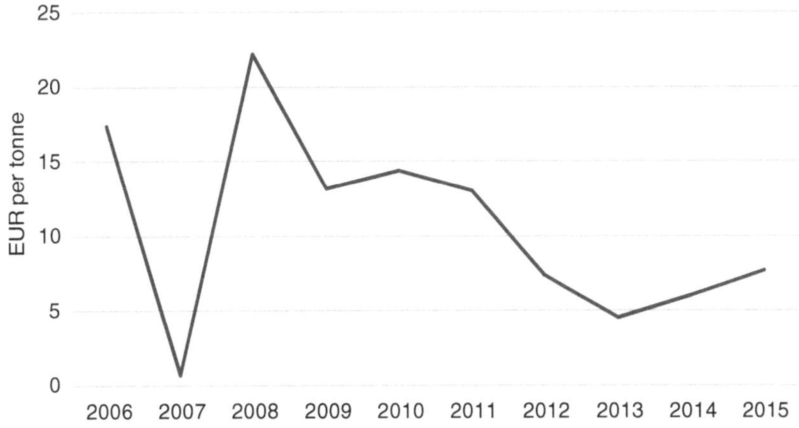

Fig. 1.7 EU ETS carbon emissions spot price. *Source* Own elaboration on Enerdata Global Energy and CO_2 Database (accessed September 2016)

Climate and Energy Framework the European Commission has brought forward proposals to address the level of oversupply in the ETS and reintroduce a meaningful carbon emissions price.[37] This reform should be seen as the crucial element towards the consolidation of the EU decarbonization path and, consequently, of the creation of a more balanced EU energy system on which renewables develop in parallel to other low-carbon and flexible solutions, such as natural gas.

1.6.2 Tightening Environmental Regulation

Considering the numerous challenges related to the development of a well-functioning carbon-pricing system at the EU level, the instrument of environmental regulation should also be exploited to rebalance the energy system along the lines of a sustainable decarbonization path. In particular, tighter emission performance standards should be applied to power plants.

In 2011, the Industrial Emissions Directive (IED)[38] came into force, updating and merging seven pieces of existing legislation, including the previously illustrated Large Combustion Plant Directive (LCPD).

The new IED places further restrictions on the level of nitrogen oxides, sulphur dioxide and particulate emissions permitted from power generators after 1 January 2016 (as until the end of 2015 the provisions of LDPD are applied).[39]

It is difficult to envisage whether these provisions will have or not a consistent impact on the European coal-fired power plants fleet. This will largely depend on the materialization of the incentive to invest in depollution equipment, a choice set to be driven by technology cost and coal pricing itself. According to Cedigaz (2014), for old coal-fired power plants there will be no incentive to invest in depollution equipment and 50–55 GW of EU coal power capacity may thus close by 2020/2023 at the latest according to the IED.[40] However, other analyses carried out by European climate think tanks suggest that a predominant share of EU coal power plants will become IED compliant, as technological changes and flexibility in IED rules will make compliance much less costly than previously estimated.[41]

The implementation of the IED should thus be followed closely, also through the system of review already adopted by the European Commission. At the same time, the EU should be ready to take further actions on environmental regulation, in order to ensure the achievement of proper environmental standards in the EU power plants fleet.

Carbon pricing and environmental regulation constitute the optimal toolset to calibrate the energy system along the lines of a sustainable decarbonization path. If correctly utilized, these tools could stimulate a further development of renewable energy sources, a greater role of natural gas in the energy mix and a reduction in the utilization of polluting coal, at one fell swoop. This readjustment seems to be the only way to make decarbonization balanced and secure up to 2030 and beyond.

Notes

1. For a wider discussion of the evolution of the EU energy and climate policy, please refer to: Simone Tagliapietra (2014).
2. 15 at the time: Austria, Belgium, Denmark, Finland, France, Germany, Greece, Ireland, Italy, Luxembourg, the Netherlands, Portugal, Spain, Sweden and the UK.
3. European Council (2002).
4. During the summit, the EU heads of State or Government discussed a plan presented by the British Prime Minister to create a true European energy policy and agreed on the need of advancing the EU action in this field. See: Tony Blair (2005).
5. European Commission (2006).
6. European Council (2006).
7. European Commission (2007).
8. European Council (2007).
9. European Commission (2011a).
10. European Commission (2011b).
11. European Commission (2014).
12. European Council (2014).
13. The package is composed by three Communications: European Commission (2015a); European Commission (2015b); European Commission (2015c).

14. European Commission (2014), *Op. Cit.*, p. 6. This attitude reflects the traditional reluctance of Member States to give up to the EU any competence concerning the composition of their own energy mixes. It also underpins the provisions on energy of the Treaty of Lisbon, contained in Article 194 TFUE. See: Official Journal of the European Union (2008).
15. European Commission (2014), *Op. Cit.*, p. 6.
16. Eurostat (2015), *Op. Cit.*
17. EDF (2015).
18. *Ibidem*, p. 1.
19. In fact, generation from wind and PV contributes to the supply of energy but their stochastic nature means that their outputs do not always coincide with periods of high demand and consequently they make a minor contribution to capacity.
20. IEA (2015), p. 112.
21. IEA (2014), *Op. Cit.*, p. 226.
22. IEA (2014), *Op. Cit.*, p. 224.
23. Official Journal (2011). The Directive aims at reducing acidification, ground-level ozone and particulates by controlling the emissions of sulphur dioxide, oxides of nitrogen and dust from large combustion plants. All combustion plants built after 1987 must comply with the LCPD emission limits. Those power stations in operation before 1987 have three options for complying: (1) by installing emission abatement equipment, e.g. flue-gas desulphurization; (2) by operating within a "National Plan" setting a national annual mass of emissions calculated by applying the emission limit value (ELV) approach to existing plants, on the basis of those plants' average actual operating hours, fuel used and thermal input, over the 5 years to 2000; or (3) by opting out of the directive. An existing plant that chooses to opt out is restricted to 20,000 total hours of operation after 2007 and must close by the end of 2015.
24. IEA (2014), *Op. Cit.*, p. 224.
25. European Commission (2014), *Op. Cit.*, p. 6.
26. In all thermal power plants, partial load operation is restricted by a minimum power generation.
27. For the complete analysis, please refer to: Eurelectric (2011).
28. IEA (2011).
29. Ibid, p. 85.

30. European Commission (2014), *Op. Cit.*, p. 14.
31. European Commission (2011a, b).
32. ExxonMobil official website: http://corporate.exxonmobil.com/en/energy/natural-gas/technology/natural-gas-as-a-transportationfuel?parentId=7bb4486e-b68e-43ee-b9fa-cff1663bd80c
33. DVN GL (2015).
34. Reuters (2014).
35. Specifically, the scheme works as follow: the overall volume of GHG that can be emitted each year by the power plants, factories and other companies covered by the system is subject to a cap set at EU level. Within this Europe-wide cap, companies receive or buy emission allowances, which they can trade if they wish. For a detailed overview, please refer to: European Commission (2013).
36. The first phase, 2005–2007: trading period used for "learning by doing". EU ETS established as the world's biggest carbon market. However, the number of allowances, based on estimated needs, turns out to be excessive; consequently, the price of first-period allowances falls to zero in 2007. The second phase, 2008–2012: the number of allowances is reduced by 6.5% for the period, but the economic downturn cuts emissions, and thus demand, by even more. This leads to a surplus of unused allowances and credits, which weighs on carbon price. See: European Commission (2013), *Op. Cit.*
37. According to the Commission (2014, *Op. Cit.*, p. 8), «the best way to achieve this is to establish a market stability reserve at the start of phase 4 trading in 2021. The market stability reserve would provide an automatic adjustment of the supply of auctioned allowances downwards or upwards based on a pre-defined set of rules and would improve resilience to market shocks and enhance market stability.»
38. Official Journal (2010).
39. In particular, in comparison with the previous LCPD, the new IED tightens emission limit values (ELVs) for sulphur dioxide from a level of 400 mg/Nm3 to a level of 200 mg/Nm3. Furthermore, power generators will have to install selective catalytic reduction from 2016 to meet the nitrogen oxides ELVs. Peaking plants (<1,500 annual operating hours) can run indefinitely, a Transitional National Plan to mid-2020 allows trading in most pollutant categories to achieve emissions reductions equivalent to the Directive's ELVs, and a derogation allows operators to run their plants for just 17,500 h after 1 January 2016 before closure, which must be before the end of 2023.

40. Cedigaz (2014).
41. In particular, Sandbag (a UK-based climate think tank) estimates that across the EU 110 out of 150 GW are or will become IED compliant. The remaining 40 GW could become compliant too if it invests in NOx abatement. According to the analysis, it had been thought that the only way to comply would be to install selective catalytic reduction that turns NOx into nitrogen and water. However, cheaper options such as selective non-catalytic reduction have become available in the meantime. See: Sandbag (2014).

References

Cedigaz. 2014. Gas and Coal Competition in the EU Power Sector. Rueil Malmaison: Cedigaz.

DVN GL. 2015. LNG as a Ship Fuel. Hamburg: DVN GL.

EDF. 2015. *Technical and Economic Analysis of the European Electricity System with 60% RES, Research and Development Division*, June 17 .

Eurelectri. 2011. Flexible Generation: Backing-Up Renewables. Brussels: Eurelectri, Tech Rep.

European Commission. *A European Strategy for Sustainable, Competitive and Secure Energy, Green Paper of the Commission.* (Brussels: 2006), COM (2006) 105 final.

European Commission. 2007.*An Energy Policy for Europe, Communication from the Commission to the European Council and the European Parliament.* Brussels: COM (2007) 1 final.

European Commission. 2011a. A Roadmap for Moving to a Competitive Low Carbon Economy in 2050, Communication from the Commission to the European Parliament, the Council, the Economic and Social Committee and the Committee on the Regions. Brussels: COM (2011) 112 final.

European Commission. 2011b. Energy Roadmap 2050, Communication from the Commission to the European Parliament, the Council, the European Economic and Social Committee and the Committee on the Regions. Brussels: COM (2011) 885 final.

European Commission. 2013. The EU Emission Trading System (EU ETS). Brussels: Climate Action Publications.

European Commission. 2014. A Policy Framework for Climate and Energy in the Period from 2020 to 2030. *Communication from the Commission*

to the European Parliament, the Council, the European Economic and Social Committee and the Committee on the Regions.Brussels. COM (2014) 15, final.
European Commission. 2015a. A Framework Strategy for a Resilient Energy Union with a Forward-Looking Climate Change Policy. Brussels, COM (2015) 80 final.
European Commission. 2015b. The Paris Protocol—A Blueprint for Tackling Global Climate Change Beyond 2020. Brussels, COM (2015) 81 final.
European Commission. 2015c. Achieving the 10% electricity interconnection target—Making Europe's electricity grid fit for 2020. Brussels, COM (2015) 82 final.
European Council. 2002. Decision Concerning the Approval, on Behalf of the European Community, of the Kyoto Protocol to the United Nations Framework Convention on Climate Change and the Joint Fulfilment of Commitments Thereunder. Brussels, 2002/358/EC.
European Council. 2006. *Presidency Conclusions*. Brussels, No 7775/1/06 REV1, March 24.
European Council. 2007. *Presidency Conclusions*. Brussels, No 7224/07, March 9.
European Council. 2014. *Conclusions on 2030 Climate and Energy Policy Framework*. Brussels, SN 79/14, October 24.
IEA. 2011. Are We Entering a Golden Age of Gas? *World Energy Outlook Special Report*. Paris.
IEA. 2015. Energy and Climate Change. *World Energy Outlook Special* Report. Paris.
Official Journal of the European Union. 2008. Consolidated versions of the Treaty on European Union (TEU) and the Treaty on the Functioning of the European Union. Brussels, OJ 2008/C 115.
Official Journal of the European Union. 2010. Directive 2010/75/EU of the European Parliament and of the Council of 24 November 2010 on Industrial Emissions (Integrated Pollution Prevention and Control). Brussels, L 334, 17.12.2010.
Official Journal of the European Union. 2011. Directive 2001/80/EC of the European Parliament and of the Council of 23 October 2001 on the Limitation of Emissions of Certain Pollutants Into the Air From Large Combustion Plants. Brussels, L 309, vol. 44, November 27.
Reuters. 2014. *Global Transport Sector Looks to Ride Natural Gas Boom*, February 5.

Sandbag. 2014. *Europe's Failure to Tackle Coal –Risks for the EU Low Carbon Transition*. London.
Simone Tagliapietra. 2014. Towards a EU Energy Union: The Need to Focus on Security of Energy Supply. Milan, Nota di Lavoro n. 95.2014, Fondazione Eni Enrico Mattei.
Tony Blair. 2005. Press Conference at EU Informal Summit Hampton Court, October 27.

Authors' Biography

Manfred Hafner is coordinator of the Energy Scenarios and Policy research programme at Fondazione Eni Enrico Mattei (FEEM), Italy. He is also Professor in international energy studies teaching at the Johns Hopkins University School of Advanced International Studies (SAIS-Europe) and at the Sciences-Po Paris School of International Affairs (PSIA). Furthermore, he is Associate Researcher at the Centre for Research on Energy and Environmental Economics and Policy (IEFE) at Bocconi University, Vice-President Consulting at Enerdata and President of International Energy Consultants (IEC). He has more then 30 years of experience analysing international energy and gas markets and has consulted extensively for industry, governments and international organizations.

Simone Tagliapietra is Senior Researcher at the Fondazione Eni Enrico Mattei (FEEM), Italy, and Research Fellow at Bruegel, Belgium. He is also Lecturer on energy economics and politics at the Universita' Cattolica del Sacro Cuore, Italy. He is an expert in international energy and climate issues, with a record of numerous publications covering the international energy markets, the European energy and climate policy and the Euro-Mediterranean energy relations, with a particular focus on Turkey. He is the author of the books 'Energy Relations in the Mediterranean: A Political Economy Perspective' (Palgrave, 2017), 'The Future of European Gas Markets' (Claeys & Casteels, 2016), 'The Geoeconomics of Sovereign Wealth Funds and Renewable Energy' (Claeys & Casteels, 2013). In the past, he also spent a year in Istanbul as Visiting Researcher at the Istanbul Policy Center at Sabanci University, and he also completed a traineeship at the Sustainable Energy Division of the United Nations Economic Commission for Europe (UNECE) in Geneva. He holds a Ph.D. in Institutions and Policies from the Università Cattolica del Sacro Cuore, Italy.

2

The Potential Role of Gas in Decarbonizing Europe: A Quantitative Assessment

Carlo Carraro, Massimo Tavoni, Thomas Longden and Giacomo Marangoni

2.1 Introduction

The dynamics of energy markets in Europe are currently experiencing a paradoxical transition. On the one hand, a revival of coal imports and a reduction of gas consumption, with an associated negative impact upon greenhouse gas (GHG) emissions in some major European economies, have been observed in recent years. On the other hand, the European Commission and all EU countries, by committing to

C. Carraro (✉)
FEEM, Venice, Italy
e-mail: carlo.carraro@feem.it

M. Tavoni · T. Longden · G. Marangoni
FEEM, Milan, Italy
e-mail: massimo.tavoni@feem.it

T. Longden
e-mail: Thomas.Longden@chere.uts.edu.au

G. Marangoni
e-mail: giacomo.marangoni@feem.it

M. Hafner and S. Tagliapietra (eds.), *The European Gas Markets*,
DOI 10.1007/978-3-319-55801-1_2

the INDC submitted at Paris COP 21, have adopted ambitious GHG emissions targets.

The European Commission itself has acknowledged the increased use of coal as a key issue for Europe, with increased CO_2 emissions being an important concern. The European Commission's contribution to the European Council of 22 May 2013 titled "Energy challenges and policy" notes that "EU consumption and imports of coal (hard coal and lignite) have increased by, respectively, 2% and almost 9% over the first 11 months of 2012, relative to the same period in 2011" (European Commission 2013).

Policies to promote the transition towards a sustainable energy system—which are likely to favour natural gas, at least in the short and medium term—have not materialized to the extent expected only a few years ago. Nevertheless, the role of natural gas as a transitional fuel within the joint climate and energy framework is an important component of the EU strategy. This was highlighted within the EU Energy Roadmap 2050, which noted that the scenarios utilized within the Roadmap "are rather conservative with respect to the role of gas … economic advantages of gas today provide reasonable certainty of returns to investors, as well as low risks and therefore incentives to invest in gas-fired power stations" (European Commission 2011).

Hence, there is a need to conduct additional analysis on the role of natural gas within the EU policy framework to address climate change. The European Union is unlikely to achieve its ambitious climate targets without relying heavily on gas rather than coal as a primary energy source. Therefore, appropriate measures need to be implemented to move energy markets in Europe closer to the optimal energy mix (where optimality obviously includes the internalization of the climate externality). Gas is likely to play a relevant role in the optimal energy mix for at least four decades (as shown within the analysis below).

To address these issues, this chapter focuses on three climate-related policy scenarios with two additional policy assumptions (two possible policy variations). In doing so, it reviews the role of natural gas within climate efforts which include the post-Copenhagen Pledges and the EU Roadmap.

It should be noted that a range of studies have focused on the impact of climate targets upon Europe, e.g. refer to Böhringer et al. (2009), Blesi et al. (2010), Capros et al. (2012a) and Bosello et al. (2013).

However, this is the first study to specifically focus on the role of natural gas across different EU climate policy scenarios. Our focus on natural gas is due to the above statement within the Energy Roadmap 2050, the current debate concerning the additional sources of gas, and the potential role of gas as a transitional fuel within the shift towards a low-carbon energy future as it provides a flexible power source which can counter the intermittency of renewables. While gas has been acknowledged to remain in the European primary energy mix within the long term (Knopf et al. 2013), the extent to which natural gas plays a role has not been given sufficient attention.

The analysis has been conducted using the World Induced Technical Change Hybrid (WITCH) model, an integrated assessment and a widely used model in the global assessment of climate and energy policies. Within the model, the main macroeconomic variables are represented through a top-down intertemporal optimal growth economic framework. This is combined with a bottom-up compact modelling of the energy sector, which details energy production and provides the energy input for the economic module and the resulting emission input for the climate module. Further information about the model is available at the website www.witchmodel.org or can be sourced from Bosetti et al. (2007), as well as in Bosetti et al. (2006, 2009).

The chapter is compiled of four sections. An introduction appears before this point, while three sections follow. Section 2.2 outlines the scenarios utilized within the analysis. Section 2.3 focuses on the main results of the analysis, with a focus on the future of natural gas within Europe. Section 2.4 concludes with a discussion of the key findings of our analysis.

As a prelude to the results of the chapter, the conclusions have been separated into three key points. The first is the importance of setting a suitable carbon price which ensures that the right incentives are given to energy markets, so that a consistent energy mix can be achieved, thus reducing the policy costs of all climate policy targets reviewed within the analysis. The second point is that natural gas is indeed a key transitional fuel for a range of climate policy targets, and therefore, policy should be very careful in designing the right incentives to sustain gas consumption, at least until intermittency remains a problem for renewables' expansion. And lastly, the importance of avoiding distortive policy instruments, e.g. subsidies, is highlighted. For example, in the near term (2020), the renewable target and related subsidies to renewables have been found to reduce carbon prices by about 10 $/tCO_2$, with clear negative impacts on incentives to adopt more energy-efficient business strategies and to invest in climate-friendly technologies and production processes. What this study shows is that a correct carbon pricing can sustain gas consumption at while transitioning coal out of the power generation mix without damaging the development of renewables, even with lower or zero subsidies.

2.2 Scenario Description

With a focus on the importance of climate policy for natural gas in Europe, we have developed a range of scenarios which capture a realistic representation of the current conditions under which policy-makers are operating. As part of this, we have implemented the scenarios presented below with underlying assumptions regarding economic growth and the expansion of nuclear power. For example, stagnant economic growth in Europe until 2020 is implemented by lowering labour productivity, and within the baseline, this results in a growth rate of approximately 0.4% per year for Europe between 2010 and 2020, increasing to approximately 1.5% per year after 2020. Table 2.1 presents the population and GDP assumptions that are implemented within the baseline scenario.

Table 2.1 Baseline demographic and economic estimations

	2010	2015	2020	2025	2030	2035	2040	2045	2050
Population (Billions)	0.513	0.520	0.525	0.528	0.530	0.530	0.530	0.528	0.526
GDP (Trillion 2005 USD MER)	15.17	15.67	16.15	17.39	18.86	20.33	21.96	23.67	25.52
GDP per Capita (2005 USD per person)	29.54	30.14	30.77	32.92	35.57	38.31	41.44	44.79	48.50

A gradual reduction of nuclear power in western Europe is also implemented across all scenarios to reflect the post-Fukushima apprehension towards the technology. Within the baseline, this results in an 8% reduction in nuclear power generation in comparison with 2010 levels at the European level for 2020, increasing to a 14% reduction in 2030.

Climate policy stringency is implemented across four different scenarios. The No Policy (No Pol) scenario is a comparative counterfactual state of the world in which no climate policy is implemented (not even in 2020) in any country in the world. As our focus is on Europe, the counterfactual nature of this scenario is clear as it does not include any of the existing policies which have already been implemented (such as the 2020 renewable and emissions target) and the main use of this scenario will be in providing a benchmark for the calculation of policy costs, including the costs of the 2020 renewable target.

The Moderate Policy (Pledge) scenario is a case where there is fragmented moderate action on climate and includes region-specific policy objectives based on the post-Copenhagen Pledges. These region-specific policy objectives include the following: (1) 2020 emission reduction targets, (2) technology-specific policies (e.g. expansion of renewable and/or nuclear) and (3) post-2020 carbon intensity targets. Within the Moderate Policy scenario, regions can trade carbon offsets internationally (for example, through a clean development mechanism type of project or via a linkage of the ETS to other regions). However, this is limited to be equivalent to 20% of abatement as at least 80% of emission reductions have to be conducted domestically. For Europe, this scenario includes the legislated 2020 targets (specifically emissions, renewables and energy efficiency) and a post-2020 extrapolation of climate policies, with a 2030 and 2050 target of 25 and 45% emissions reductions with respect to 2005.

The Stepped up Policy (Pledge+) scenario replicates much of the settings of the Moderate Policy scenario, except that the level of ambition is stepped up in 2020 and beyond within all regions. This scenario mimics the implications of the Paris agreement for the EU. This results in a tightening of the supply of emission carbon offsets up to and including 2020 (or equivalently, this can be interpreted as having raised the ambition of emissions reductions in 2020 to 30% wrt to 1990). For 2030 and 2050, emission reductions would be 40 and 60% wrt 2005, respectively.

The 2 °C Policy (2°) scenario moves away from the fragmented representation of climate policy and captures a situation where the Durban

Action Platform delivers a binding international climate treaty entering into force in 2025 with the aim of maintaining global temperature increase below 2 °C with sufficiently high probability. It is important to remark that since the model has a global scope, each policy scenario has a detailed formulation for all the regions of the model (13 regions), and not just for Europe.

Two additional policy assumptions are then imposed on top of the implementation of the level of climate policy stringency with the Base case, being the standard representation of the policy. Note that for Europe, this means that the Base case includes the legislated 2020 targets (specifically emissions, renewables and energy efficiency) in all scenarios, except for NoPol. The first additional policy assumption that is implemented is the no renewable target (No RET), where the 20% renewable target (as a share of final energy) in Europe for the year 2020 and beyond is not activated. This allows disentangling the impact of the renewable target upon Europe—its cost for the EU in particular—in comparison with the alternative cases.

The second additional policy assumption is a case where Europe pursues energy efficiency policies in 2020 and beyond. This, in turn, stimulates high energy efficiency (HEE) where demand stays relatively flat between 2010 and 2050. The implementation of the HEE scenario has been separated into two potential options for policy design and implementation. The first of which is an energy intensity (HEE_I)-based policy where technical change improves energy efficiency. The second is where the policy is imposed as a target on energy demand (HEE_D) and can be achieved by reducing energy demand, rather than through energy intensity. As will be discussed in Sect. 3.1, the distinction is important with respect to policy design and policy costs but is irrelevant with respect to the energy mix. Thus, the distinction will be retained only when presenting carbon prices and policy costs.

2.3 Main Results

Before focusing upon Europe, it is important to briefly review the overall climate policy framework that is being implemented in all world regions as part of the same scenarios. Figure 3.1 reviews the impact of the climate policy stringency scenarios upon global greenhouse gas emissions between 2010 and 2050. The Pledge and Pledge+ policies lead to a peak of global emissions by 2050 and 2040, respectively (and decline thereafter), whereas the 2Deg policy moves this peak back to 2020. The graph highlights the growing global gap in emissions between the case in which no action on climate is undertaken (NoPol) and the different climate policy scenarios. If emissions continued to grow unabated, in line with historical trends, the effects of climate change would be potentially significant, with a global increase in temperature by the end of the century estimated around a mean of 4 °C. On the other hand, the three policies analysed in this chapter have the potential to reduce the temperature increase, depending on the stringency of emission reductions.

Greenhouse gas emissions of selected major regions for the Pledge and Pledge+ policies, reflecting the commitments made within the Copenhagen Pledges, are shown in Fig. 2.1.

In these fragmented policy scenarios, OECD countries would reduce emissions, while emissions in China and India increase before 2030. In the case of China, emissions level off in 2030 and decrease thereafter, as decided at Paris COP 21, thus reflecting a firm commitment towards climate and air pollution reduction objectives, while emissions in India continue to increase up until 2050, given the different stages of

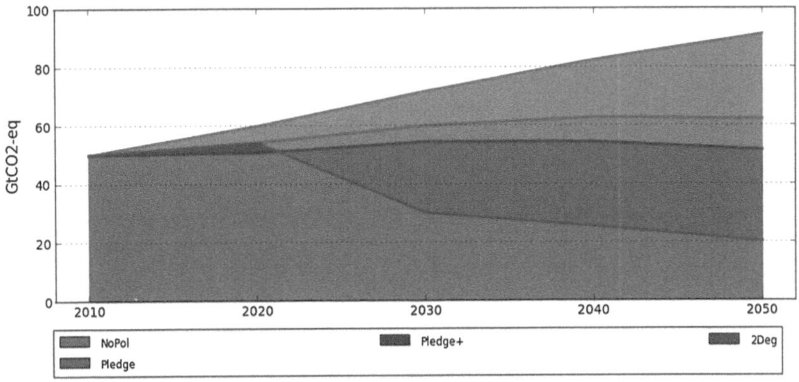

Fig. 2.1 Global greenhouse gases by scenario

economic development. In the case of China, CO_2 emissions in 2010 are 22.7% of the global total and peak at 30.1% in the pledge case in 2030, decreasing to 26.7% in 2050. This is in comparison with 31.5% of global emissions in 2030 and 30.8% in 2050 within the no policy scenario.

Let us now pay attention to the level of action by Europe across the scenarios presented in Fig. 2.2. CO_2 emissions associated with Europe were 12.1% of the global total in 2010, and under the Pledge scenario, this would decrease to 6.6% in 2050 (in comparison with 8.7% in the no policy baseline). In terms of abatement, in 2050 Europe would be responsible for 13.6% of global emission reductions in the Pledge scenario, which decreases to 11.6% in Pledge+ and 8.4% with a unilateral focus on achieving 2Deg. Note that the percentage of emissions/abatement differs based on the level of commitment by regions outside Europe and the overall worldwide emissions in total.

Before reviewing the role of natural gas, it is important to evaluate the climate policy stringency targets for Europe. Figure 2.3 shows the European greenhouse gas targets for the Pledge and Pledge+ scenarios with a comparison between emissions with respect to the NoPol case. Note that Fig. 3.3 makes a distinction between the allowance allocation of emissions and the total amount of emissions that occur within Europe, once international carbon offsets have been accounted for. As

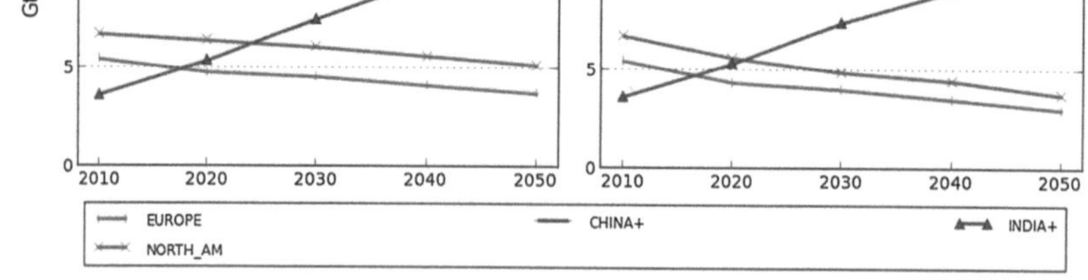

Fig. 2.2 Greenhouse gases by selected major region—Pledge and Pledge+

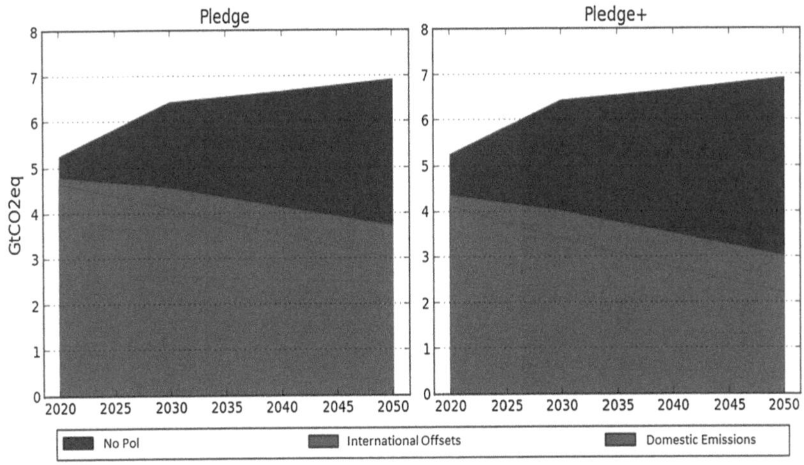

Fig. 2.3 European greenhouse gas targets—Pledge and Pledge+

already implemented today, Europe is allowed to fulfil a fraction of its domestic emissions reductions targets by buying a certain amount of emission permits outside the region, most notably in the developing countries where abatement opportunities are cheaper.

As previously noted, the two policies considered foreseeing a gradual reduction in emissions in Europe, with emission reduction targets in 2030 of 25 and 45% (with respect to 2010) for the Pledge and Pledge+ policy scenarios, respectively. These targets would increase to 45 and 60% by 2050, with a rather linear schedule.

2.3.1 Power Generation Within Europe

We start by providing an overview of the welfare-maximizing power generation mix for coal, gas, nuclear and non-biomass renewables across the Pledge and Pledge+ scenarios and the additional policy assumptions. These are shown in Figs. 2.4 and 2.5. The general trend in power generation for the Pledge and Pledge+ policy scenarios is a reduction in coal and an increase in gas and renewables, as well as a decreasing role for nuclear due to the inclusion of the potential impact of post-Fukushima apprehension within western Europe. These trends are robust across the different policies.

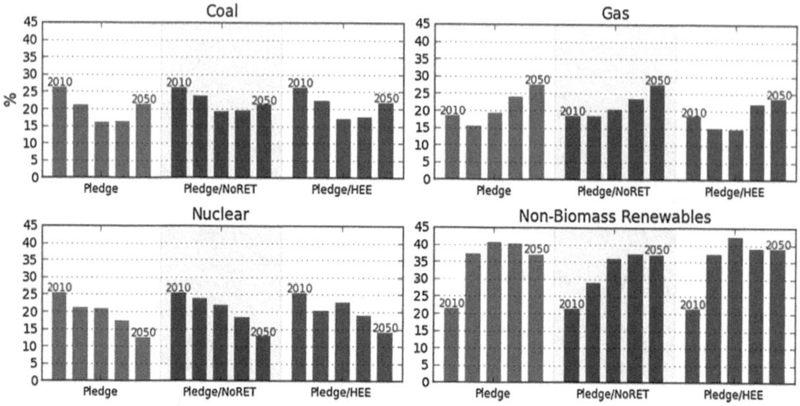

Fig. 2.4 Power generation shares by fuel—full range of Pledge scenarios, from 2010 to 2050

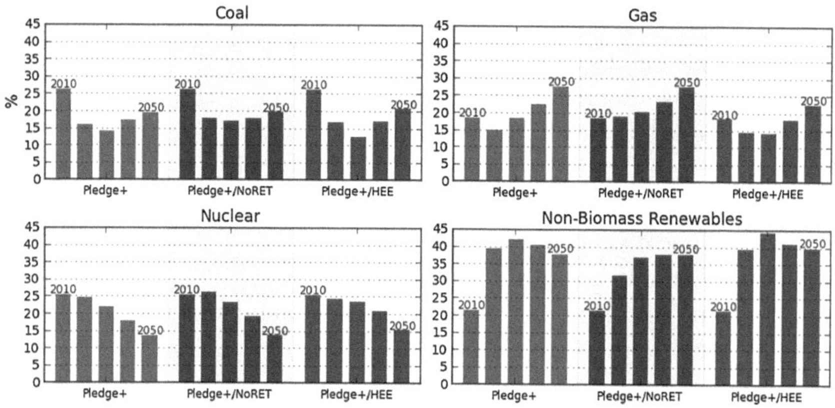

Fig. 2.5 Power generation shares by fuel—full range of Pledge+ scenarios, from 2010 to 2050

In all scenarios, coals lose 10% of market share by 2030, recuperating slightly thereafter due to the deployment of CCS technology. Gas gains 10–15% points, after an initial reduction in 2020 over 2010 due to the economic recession. Renewables show a fast growing pattern in the short term, spurred to a large extent by existing incentives, but also a long-term saturation, due to increase in system integration costs.

Specifically, the power generation shares for Europe within Pledge in 2020 are 21% for coal, 16% for natural gas, 37% for non-biomass renewable and 21% for nuclear, in comparison with shares of 25, 17, 27 and 25% in 2015. The removal of the renewable target for 2020 results in power generation shares for Europe within Pledge/NoRET in 2020 of 24% for coal, 18% for natural gas, 29% for non-biomass renewable and 24% for nuclear, with an additional 5% decrease in total electricity demand.

In the case of Pledge+, the power generation shares for Europe in 2020 are 16% for coal, 15% for natural gas, 39% for non-biomass renewable and 23% for nuclear, in comparison with shares of 25, 17, 27 and 25% in 2015. The removal of the renewable target for 2020 results in power generation shares for Europe within Pledge+/NoRET in 2020 of 18% for coal, 19% for natural gas, 32% for non-biomass renewable and 26% for nuclear, as well as a 6% decrease in total electricity demand.

Underlying a review of Europe which focuses on 2020, as done above, are the issues of low economic growth and the impact of the renewable target. Figures 3.4 and 3.5 show indeed that natural gas within Pledge and Pledge+ is expected to slightly decline in 2020 wrt 2010 and this is related to the slow demand growth in total electricity. However, the impact of the renewable target is notable with no contraction in the share of natural gas occurring within the NoRET cases.

Irrespective of the impact of the renewable target, after 2020 both the Pledge and Pledge+ climate policies induce gas to increase significantly and coal to continue decreasing (until it is somewhat revived when coupled to CCS by mid-century). Figure 2.6 provides the changes in natural gas from electricity in terms of the level of production. The chart indicates that natural gas would eventually increase its contribution to the power mix in a significant way, with an expected generation by mid-century of 1000–1200 TWH, which roughly corresponds to a doubling from today's levels.

The exact timing of the increase in the use of gas depends on assumptions about the economic recovery and the set of policies in place after 2020. As evidenced from Fig. 2.6, the impact of the renewable

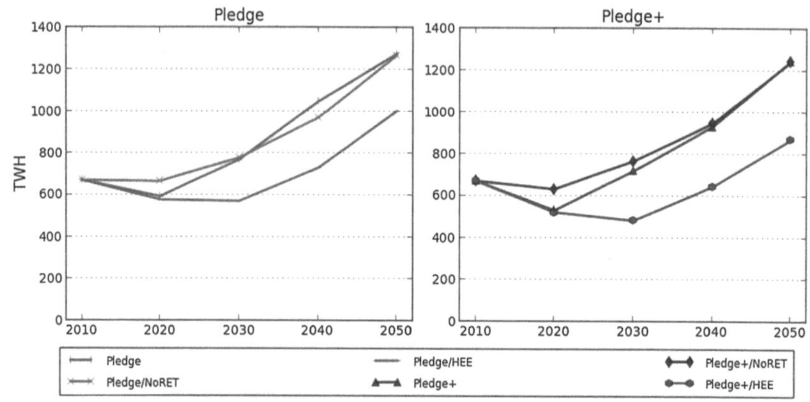

Fig. 2.6 Natural gas electricity—level of power generation

target upon the amount of gas within power generation is visible only in 2020. The impact of the renewable target in 2020 vanishes after that due to an increased role played by renewable energy sources in the long term across all of the additional policy assumptions due to the level of carbon prices in the market. On the other hand, strong post-2020 legislation on energy efficiency is shown to have a sizeable impact on the prospects of natural gas, as a result of lower electricity demand due to increased savings.

Underlying results that have been shown in this section are changes in the investments related to providing the capacity for the power generation options reviewed. Focusing on the Pledge and Pledge/NoRET scenarios, Fig. 2.7 focuses on the impact of the renewable target on investments across coal, natural gas and modern renewables.

The chart shows two contrasting trends for coal and gas on one side, and renewables on the other. Investments in both coal and gas are expected to grow over time, in the range of 100–300 USD billions per decade, but only after the post-2020 economic recovery. Despite its decreasing role in the power mix, investments in coal remain substantial, due to the higher overnight capital costs of coal power, and the fact that after 2030 the majority of coal is equipped with carbon capture and storage (CCS) technology. Indeed, for coal

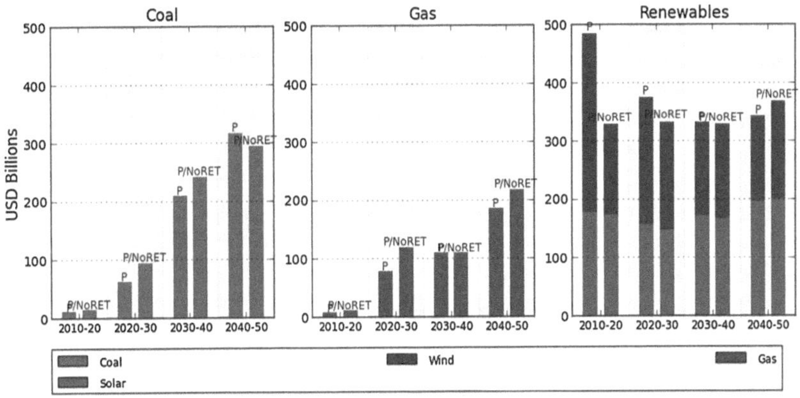

Fig. 2.7 Decadal investments across key power generators

to remain in the optimal energy mix, and still enable the achievement of emissions consistent with Copenhagen Pledges, coal needs to be equipped with CCS after 2030. As a comparison, natural gas is also coupled with CCS; however, this occurs after 2040 within the Pledge scenario. Despite providing a much larger electricity share, investments in gas are smaller, due to the low overnight capital costs assumed for CCGT technologies.

For renewables, investments on the other hand slightly decrease after 2020, due to the improved economics of renewables, as well as a saturation of their contribution due to the already highlighted system integration constraints. In 2020, policies supporting renewables increase investments by about 50%. Between 2010 and 2030, the Pledge scenario corresponds with investments in modern renewables, being 55% of total investments related to the supply of electricity. In terms of capacity, this equates to 65% of new power capacity between 2015 and 2030. Note that projections completed by Bloomberg New Energy Finance forecast that renewables will account for between 69 and 74% of new power capacity added between 2012 and 2030 at the global level.

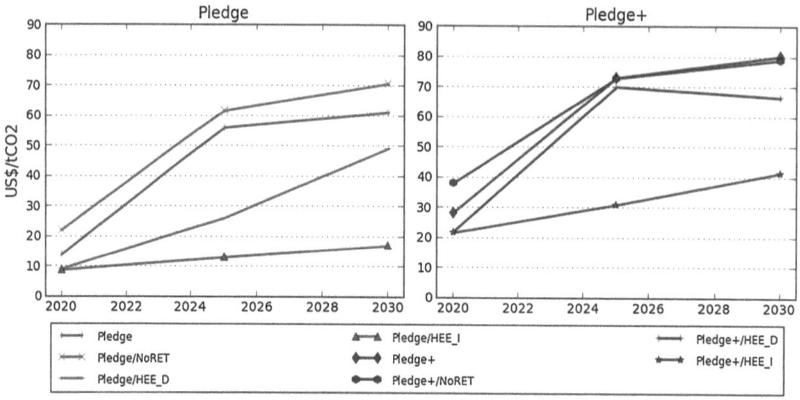

Fig. 2.8 Carbon prices—full range of Pledge and Pledge+ scenarios

2.3.2 Carbon Market and Policy Costs

We now turn to the economic implications of the economic, energy and climate scenarios analysed within this chapter. We begin by looking at an important indicator, namely the carbon prices which emerge from the EU carbon market, see Fig. 2.8. The chart highlights the expected fact that carbon prices grow in the stringency of the emissions reduction target, both over time (by about 5 $/t$CO_2$ each year) and across the policies (with Pledge+ adding 10–15 $/t$CO_2$ to the Pledge case).

Carbon prices in 2020 for the cases where the renewable target is implemented are 9–14 $/t$CO_2$ in the Pledge policy and 22–28 $/t$CO_2$ in the Pledge+ policy scenario, depending on the impact of high energy efficiency. However, the carbon price without the renewable target imposed would be 22 $/t$CO_2$ in Pledge/NoRET and 38 $/t$CO_2$ in the Pledge+/NoRET. This indicates that the renewable target suppresses carbon prices in 2020 by approximately 10 $/t$CO_2$. The importance of the differences in carbon prices lies in the need to provide clear incentives to energy markets—indeed, a stable and long-term signal which increases over time would prevent the recent expansion of coal within Europe which was noted within the introduction.

In addition, if full auctioned, the sales of permits have the potential to generate significant fiscal revenues, which are important at times of

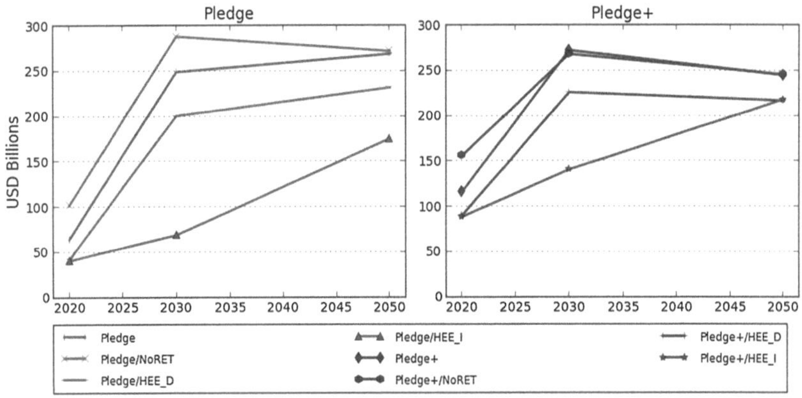

Fig. 2.9 Fiscal revenues from the carbon market

consolidation of public debt. We estimate that public revenues with Pledge and Pledge+ are associated with potential revenues of 65–166 billion USD and exceed 200 billion USD after 2030 (refer to Fig. 2.9). In 2020, the renewable target would reduce revenues by almost 40 billion USD irrespective of whether Pledge or Pledge+ is followed. This highlights that subsidies and/or incentives for modern renewables, in addition to being costly, also reduce the revenues from issuing emission permits.

Figures 2.8 and 2.9 also show the carbon prices and permit revenue associated with two different approaches to implement the same energy efficiency improvements—that being either through energy intensity improvements with technical change (HEE_I) or through energy demand reductions (HEE_D). Between these two scenarios, the differing impact of the imposition of the energy efficiency improvements is highlighted with energy intensity improvements through technical change reducing the burden of emission reductions which occur within the economy and hence have a downward impact upon the amount of carbon offsets which are sourced by Europe from abroad.

Carbon prices are imperfect indicators of macroeconomic costs. Hence, we assess these costs—as measured by GDP losses—separately in Figs. 2.10 and 2.11. Policy costs in the Pledge scenario are found to be in the order of 0.5% GDP loss in 2020, growing to 1.5% by

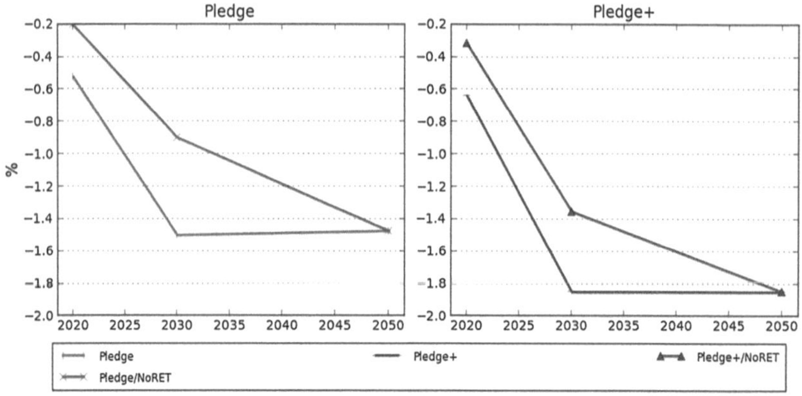

Fig. 2.10 Policy costs in comparison with the no policy scenario—selection of Pledge and Pledge+ scenarios

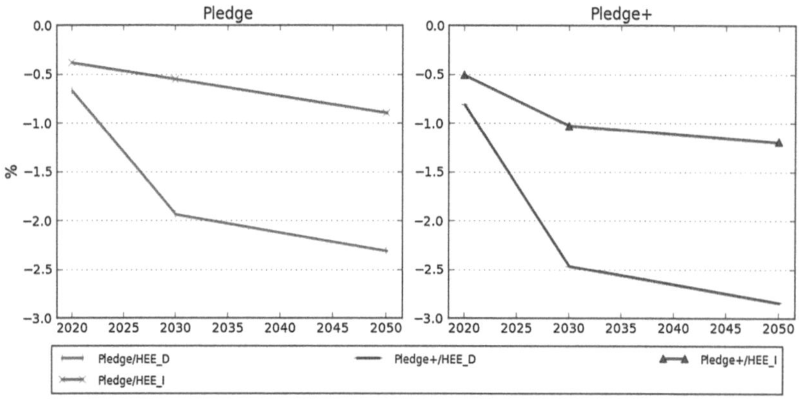

Fig. 2.11 Policy costs in comparison with the no policy scenario—focus on high energy efficiency (HEE) Pledge and Pledge+ scenarios

the mid-century. The renewable target is responsible for a considerable fraction of short-term costs, more than doubling 2020 policy costs; however, these converge over time once the impact of the 2020 renewable target disappears. The Pledge+ policy induces moderately higher costs—0.6 and 0.3% for the Base case and NoRET, respectively. Note that upon adjusting their analysis for an economic recession,

Bosello et al. (2013) find a similar level of policy costs for a scenario similar to Pledge using the ICES model (another integrated assessment model developed and used at FEEM), with a policy cost of 0.5% for the EU when implementing its energy and climate policy unilaterally.

Figure 2.11 also shows policy costs associated with the two different approaches to implement the same energy efficiency improvements—that being either through energy intensity improvements (HEE_I) or through energy demand reductions (HEE_D). In 2020, the difference in policy cost is limited as the difference in energy demand with respect to the baseline is small due to the assumption of suppressed economic growth. However, over time the level of electricity demand within both of these scenarios is notable (20% lower in 2050) with policy costs between HEE_I and HEE_D differing by approximately 1.5% of GDP. Indeed, the changes over time show that the costs of the HEE scenarios crucially depend on policy design and implementation. If the energy efficiency target is designed as energy intensity improvements and implemented as increased technological change, then costs are lower than in the other scenarios.

However, if the energy efficiency target is designed as a target on energy demand reduction (as done in the EU Energy Efficiency Directive), then costs and the demand for offsets are notably higher. In reality, the response to a target would likely be made up of a mixture of energy efficiency improvements and reduced energy demand. However, the policy costs shown within Fig. 2.11 highlight the importance of providing an incentive for a mixed response to a given target. Whether the current European target within the Energy Directive is based on energy demand is suitable will be contingent on the response of industry and consumers, rather than being driven by policy design.

2.3.3 2 °C Durban Action Policy

Let us now turn to how these scenarios differ to a situation where the Durban Action Platform delivers a binding international climate treaty entering into force in 2025 with the aim of ensuring that the

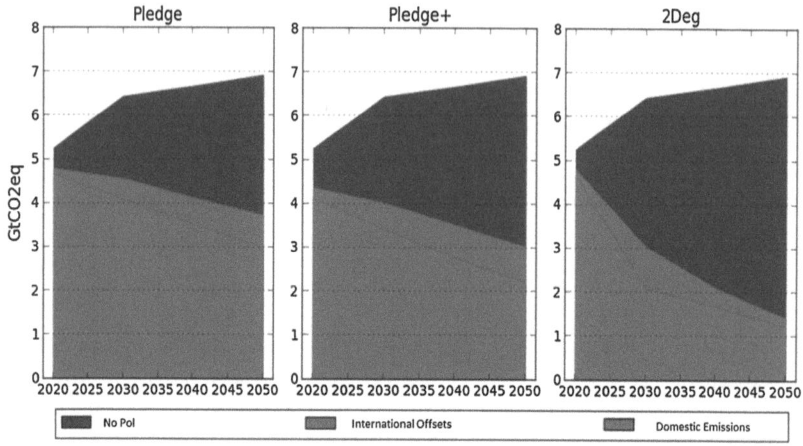

Fig. 2.12 European greenhouse gas targets—Pledge, Pledge+ and 2Deg scenarios

2100 global temperature increase is below 2 °C with sufficiently high probability.

Figure 2.12 updates the European greenhouse gas targets for the Pledge, Pledge+ scenarios, including also the case of 2Deg. Under the 2Deg policy, emissions in Europe would need to be cut significantly more than in the Pledge and Pledge+ policies, by 60% in 2030 and 80% in 2050. This result is consistent with the emission reductions specified within the EU 2050 Roadmap.

The power generation shares for Europe in the 2Deg policy scenario are shown in Fig. 2.13. In 2030, the power generation shares are 11% for coal, 19% for natural gas, 38% for non-biomass renewable and 26% for nuclear, in comparison with Pledge shares of 16, 19, 41 and 21%, respectively. Natural gas maintains a similar (albeit slightly lower) share in the power mix than in the moderate and stepped up policies (i.e. Pledge and Pledge+). Underlying these numbers are strong energy efficiency improvements with 2Deg in 2030, having a 10% reduction of total electricity demand in comparison with the Pledge case which is almost equivalent to the high energy efficiency scenarios reviewed within the fragmented policies. The strength of the reduction in energy demand results in the spike for nuclear within Fig. 2.13 in 2030 as the

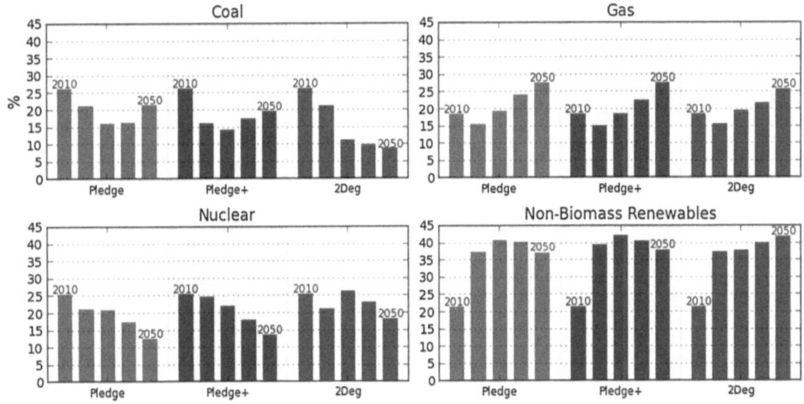

Fig. 2.13 Power generation shares by fuel—Pledge, Pledge+ and 2Deg scenarios

capacity of nuclear has been fixed to reflect a partial phase out nuclear of within western Europe.

In terms of TWh, increased demand for gas wrt 2010 tends to occur in all but the HEE scenario and when the renewable target has an impact (i.e. 2020 within 2Deg, but not within 2Deg/NoRET). In comparison with the Pledge and Pledge+ cases, there is a lower demand for natural gas with the 2050 amount in 2Deg being 976 TWh in comparison with 1276 TWh in Pledge and 1245 TWh in Pledge+. Policy costs within the 2Deg scenario are significant irrespective of global action, and in 2050, costs are over three times larger than in the other policies considered (6.27% in comparison with 1.47% in Pledge and 1.85% in Pledge+).

2.4 Conclusions

This chapter used WITCH, an integrated assessment energy-economic model, to assess a range of energy and climate policy scenarios, as a way to pin down the prospects for natural gas within the welfare-maximising energy mix in Europe for the next four decades. In doing so, it reviewed the role of natural gas within various climate efforts and policy schemes. Two main conclusions can be highlighted. The first is the importance of setting a suitable and sustained carbon price, which ensures that the right incentives are given to energy markets so that the welfare-maximising energy mix can be achieved. This would also reduce the policy costs related to all of the climate policy targets reviewed within the analysis. The second is that natural gas is very likely to be the key transitional fuel within the cost-effective achievement of a range of climate policy targets.

2.4.1 Carbon Pricing

In this chapter, we have shown that even a moderate and fragmented climate policy is sufficient to provide the appropriate incentives for realigning energy markets dynamics with climate objectives. This would require a carbon price of above 15 $/t$CO_2$ which grows to 60–70 $/t$CO_2$ over time. This can be achieved at moderate economic cost by a 2030 emission reduction target in the range of 25–35%, and a 2050 target of 40–60% (all relative to 2005).

The 2050 Energy Roadmap (reduction targets of 60% in 2030 and 80% in 2050 which are consistent with a global objective of 2 °C in 2100) would have significantly higher economic impacts (much higher GDP losses) than the fragmented carbon policy scenarios identified as Pledge and Pledge+, even with global action consistent with the Durban Action Platform.

In relation to providing appropriate incentives for energy markets via a carbon price, it is important to note that modern renewables, such as solar and wind, are becoming competitive due to the existing targets and incentives. Modern renewables would continue to play an important role after 2020 as long as carbon prices are sufficiently high (e.g. 20–50 $/t$CO_2$), and this will occur even without additional incentives or subsidies.

Energy efficiency regulation could play an important role by reducing overall electricity demand; however, the policy design will matter with a notable impact in terms of policy costs, depending on whether it is implemented through improved intensity or reduced demand. Indeed, if the energy efficiency target is designed as a target on energy demand reduction (as done in the EU Energy Efficiency Directive), then costs and the demand for offsets are notably higher.

2.4.2 Gas as a Transition Technology

Due to slow growth in demand and the growing role of renewables which has been induced by the EU target and related incentives/subsidies, natural gas use in power generation is expected to slightly decline until 2020 (unless important changes in gas supply related to shale gas production occur).

Irrespective of a decrease in the share of natural gas until 2020 due to the renewable target, the share of natural gas rises after 2020 and an increase in gas is consistent with the cost-effective achievement of a range of climate targets—refer to the discussion surrounding Fig. 2.6 for further details. In other words, although natural gas's share falls through 2020, it will rise after 2020 if climate targets are to be met cost effectively.

After 2020, both the Pledge and Pledge+ climate policies would induce an increase in gas consumption, while the use of coal decreases. After 2020, increases in gas consumption and a phase out of coal would be enhanced by promoting climate policies which sustain carbon prices above 15 $/tCO_2$ and up to 50–70 $/tCO_2$ in the following decades.

Gas demand would increase after 2020 in all simulated policy scenarios, including the 2Deg scenario through linkages to CCS. The growth of renewables is likely to slow down after 2020 due to limitations of system integration. This will enhance the role of gas as a transition fuel. However, to achieve the 2 °C target, a further development of renewables is required, even at high electricity storage costs, which explains the high policy cost of the 2Deg scenario.

References

Blesl, Markus, Tom Kober, David Bruchof, and Ralf Kuder. 2010. Effects of Climate and Energy Policy Related Measures and Targets on the Future Structure of the European Energy System in 2020 and beyond. *Energy Policy* (The socio-economic transition towards a hydrogen economy—findings from European research, with regular papers), 38 (10): 6278–6292. doi:10.1016/j.enpol.2010.06.018.

Böhringer, Christoph, Thomas F. Rutherford, and Richard S. J. Tol. 2009. The EU 20/20/2020 Targets: An Overview of the EMF22 Assessment. *Energy Economics* (International, U.S. and E.U. Climate Change Control Scenarios: Results from EMF 22), 31 (Supplement 2, December): S268–S273. doi:10.1016/j.eneco.2009.10.010.

Bosello, Francesco, Lorenza Campagnolo, Carlo Carraro, Fabio Eboli, Ramiro Parrado, and Elisa Portale. 2013. Macroeconomic Impacts of the EU 30% GHG Mitigation Target. FEEM Working Paper 2013.028, Fondazione Eni Enrico Mattei. http://www.feem.it/userfiles/attach/201349177484NDL2013-028.pdf.

Bosetti, Valentina, Carlo Carraro, Marzio Galeotti, Emanuele Massetti, and Massimo Tavoni. 2006. WITCH: A World Induced Technical Change Hybrid Model. *The Energy Journal* (Special Issue on Hybrid Modelling of Energy-Environment Policies: Reconciling Bottom-up and Top-down) (01): 13–38. doi:10.5547/ISSN0195-6574-EJ-VolSI2006-NoSI2-2.

Bosetti, Valentina, Emanuele Massetti, and Massimo Tavoni. 2007. The WITCH Model: Structure, Baseline, Solutions. FEEM Working Paper 2007.010, Fondazione Eni Enrico Mattei. http://www.feem.it/userfiles/attach/Publication/NDL2007/NDL2007-010.pdf.

Bosetti, Valentina, Massimo Tavoni, Enrica De Cian, and Alessandra Sgobbi. 2009. The 2008 Witch Model: New Model Features and Baseline. FEEM Working Paper 2009.085, Fondazione Eni Enrico Mattei. http://www.feem.it/userfiles/attach/2009111910584485-09.pdf.

Capros, Pantelis, Nikolaos Tasios, Alessia De Vita, Leonidas Mantzos, and Leonidas Paroussos. 2012a. Transformations of the Energy System in the Context of the Decarbonisation of the EU Economy in the Time Horizon to 2050. *Energy Strategy Reviews* (European Energy System Models), 1 (2): 85–96. doi:10.1016/j.esr.2012.06.001.

Capros, Pantelis, Nikolaos Tasios, and Adamantios Marinakis. 2012b. Very High Penetration of Renewable Energy Sources to the European

Electricity System in the Context of Model-Based Analysis of an Energy Roadmap towards a Low Carbon EU Economy by 2050. In *2012 9th International Conference on the European Energy Market*, 1–8. doi:10.1109/EEM.2012.6254669.

European Commission. 2011. Energy Roadmap 2050. http://ec.europa.eu/energy/energy2020/roadmap/doc/com_2011_8852_en.pdf.

European Commission. 2013. Energy Challenges and Policy—Commission Contribution to the European Council of 22 May 2013. http://ec.europa.eu/europe2020/pdf/energy2_en.pdf.

Knopf, Brigitte, Yen-Heng Henry Chen, Enrica De Cian, Hannah Förster, Amit Kanudia, Ioanna Karkatsouli, Ilkka Keppo, Tiina Koljonen, Katja Schumacher, and Detlef P. Van Vuuren. 2013. Beyond 2020—Strategies and Costs for Transforming the European Energy System. *Climate Change Economics* 4 (supp 01): 1340001.

Authors' Biography

Carlo Carraro is Professor of Environmental Economics at the University of Venice. He holds a Ph.D. from Princeton University. He was President of the University of Venice from 2009 to 2014 and Director of the Department of Economics from 2005 to 2008. Since 2008, he is Vice-Chair of the Working Group III and Member of the Bureau of the Intergovernmental Panel on Climate Change (IPCC). He is the Scientific Director of the Fondazione Eni Enrico Mattei (FEEM), Italy and Director of the International Centre for Climate Governance. He was elected President of the European Association of Environmental and Resource Economists (EAERE) for 2017–18 and Fellow of the Association of Environmental and Resource Economists (AERE).

Massimo Tavoni is associate professor at the School of Management of Politecnico di Milano, and coordinator of the Climate Mitigation research programme at Fondazione Eni Enrico Mattei (FEEM), Italy. He has been fellow at the Center for Advanced Studied in Behavioural Sciences (CASBS) at Stanford University, and post doctoral fellow at Princeton University. His research is about climate change mitigation policies, and has appeared in major scientific journals. He was lead author of the 5th assessment report of the IPCC, co-directs the International Energy Workshop and is deputy editor for the journal 'Climatic Change'. He is the recipient of a grant from the European Research Council.

Thomas Longden is a Senior Research Fellow at the Centre for Health Economics Research and Evaluation (CHERE) at the University of Technology Sydney (UTS). Before joining CHERE in 2016, Dr Longden was based at Macquarie University, the University of Sydney and Fondazione Eni Enrico Mattei (FEEM), Italy. Thomas holds a PhD from the University of New South Wales (UNSW) and his main areas of research interest are applied econometrics, health economics, environmental economics and energy economics. Thomas was a Contributing Author on the latest AR5 WGIII Intergovernmental Panel on Climate Change (IPCC) report titled Mitigation of Climate Change.

Giacomo Marangoni has been researching at Fondazione Eni Enrico Mattei (FEEM), Italy since 2011 on the topics of climate change mitigation and adaptation policy assessment. His publications focus mainly on future transformations of the energy system and the economy, under short- and long-term carbon-constraining climate and energy policies, both at a global and regional level. He holds a PhD in Management Engineering from the Polytechnic University of Milan, Italy. He was also a Contributing Author of the chapter "Assessing transformation pathways" by Working Group III in the Fifth Assessment Report of the Intergovernmental Panel on Climate Change.

3

What Outlook for European Gas Demand? An Overview of Possible Scenarios

Anouk Honoré

3.1 Setting the Context

3.1.1 Natural Gas in Europe: A Story of Success…

Natural gas consumption in Europe[1] had been a story of success since its early developments in the 1960s. Northern European markets, closest to the onshore Dutch discoveries and those offshore the UK, Norway and Denmark were the first to develop at scale, displacing coal and oil products in the space heating and industrial sectors. Since the mid 1990s, the widespread adoption of the highly efficient combined cycle gas turbine in liberalised markets with ambitions on curbing carbon dioxide (CO_2) emissions provided an additional spur for European gas demand in the power sector, again replacing coal and oil products.[2] The market share of natural gas has increased rapidly from less than 10% of the total primary energy supply (TPES) in the early 1970s to about 23% in 2015 (Fig. 3.1).

A. Honoré (✉)
Oxford Institute for Energy Studies, Oxford, UK
e-mail: anouk.honore@oxfordenergy.org

M. Hafner and S. Tagliapietra (eds.), *The European Gas Markets*,
DOI 10.1007/978-3-319-55801-1_3

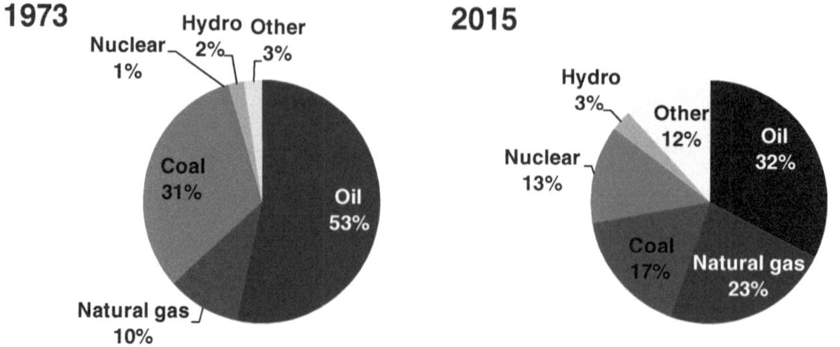

Fig. 3.1 Total primary energy supply in Europe, by fuel in 1973 and 2015 *Source* Based on International Energy Agency data from International Energy Agency, *Natural Gas Information 2016 Edition* (Paris: OECD, 2016), III.17

3.1.2 ... at Least Until the Mid 2000s

Rapid gas demand growth slowed in the mid-2000s as a result of a maturing market, low population growth, higher gas prices (in large part due to the oil price linkage in much of its contracted imports), growing competition in the energy mix and the migration of manufacturing industry to other world regions. Weather corrected data shows that gas demand peaked in 2008, although observed gas demand peaked in 2010 as a result of especially cold temperatures that year. The key message is that the fundamentals, which had been historically driving gas demand up, had already changed when the effects of the economic recession started to be felt. Therefore, the 2008/2009 financial crisis and subsequent recession happened in a context of already moderating gas demand growth in Europe.

Contrary to earlier scenarios,[3] gas demand fell in the early 2010s (between 2010 and 2014) to levels not seen since the late 1990s.[4] Most of the sectors (except transport) were hit by the combined effects of slow economic growth, improvements in efficiency measures, relatively high gas prices (especially to coal prices) and the development of renewable energy. Total gas demand picked up year on year in 2015, mainly thanks to colder temperatures in the early months, and reached 472 billion cubic metres (bcm),[5] still 96 bcm below its record level of 2010.[6]

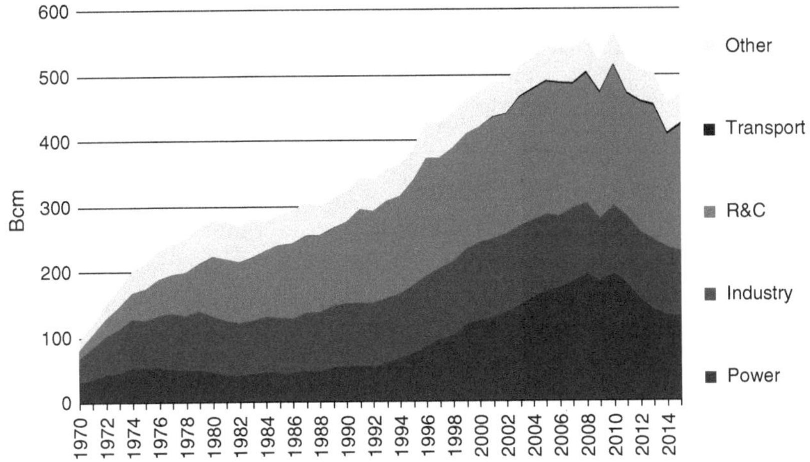

Fig. 3.2 Natural gas demand in Europe, by sector (1970–2015). *Source* Based on International Energy Agency data from International Energy Agency, *Natural Gas Information,* several editions (Paris: OECD); Data for 2015: author's estimates, in Anouk Honoré, "Demand production vs demand destruction" (presentation at the Flame Conference, Amsterdam, 11 May 2016)

The most impressive evolution happened in the power sector, which lost about a third of its gas demand in 2010–2014 (Fig. 3.2). A combination of factors explain this: the economic slowdown restricted power demand growth which, combined with the fast increase of renewable energy,[7] left little room for other fuels in the generation mix. A sharp drop of coal prices since 2011 made coal more competitive than gas over the period,[8] a situation reinforced by the parallel decline of the price of carbon in the European Union Emissions Trading Scheme (EU ETS).[9] As a result, from early 2012, spark spreads were negative, or at least well below dark/clean dark spreads, in most European countries. This unprecedented evolution casted a shadow of uncertainty on prospects for a continued future gas demand growth in Europe.

3.1.3 Uncertainties on Future Gas Demand Growth

In the early 2000s, expectations of growing European gas demand were still largely undisputed, when natural gas was the fuel of choice for new power generation capacity and was seen as the key driver for additional demand in the next two to three decades at least. The development of low carbon policies created slow changes in the energy mix but without too much effects on the gas industry thanks to energy and power demand growth, which left enough room for gas even in a growing competitive environment. Nonetheless, scenarios taking into account a more optimistic development of environmental policies, with fast growing renewables and the maturity of the older gas markets, started to question the linear trajectories of gas demand growth.[10] Scenarios were still showing growth, but expectations were revised down. For example, the IEA expected 868 bcm in 2030 in its World Energy Outlook (WEO) published in 2002, but revised its expectations to 615 bcm in 2030 in its WEO 2010 and to 520 bcm in 2030 in its WEO 2015.[11]

The energy world and of course the gas industry are not isolated from what happens in the rest of the economy, and the impacts of the economic crisis of 2009 came as a shock for many gas players. Gas demand in Europe lost more than 33 bcm in 1 year alone or about 6% of total gas consumption. More importantly, the following years reminded everyone that natural gas does not have a captive market, and its market share can evolve quickly, especially in the power sector (21% of total gas demand in 2015[12]). The COP21 meeting held in Paris in December 2015 resulted in an agreement between 195 governments to cooperate to hold global temperature increase below 2 °C. To achieve this target, a global peaking of Green House Gas (GHG) emissions and emissions neutrality after 2050 are needed. Policy measures in the European Union and in various Member States also focus on reducing emissions progressively up to 2050. Although all sectors are expected to contribute, the power sector is seen as the biggest potential for cutting emissions. Electricity is expected to come from renewable sources like wind, solar, water and biomass or other low emission sources such as nuclear power plants or even fossil fuel power stations equipped with Carbon Capture and Storage (CCS) technology. All these factors, both at the regional and at the national levels, have created a level of confusion as to the future of gas in Europe that is unprecedented, and even raises the possibility of a peak demand for gas having occurred in 2008/2010.

3.2 Future Gas Demand Growth in the Power Sector?

3.2.1 Eroding Role of Natural Gas in the Power Generation Mix

In the first half of the 2010s, not only has gas demand for power generation declined faster than total gas consumption[13] but gas has also lost market share to other fuels (Fig. 3.3).

In the power sector, electricity produced from gas can be substituted by electricity produced by other readily available fuels. The main option, one which can be done quickly (and on a much larger scale than by using oil products), is to use additional coal, although this would add significantly to carbon emissions (and potentially create other environmental problems). Switching from gas to coal

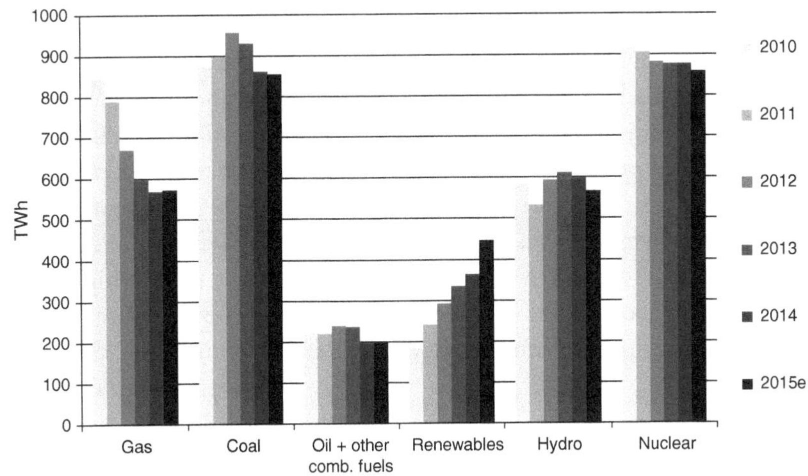

Fig. 3.3 Power generation mix in Europe, by fuel (2010–2015). *Source* Based on International Energy Agency data from International Energy Agency, *Electricity information,* several editions (Paris: OECD); Data for 2015: author's estimates, in Anouk Honoré, "Demand production vs demand destruction"

happened in 2010–2012 due to competitive coal prices relative to gas and low carbon prices in the EU ETS system. Coal continued to be an important competitor to gas, but post-2013, the share of coal in the mix also started to decline due to various coal plants closures as a result of the Large Combustion Plant Directive (LCPD). In 2015, the share of coal in the mix was slightly below its level in 2010, certainly not higher.

Nonetheless, the decline in the share of electricity generated from gas continued. This was due to flat power demand and rapid growth in renewables such as wind and solar, which contributed to just about 13% in 2015 (compared with 5% in 2010).[14] Between 2010 and 2015, the share of natural gas in total generation mirrored the evolution of the share of renewable, but in the opposite direction as the share dropped from 23 to 16%—not far from its level in 2000 (15.7%).

While renewables benefit from priority dispatch in power generation,[15] wind and solar—the two fastest growing renewable energy sources in Europe—are both intermittent and unpredictable. Their availability depends on external factors such as sunshine and wind. They cannot be switched on and off as needed, unlike other power plants. As a result, direct substitution of gas plants by renewables is limited. But their growing share in the mix, and flat growth of power demand post-recession, has had a major impact on power generation from gas.

In 2015, natural gas was used mostly in applications (such as combined heat and power plants) which must run or when gas-fired stations are needed to meet short-term capacity (such as peak shaving, which does not involve large gas volumes).[16] Gas for power demand showed some signs of recovery in 2016 thanks to lower gas prices, higher coal prices in the second half of the year and continued closures of coal plants, which triggered some hope of small improvement this side of the 2020s.

3.2.2 Drivers and Constraints: Conflicting Factors

There are numerous factors that influence gas consumption, and these differ from one country to another depending on the national characteristics such as indigenous resources, supply contracts, transportation capacity (and access to it), power generation mix, policies, taxes and financial support mechanisms. Economic growth and gas price competitiveness also remain key elements in the evolution of gas consumption. However, the future role of gas in Europe will increasingly be a consequence of what kind of energy policies and environmental measures are put in place both at the regional and at the national levels—and also their affordability, which might limit their ultimate scale.

The role of gas in the energy mix is driven by the overall consumption of energy, which is a function of gross domestic product (GDP) and energy efficiency, and by competition with other fuels in the different sectors. The European Union has made the mitigation of climate change one of its key priorities. The Kyoto protocol, the 20-20-20 targets, the 2030 framework (Fig. 3.4) and the 2050 roadmap all propose to transform the region into a low carbon economy via three main measures: increasing renewables, a CO_2 cap and price and better efficiency. Despite the ambitious policies and targets proposed at the regional level, the measures decided by the European Union mainly set a common structure, as energy remains an important economic and strategic challenge for national governments. Energy policies are still very much a matter of national interest albeit inside the regional framework of the decarbonisation of the economy. It is a complicated and questionable task to try to sum up the policies of the very different European countries, but it is fair to say that natural gas has not been getting much attention in energy policy, or better said, the focus is almost always entirely placed on other technologies or objectives that will, on a short-term or long-term basis, have an impact on the use of gas. This creates a high degree of uncertainty in the gas industry. Will gas for power demand ever recover? This will depend on several—sometimes conflicting—factors.

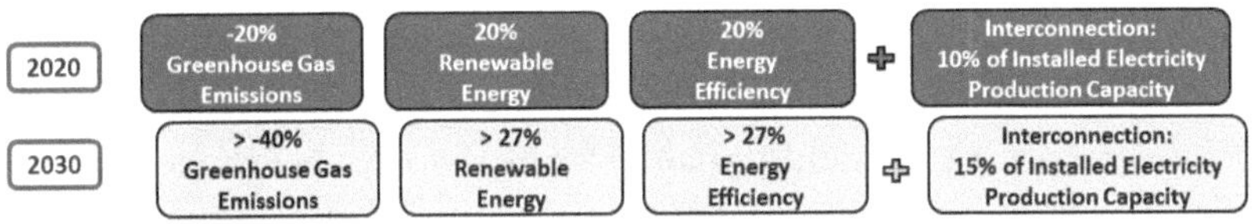

Fig. 3.4 EU framework for climate change and energy, targets for 2020 and 2030. *Source* Author elaboration from http://ec.europa.eu/clima/policies/strategies

- First, the continued policies aimed at improving energy efficiency mean that even with some GDP growth,[17] the effects on energy and power demand are likely to be flattened by these measures.[18]
- Second, renewables are at the centre stage of the European energy policy framework. As part of the EU framework for climate change and energy, renewable energy is expected to continue to rise albeit at a slower pace than seen since the early 2000s due to the downward revision of support schemes across Europe. Upgrades and better interconnections between countries will also contribute to sustaining the role of renewables, in order to meet the EU 2020 and 2030 targets.
- Third, the competitiveness of gas prices against coal prices is not expected to change dramatically over the period (with the expection of the second half of 2016). This is because coal prices are expected to remain low, and the pricing of carbon within the EU ETS is unlikely to climb high enough to make a difference in the dispatch order, despite a series of measures envisaged by the European Commission.[19] The low(er) liquefied natural gas (LNG) prices expected in the second half of the 2010s and early 2020s will create some demand in the power sector, but gas prices will need to drop to very low levels to start making a real difference in the regional mix. It is impossible to give a “magic price” (of gas, coal, or EU ETS) at which coal-to-gas switching would start to happen in the whole region, due to the wide heterogeneity of the market.[20] As an illustration, in a market with spare capacity to be used and highly efficient gas and coal plants, at a gas price of \$4/MMBtu and a coal price of \$50/t, switching may happen at a carbon price of about 20 €/tCO_2.[21]

It is hard to imagine any of those factors being reversed anytime soon, at least this side of 2020, but one can be cautiously optimistic. Lower gas prices thanks to a global LNG glut[22] are raising new hopes of gas demand recovery, but additional measures would be needed to trigger a shift to gas in the short term, like in the UK for instance. In this country, the carbon price floor (a national measure which comes on top of the EU ETS price and which reached £18 in April 2015[23]) changed the relativity of the spark and clean dark spreads, with more gas being used in the system and even some days in 2016 with no coal at all in the mix (an event that has not happened in over 100 years). The UK example will not be easily replicated as the special characteristics of the market have contributed to this evolution[24] such as the type of gas plants in the market (mostly Combined Cycle Gas Turbines—CCGTs—belonging to utilities which are more reactive to price changes), the closure of several gigawatts (GW) of coal capacity due to the LCPD, the still relatively low level of renewables (including hydro), and maybe even more importantly, the limited interconnections with the rest of the European system. Exactly how this model can be replicated in other markets is uncertain due to differences between markets, but it has the merit to offer a concrete example that something can be done with higher CO_2 prices.

3.2.3 Role Still Has a Role to Play… in Theory

One should bear in mind that a return to the situation as seen in the 2000s is impossible. The power sector lost about 65 bcm between 2010 and 2015 in Europe, and even in a theoretical scenario of a return to the level of competitiveness between coal and gas seen in 2010, this would probably translate into only about 30 bcm of additional gas demand due to the rise in renewable generation in the mix in the meantime,[25] all other factors being equal.

However, "all things will not remain equal" and some existing capacity will close down in the time frame, due to EU Directives, national measures, old age and inefficiency.[26]

- Some plants have been pushed out of the system already, as a result of the LCPD (around 55–60 GW by the end of 2015),[27] and some additional ones will also opt out of the Industrial Emissions Directive (IED) and will close down by the first half of the 2020s.[28] Although it was too soon to tell at the time of writing, as many generators decided to include their plants in the National Transition Plans which gives them time to decide to invest or not to comply with the directive,[29] the IED and additional measures on GHG emissions will probably lead to the closure of between 50 and 100 GW by the mid 2020s.[30]

- Finally, some nuclear capacity will shut down due to economics but also political decisions to either phase out nuclear or to decrease the role of nuclear in the mix.[31] The removal of this (large amount of firm) capacity will create a gap between the need for power generation and the capacity in place.

As a result, the gap will rise between power demand and how much renewables can fulfil (even with flat power demand and rising renewables in the mix). Much will depend on how big the gap is and how it is filled. The number of new coal plants will be limited as investment decisions are complicated by low baseload electricity prices and the difficulty of obtaining approval for construction due to environmental regulations (mostly in Western Europe). There is no coherent strategy on nuclear power in Europe,[32] but the nuclear generating fleet is ageing and limited new capacity is also to be expected.[33] It would be optimistic to see any substantial increase in nuclear power production in Europe post-2020. The main issue for nuclear power is, rather, prolonging the operating life of existing stations beyond original design and acceptance by the population. The gas-fired generation will benefit from expected closure of firm capacity (coal and nuclear), especially in the 2020s.

Although the exact amount of retiring plants is yet unknown, this will leave some space for gas in the mix as renewables will not be able to compensate the entire loss, and new coal, nuclear and even large hydro power plants will be limited. Even in a world of tighter and lower carbon emissions, there is a possible role for gas in the generation mix in Europe, but this will require that enough gas plants are in place and ready to be used (which is yet uncertain) and more importantly that the gas industry manages its high-carbon status in the 2020s and beyond by developing power stations equipped with CCS technology (the timing of which is also yet uncertain).

The main unknown is whether gas will be able to play its role when the time comes. About 50 GW of gas-fired plants were closed down or mothballed in the first half of the 2010s[34], and very few new non-renewable plants have taken final investment decision (FID) since the 2008/2009 crisis—apart from Turkey (as of mid-2016).[35] If capacity mechanisms are put in place in an efficient and timely manner, then gas used for power may start to recover slowly in the late 2010s and post-2020, when much nuclear capacity gets retired and coal starts—hopefully—to decline in the mix.[36]

In this scenario, natural gas demand for power would remain modest and up to 135 bcma by 2020, but grow more rapidly thereafter to reach 140–160 bcma in 2030, depending on how renewable policies, interconnections and coal plants closures evolve.[37] In a theoretical scenario with more ambitious goals on de-carbonisation and assumptions of no power demand increase (thanks to energy efficiency and energy saving plus 2020 and 2030 renewable targets met), gas demand for power would probably remain flat up to 2020 and decline to about 115 bcma in 2030.[38]

3.3 Non-Power Sectors: Limited Expectations

3.3.1 Residential and Commercial Sector

The residential and commercial sector is the largest consuming sector in Europe (41% of total gas demand in 2015[39]). This sector is traditionally less influenced by the economic situation in the short term, but rather by cold temperatures in winter when gas is used for space heating.

However, since 2012, it seems that the reaction to cold spell has been more cautious than just a few years before. Cold temperatures have not raised gas demand by as much as they would have done in the past.[40] This result could be explained by a change of attitude, with people starting their boilers later in the year and/or switching them off earlier at the end of the heating season in addition to lowering the thermostat.

Better insulation of new (and old) houses and metering systems will help to keep gas demand growth in this sector at a low rate.

Small-scale generation from renewable energy sources with solar roof panels, small turbines or heat pumps in the garden for heating and cooling is also being developed in the residential and commercial sector. Self-generated and self-consumed power will have an impact on the gas consumed in this sector but it will also reduce the need for centralised generation and therefore gas for power demand.

The processes using gas for heating could be replaced by processes using electricity such as the use of heat pumps in new buildings (the best example) and also direct heating and heat storage systems. In a low- or zero-energy house, all heating might be covered by the exhaust heat of electric appliances. Both solutions have been used in new buildings in Germany over the last 10 years.

Electrification of heating systems could have important consequences in countries with large changes in temperatures influencing the level of power demand which will peak when temperatures rise or drop to their extremes. This usually happens at times of high pressure, and therefore, when there is little or no wind. Sensitivity to a change in temperatures and additional variations in power demand will create a new need for CCGTs in order to respond rapidly to these changes. For example in France, which has electrified its heating system more than any other country in Europe (aside from Sweden), CCGTs are mostly used to cope with winter variations. These rapid changes in power demand can be enormous as one-degree drop in the mean temperatures creates an extra 2.3 GW load on the system, out of about 5 GW extra load in Europe for every one-degree drop in temperature. As a matter of comparison, the temperature sensitivity of power demand in Germany, the largest European electricity market, is closer to 0.5 GW (0.6 GW in the UK and 0.3 GW in Italy according to the French TSO RTE[41]). The electrification of the heating system could potentially create another niche market for the CCGTs during winters, albeit with low utilisation rates throughout the year and therefore limited impacts on annual demand for gas, except maybe in exceptionally cold winters. Interestingly, higher electricity penetration also means that the higher energy efficiency targets may not necessarily mean lower power consumption.

Governments are also looking at increasing the generation of heat (and cooling) in buildings from renewable energy sources rather than fossil fuel (including natural gas) systems, as stated in their Renewable Energy Action Plans.[42]

As a result, gas consumption growth in the residential and commercial sector is anticipated to be slow despite some switching to gas from oil in the heating sector. This is due to the near saturation of the sector in most European countries, apart maybe from Turkey because of its population growth and its rapid urbanisation and in some smaller markets. However, seasonal variations are expected to remain, with peaks progressively appearing also in the summer.

3.3.2 Industry Sector

Natural gas has a multitude of industrial uses, including providing the base ingredients for various products such as plastic, fertiliser, anti-freeze, and fabrics, and in Europe, it is consumed primarily the chemical industry, followed by non-metallic mineral products, food processing beverages and tobacco and many others, and new applications are being developed frequently. The fall in energy intensity thanks to a shift to light industry instead of heavy industry (shift to less gas-intensive sectors), technological improvements (less gas used in production process) and high gas prices, especially post-2003, are the main factors for gas demand decline in this sector, which still represented about 21% of total gas demand in Europe in 2015.[43]

Still, the sector suffered a significant hit with the 2008/2009 economic crisis. Recovery has been slow, and industrial production was still not back to pre-crisis levels in 2016. Permanent gas demand destruction is likely to have happened in this sector due to factory closures. Competition from regions with lower gas prices, namely North America and the Middle East, means that it is not certain that even in the case of an economic recovery in Europe, industrial gas demand will reach previous levels. Even with lower gas prices expected thanks to the wave of LNG in the second half of the 2010s, it may not be enough to change the relative competitiveness with other regions. Apart from Germany and some Eastern/Southern European countries (including Turkey), no growth is expected in this sector as a result of lower gas prices.

Additional growth will also be curtailed by efficiency policies, even if some sectors such as fertilisers have already implemented energy measures and may not be able to make their production much less energy intensive. Over time, improvements in energy efficiency become more difficult once the "low hanging fruit" has been harvested. As a result, the markets located in Northwestern Europe, for instance, will have limited possibilities to lower gas consumption via improved energy efficiencies, due to past investments and technologies already in place, compared with other regions of Europe.

Another evolution to be expected is the development of decentralised generation by the industries, even small ones. This will not impact the statistics on gas demand for the industrial sector per se, but will change the need for centralised generation, and as a consequence, may lower the future gas needs in power.

3.3.3 Transport Sector

Gas for transport, about 1% of total demand, is expected to be the next key driver for additional demand in Europe. As with the other sectors, natural gas does not have a captive market, and if gas for transport is to grow in Europe or at least in some countries, it will be a policy-driven evolution and as a consequence, may be concentrated in some national markets only, rather than being a true European revolution. While this new market has some potential, demand will not be in the range of the other major sectors such as power, industry or residential and commercial, at least in the period considered in this chapter.[44]

3.4 Conclusions

Gas demand expectations in Europe have been revised downward since the early 2000s. The impacts of energy policies and the effects of the financial and economic crisis of 2009 were largely underestimated. The power sector, "the former key driver for additional demand", has been the main driver for demand decline in the first half of the 2010s as gas-fired plants were sqeezed out of the generation mix, a very different picture from the pre-crisis scenarios.

Natural gas demand in Europe fell from 567 bcm in 2010 to just over 472 bcm in 2015. The power sector is no longer synonym of certain additional gas demand as previously thought in the scenarios designed in the 2000s, and in a majority of countries, it will not be. Nonetheless, most scenarios still expect some additional demand, even in a context of slow economic growth and decarbonisation of the energy sector.

A large share of the renewables in the energy mix needs to be seen as a longer term enduring change even with the revision of support schemes for renewable energy which is happening across Europe. The more the renewables, the less the annual average load factor of thermal generation, especially if electricity demand growth does not pick up again rapidly. Reduced operating hours and an increasing number of plant start-ups and shutdowns in order to balance renewable energy supply is rather new, but power generators with gas-fired capacity will need to adapt to the new role of gas in power generation and create a new business model.

Gas demand growth in the power sector will only happen if (much) more competitive gas prices can help it compete with coal (as the place of nuclear and renewables in the merit order is not affected by changes in fossil fuel prices). The competitiveness of gas prices against coal prices is not expected to change dramatically over the period considered in this paper despite the episode of higher coal prices seen in the second half of 2016. Coal prices are expected to remain low, and the pricing of carbon within the EU ETS is unlikely to climb high enough to make a difference in the dispatch order. Lower gas prices (potentially up to the early 2020s thanks to a global LNG glut) are raising new hopes of gas demand recovery, but to be competitive gas prices would need to drop below $3.5–4/MMBtu in (Western) Europe before switching starts to happen for baseload generation. Additional measures would be needed to trigger a shift to gas in the short term, like in the UK for instance.

However, as coal plants are retired due to the LCPD/IED/EU ETS, this coal-gas prices relationship becomes less and less relevant in most countries in the 2020s. About 50–100 GW of the existing baseload capacity will face closure in our time frame. Much will depend on how the gap between power demand and renewables is filled (and how big the gap is), but this is a possible sign of improvement for gas in the power generation sector on the condition that gas manages its high carbon status and start developing CCS technology in the (close) future.

The most interesting result shown by these scenarios is that European gas demand does not seem to be doomed, however it does not return to 2010 levels: this author expects total gas demand to rise from 472 bcm in 2015 to 482 bcm in 2020 and 512 bcm in 2030.[45] More ambitious decarbonisation policies may limit gas demand growth further and probably keep it close to 460 bcm in 2030.[46] However, at the time of writing, there were no strong signs for this very ambitious scenario to be realised in the time frame considered (Fig. 3.5).

The scenarios represent this author's views at the time of writing the paper (mid- to late 2016, with information available at the time).[47] They will need to be updated as policies/prices/generation mix evolve, but the main conclusion of this research is that the outlook for the gas industry includes the potential for modest future growth at least up to 2030.

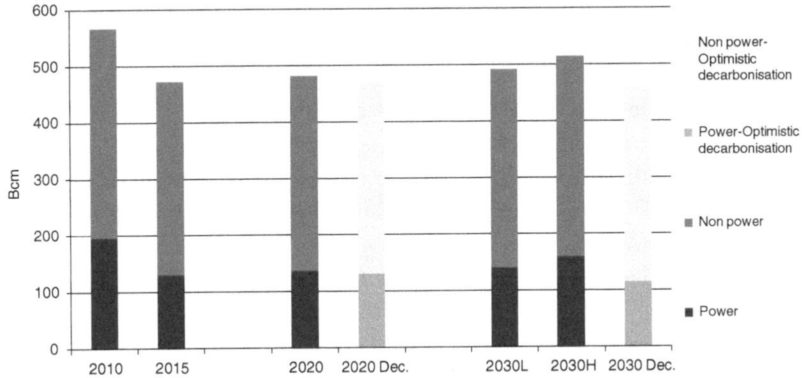

Fig. 3.5 Scenarios for natural gas demand in Europe, by sector (2020 and 2030). *Source* Data for 2010 and 2015: Based on International Energy Agency, *Natural Gas Information, various issues*; Data for 2020 and 2030: Anouk Honoré, "Looking further ahead—What is the outlook for European Gas from 2020–2030?", presentation at Platts conference, Dusseldorf, 28 September 2016

Notes

1. In this chapter, the definition of "Europe" means OECD-Europe, a region which comprises Austria, Belgium, Czech Republic, Denmark, Estonia, Finland, France, Germany, Greece, Hungary, Iceland, Ireland, Italy, Luxembourg, the Netherlands, Norway, Poland, Portugal, Slovak Republic, Slovenia, Spain, Sweden, Switzerland, Turkey and United Kingdom.
2. See Anouk Honoré, *European Natural Gas Demand, Supply and pricing, cycles, seasons and the impact of LNG price arbitrage* (Oxford: Oxford: Oxford University Press, 2010), Chap. 1, Sect. 1, for additional information (Honoré 2010).
3. See International Energy Agency, *World Energy Outlook,* several editions (Paris: OECD), for an example.
4. For more details, see Anouk Honoré, *The Outlook for Natural Gas Demand in Europe,* NG87 (Oxford: Oxford Institute for Energy Studies, 2014) (Honoré 2014).
5. International Energy Agency, *Natural Gas Information 2016 Edition* (Paris: OECD, 2016), III.20 (International Energy Agency 2016).

6. 85 bcm below 2008 levels. Source: International Energy Agency, *Natural Gas Information 2016 Edition.*
7. Thanks to the support schemes put in place to reach the EU 2020 targets, more than 50% of the new generation capacity since 2000 has been in some form of renewable energy; this has near zero marginal costs, priority dispatch, and guaranteed access to the grid. Data source: European Wind Energy Association. *Wind in Power, 2015 European Statistics* (2016).
8. In early January 2011, coal prices (CIF ARA) were \$131/mt. In early January 2016, coal prices were \$44.5/mt, but had climbed to \$66.5/mt in September. Source: Platts, Power in Europe, 10 January 2011, 18 January 2016 and 26 September 2016.
   For more information on coal, EU ETS, and gas prices in the early 2010s, see Anouk Honoré, *The Outlook for Natural Gas Demand in Europe.*
9. The carbon price in the EU ETS has declined from above €25/t in 2008 to €4–8/t since 2013 (time of writing: August 2016).
10. See Anouk Honoré, *European Natural Gas Demand, Supply and Pricing, cycles, seasons and the impact of LNG price arbitrage*, Chap. 3 for more information.
11. New Policy Scenarios for OECD-Europe, in International Energy Agency, *World Energy Outlook 2002* (Paris: OECD, 2002), International Energy Agency, *World Energy Outlook 2010* (Paris: OECD, 2010), International Energy Agency, *World Energy Outlook 2015* (Paris: OECD, 2015) (International Energy Agency 2002, 2010, 2015).
12. International Energy Agency, *Natural Gas Information 2016 Edition*, III.20.
13. Natural gas demand for power has declined faster than total consumption since 2010. In 2015, gas demand for the power sector was almost 40% below 2010—equalling levels not seen since the late 1990s—while total demand declined by 16% over the same period. Sources: International Energy Agency, *Natural Gas Information 2016 Edition* for 2010 data and author's estimate for 2015 data.
14. Data for renewables generation without counting hydro power generation. Source: International Energy Agency, *Electricity information,* several editions (Paris: OECD) and author's estimate for 2015 data in Anouk Honoré, "Demand production vs demand destruction"

(presentation at the Flame Conference, Amsterdam, 11 May 2016) (Honoré 2016).

15. Renewables also benefit from interdiction of significant curtailment, see European Union, "Directive 2009/28/EC, Promotion of the use of energy from renewable sources" (2009) for more information (European Union 2009).
16. With the notable exception of the UK.
17. GDP growth is not expected to be impressive in the period considered in this book, and forecasts in the early 2010s have been revised downwards several times. See Organisation for Economic Co-Operation and Development, *Interim Economic Outlook*. (Paris: OECD, 2016) for additional information (Organisation for Economic Co-Operation and Development 2016).
18. The EU energy savings expected thanks to the (non-binding) 2020 targets will probably reach 17% (maybe as high as 18–19%) by 2020, mostly thanks to the impacts of the financial crisis—and will miss the 20% target by 1–2%. In 2014, the EU countries agreed on a new energy efficiency target post-2020 of 27% (or greater) by 2030.
19. For more information on the EU ETS, see European Commission, "The EU Emissions Trading System (EU ETS), Climate Action", accessed 31 August 2016, http://ec.europa.eu/clima/policies/ets/index_en.htm (European Commission 2016).
20. This would depend on various factors including plant efficiencies, the type of plants in the market (combined cycles, combined heat and power, heat plants), the national mix, the available capacity and interconnections with other markets, and any measures affecting the relationship between these prices—such as the carbon price floor in the UK for instance. For more information on the gas/coal/EU ETS price relationship and additional details at the national level in Europe, see Anouk Honoré, *The Outlook for Natural Gas Demand in Europe*.
21. With highly efficient gas plants and less efficient coal plants, switching may happen at a much lower carbon price (about €/5t CO_2). Source: Anouk Honoré, "Demand production vs demand destruction".
22. See Anne-Sophie Corbeau and David Ledesma, *LNG market in transition: the great reconfiguration* (Oxford: Oxford University Press, 2016) for further details (Corbeau and Ledesma 2016).
23. For more information on the UK carbon price floor, see UK Government. "Carbon price floor: reform". Business tax—policy paper.

Accessed August 31, 2016. https://www.gov.uk/government/publications/carbon-price-floor-reform (UK Government 2016).

24. Anouk Honoré, "Looking further ahead—What is the outlook for European Gas from 2020–2030?", (presentation at the Platts Conference, Dusseldorf, 28 September 2016) (Honoré 2016).
25. Renewable generation is dispatched first, reducing the need for other types of generation.
26. About 320 GW are older than 30 years, 60% of which are fossil fuelled (mostly coal and oil), but there are (in the UK for instance) also some old nuclear plants that may shutdown. Source: Anouk Honoré, *The Outlook for Natural Gas Demand in Europe*.
27. Data source: author's estimates. Source: Anouk Honoré, "Looking further ahead—What is the outlook for European Gas from 2020–2030?". For more details on the LCPD, see European Commission. "Large Combustion Plant Directive". Accessed August 31, 2016. http://ec.europa.eu/environment/archives/industry/stationary/lcp/implementation.htm (European Commission 2016).
28. If the plants are opted out, they will be allowed to run a maximum of 17,500 h between 2016 and 2023 without complying with the new emission limit values, and will then need to be retired.
29. For more details on the IED, see European Commission. "The Industrial Emissions Directive". Accessed 31 August 2016. http://ec.europa.eu/environment/industry/stationary/ied/legislation.htm (European Commission 2016).
30. Author's estimates, see Anouk Honoré, "Demand production vs demand destruction".
31. A phase-out has been decided in Germany (2022), Belgium (2025) and Spain (2028). The role of nuclear is expected to decrease from about 75–50% in the generation mix in France by 2025.
32. Several important gas markets such as Italy, Turkey and Austria do not have nuclear in their energy mix, while some countries (Germany by 2022, Belgium by 2025, Spain in 2028 and Switzerland in 2035) have decided to phase out nuclear. In other countries (for example the UK), plants will be closed, having reached the end of their operating lives. The use of existing plants may also be curtailed following political decisions—such as the position in France where the share of nuclear production in total power generation should be reduced to 50% by 2025 compared to about 75% in 2014.

33. There are only four new reactors under construction in Europe: one in Finland, one in France (both are EPRs of 1600 MW which are experiencing budget and time overruns) and two in Slovakia (each 440 MW). Several countries have expressed interest in building new reactors (for instance the UK, Netherlands, and Sweden) or in introducing nuclear in their mix (Poland, Turkey). However, due to construction lead times, no new reactors (apart from those already under construction) will be operational prior to 2020.
34. This is difficult to estimate as some plants formerly declared mothballed were only shut down for the summer months; others only mothballed part of their total capacity, and some mothballed plants will reopen (such as the SSE Keadby gas-fired power station in the UK for instance).
35. Uncertainty on future load factors and revenues means investment decisions are more difficult; there will be no new conventional thermal plant while there is a merchant risk and zero long-term price visibility.
36. See Anouk Honoré, *The Outlook for Natural Gas Demand in Europe* for details on assumptions and country-by-country scenarios.
37. Anouk Honoré, "Looking further ahead—What is the outlook for European Gas from 2020–2030?".
38. Anouk Honoré, "Looking further ahead—What is the outlook for European Gas from 2020–2030?".
39. Anouk Honoré, "Demand production vs demand destruction".
40. See Anouk Honoré, *The Outlook for Natural Gas Demand in Europe* for more information.
41. For more information, see Réseau de Transport d'Électricité. Accessed August 31, 2016. http://www.rte-france.com/en (Réseau de Transport d'Électricité 2016)
42. European Commission. "renewable energy Action Plans". Accessed 31 August 2016. http://ec.europa.eu/energy/renewables/action_plan_en.htm (European Commission 2016).
43. Anouk Honoré, "Demand production vs demand destruction".
44. For more information, see Christopher Le Fevre, *The prospects for natural gas as a transport fuel in Europe,* NG84 (Oxford: Oxford Institute for Energy Studies, 2014) (Le Fevre 2014).
45. Anouk Honoré, "Looking further ahead—What is the outlook for European Gas from 2020–2030?".

46. Anouk Honoré, "Looking further ahead—What is the outlook for European Gas from 2020–2030?".
47. The main assumptions behind this are no additional power demand between 2015 and 2030 thanks to energy efficiency and energy savings and renewable targets are met.
48. While the region is on the road to the energy transition to a low carbon economy, only the consequences of existing or future policies or measures that can be reasonably expected are considered. Other assumptions involve primarily economic growth, market structure, the competitiveness of the European industry, the competitiveness of gas versus coal in the power sector, available generation capacity and the evolving mix. It is not easy to know which one(s) will be the most important and it has been a different story in each of the various markets. This method created annual scenarios for each sector in each market, and this patchwork was then combined to create a bottom-up regional scenario to the 2030 horizon.

Bibliography

Corbeau, Anne-Sophie, and David Ledesma. 2016. *LNG Market in Transition: The Great Reconfiguration*. Oxford: Oxford University Press.

European Commission. 2016a. The EU Emissions Trading System (EU ETS), Climate Action. http://ec.europa.eu/clima/policies/ets/index_en.htm Accessed 31 Aug 2016.

European Commission. 2016b. Large Combustion Plant Directive. http://ec.europa.eu/environment/archives/industry/stationary/lcp/implementation.htm. Accessed 31 Aug 2016.

European Commission. 2016c. The Industrial Emissions Directive. http://ec.europa.eu/environment/industry/stationary/ied/legislation.htm. Accessed 31 Aug 2016.

European Commission. 2016d. Renewable Energy Action Plans. http://ec.europa.eu/energy/renewables/action_plan_en.htm. Accessed 31 Aug 2016.

European Union. 2009. Directive 2009/28/EC, Promotion of the Use of Energy From Renewable Sources. 2009. http://eur-lex.europa.eu/legal-content/EN/TXT/PDF/?uri=CELEX:32009L0028&from=en. Accessed 5 Dec 2016.

European Wind Energy Association. 2016. *Wind in Power, 2015 European Statistics*. 2016. http://www.ewea.org. Accessed 31 Aug 2016.

Honoré, Anouk. 2010. *European Natural Gas Demand, Supply and Pricing, Cycles, Seasons and the Impact of LNG Price Arbitrage*. Oxford: Oxford University Press.

Honoré, Anouk. 2014. *The Outlook for Natural Gas Demand in Europe*. NG87. Oxford: Oxford Institute for Energy Studies, 2014. https://www.oxfordenergy.org/wpcms/wp-content/uploads/2014/06/NG-87.pdf. Accessed 31 Aug 2016.

Honoré, Anouk. 2016. Demand Production vs Demand Destruction. Presentation at the Flame Conference, Stream D: Outlook from Gas to Power, Amsterdam, May 11.

Honoré, Anouk. 2016. Looking Further Ahead—What is the Outlook for European Gas from 2020–2030? Presentation at the Platts Conference, Dusseldorf, September 28.

International Energy Agency. *Electricity Information*. Several editions. Paris: OECD.

International Energy Agency. 2016. *Natural Gas Information 2016 Edition*. Paris: OECD.

International Energy Agency. 2002. *World Energy Outlook 2002*. Paris: OECD.

International Energy Agency. 2010. *World Energy Outlook 2010*. Paris: OECD.

International Energy Agency. 2015. *World Energy Outlook 2015*. Paris: OECD.

International Energy Agency. *World Energy Outlook*. Several editions. Paris: OECD.

Le Fevre, Christopher. 2014. *The Prospects for Natural Gas as a Transport Fuel in Europe*. NG84. Oxford: Oxford Institute for Energy Studies. https://www.oxfordenergy.org/wpcms/wp-content/uploads/2014/03/NG-84.pdf. Accessed 31 Aug 2016.

Organisation for Economic Co-Operation and Development. 2016. *Interim Economic Outlook*. Paris: OECD. http://www.oecd.org/economy/oecd-warns-weak-trade-and-financial-distortions-damage-global-growth-prospects.htm. Accessed 31 Aug 2016.

Réseau de Transport d'Électricité. 2016. http://www.rte-france.com/en. Accessed 31 Aug 2016.

UK Government. Carbon Price Floor: Reform. Business tax—Policy Paper. https://www.gov.uk/government/publications/carbon-price-floor-reform. Accessed 31 Aug 2016.

Author Biography

Anouk Honoré joined the Oxford Institute for Energy Studies in 2004. At the OIES she focuses on the European and South American natural gas markets, building scenarios on demand and supply balances in various countries. She focuses on gas issues with particular emphasis on market fundamentals and policies. Her research also covers power generation. Before joining the OIES, she worked at the International Energy Agency where she focused on China, Latin America and later the IEA members' gas markets.

4

Impact of German "Energiewende" on Gas: Responding to Marginalisation in a Politicised Environment

Ludwig Moehring

4.1 Setting the Context

4.1.1 Climate Protection as a Major Disruption for Modern Energy Policy and Global Economies

The Paris Climate Conference at the end of 2015 (COP21) has a good chance to be considered a turning point in modern energy policy. It appears that a huge majority of countries supports a rigid and rapid global reduction of greenhouse gas (GHG)[1]—aiming to hold "the increase in the global average temperature to well below 2 °C above pre-industrial levels and to pursue efforts to limit the temperature increase to 1.5 °C above pre-industrial levels" (Art. 2, 1. (a) of the Paris Agreement)—and the signing countries are aware that the current commitments of GHG reduction will not be sufficient to meet these targets,

L. Moehring (✉)
Wingas, Kassel, Germany
e-mail: ludwig.moehring@wingas.de

M. Hafner and S. Tagliapietra (eds.), *The European Gas Markets*,
DOI 10.1007/978-3-319-55801-1_4

which are supposed to provide "a bridge between today's policies and climate-neutrality before the end of the century".[2]

In early September 2016, also China and the USA, the largest emitters of GHG emissions worldwide announced to ratify the Paris Agreement, which is considered a massive step in the battle against global warming. Of course, it remains to be seen whether the states who (will) have ratified the so-called Paris Agreement will actually provide the funds committed and deliver on their promises. Yet, the broad acknowledgement to reduce carbon emissions will in any case put severe pressure on governments across the globe to actively contribute to GHG reduction—however challenging the targets for global GHG reduction may be. And, above all, there is a very relevant monetary dimension to all of this: will the signing states remain committed to their GHG reduction effort, if they are confronted with a short- or medium-term negative macroeconomic impact?

Nevertheless, the global climate protection efforts will certainly put a lot of stress on the global oil and gas markets (demand and supply) which, in turn, will impact many economies who will have to find their respective places in a new climate-driven equilibrium. Next to the digitalisation of consumer and industrial lives, the march towards the low-carbon world can be seen as one of the major disruptions of our time.

The reduction of GHG will be ever more challenging given the increasing global demand for energy—this, in brief, resulting from the major growth of global population as well as increasing industrial production and growing mobility in particular in China and India. Even with optimistic assumptions of energy efficiency, it is hard to believe how global energy consumption might go down compared to current projections with an expected growth of energy consumption of up to 20% within the next 25 years (See Fig. 4.1, below).

However, growth projections for global gas demand may indeed come under pressure. In particular, in Europe we have seen stable if not receding gas demand (corrected for weather) in recent years. And whilst natural gas had been considered a global fuel of choice for many years and the IEA stipulated the "Golden Age of Gas" just a few years ago, given the effort to reduce CO_2 this may have been a premature conclusion. Many believe that carbon fuels may not be needed anymore

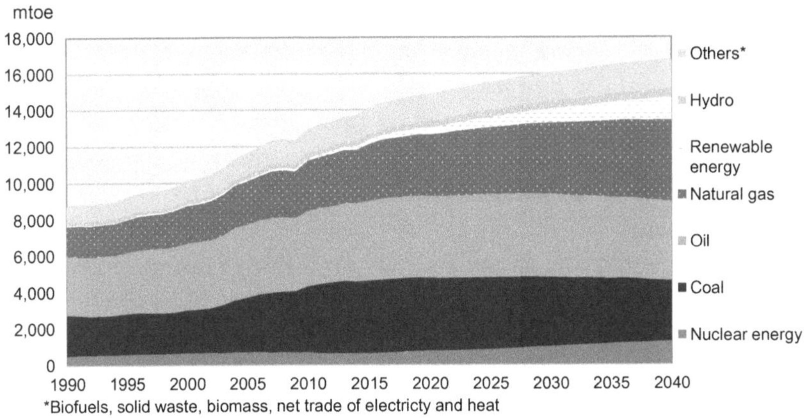

Fig. 4.1 Global Energy Demand 1990–2040. *Source* author's elaboration on IHS Markit, Rivalry outlook (2016)

at some stage of this century and feel encouraged through the Paris Agreement that this will indeed be the case.

4.1.2 Germany's Energy Transformation (Energiewende) Has Resulted in a Major Exposure for Gas

At the global level, Germany is the 6th largest emitter of CO_2. And apparently it has understood the challenge on climate change: it perceives itself as a global front runner when it comes to climate protection and the transformation of the German power generation sector: from nuclear and coal to renewables, aiming at reducing CO_2 emissions by 80–95% compared to 1990 and produce 80% of renewable power consumption, all of this by 2050.[3] In such context, policy makers increasingly demand an exit strategy from fossil fuels, which manifests itself not just in the power generation sector, but also in the heating and mobility sector: ideas have been floated at the political level to, e.g. disallow (or make unattractive) new gas- or oil-fired heating systems as of 2030 or even earlier, or to grow electric mobility through massive public subsidies. This approach on the political side receives a lot of support by leading media as well as other pressure groups consisting of, e.g. NGOs as much as of the two major Christian churches who have shown a strong support of the *Energiewende*.[4]

Natural gas is at best tolerated as unavoidable in such context: to support power generation when fluctuating renewable power is not available in case of no wind and/or no sun, or as a phasing out fuel for heating purposes.

Coal on the other hand can still rely on influential supporters, which became very visible in the process towards the so-called coal compromise ("Kohlekompromiss") in 2015, which prevented the envisaged levy on coal-fired power plants in an attempt to reduce GHG emissions in the power sector by an additional 22 mln tons/y—against the background of heavy protests related to losing (!) the lignite production in Germany.

The usual line of argumentation around CO_2 advantages of gas or—in the mobility sector—also around NOX and particles goes largely unnoticed and gets ignored in a politically driven public discussion, which increasingly pursues also the vision of a manoeuvering all areas of energy consumption into a renewable world. A very good reference point for such considerations is the Climate Action Plan 2050 ("Klimaschutzplan 2050") of the Ministry of Environment, which was released in a first draft in June 2016, with a second draft in early September 2016 and ultimately agreed within the government in November 2016—following many discussions with other stakeholders within and outside the government.[5] This energy landscape would be based on massively increasing use of electricity also for transport, housing and industry[6] (which in the first two drafts was called "electrification strategy"; this term was no longer used in the final version without the underlying strategy being adjusted). Such approach would not just aim at the GHG emission-free society by mid of this century, it would also rely on an "all-electric" energy demand.[7] In such world, also gas is supposed to be put on the sideline sooner rather than later despite its economic and environmental advantages compared to other options for CO_2 reduction.

As much as it is useful to develop and pursue radical visionary thinking on the political side how the low-carbon world can be achieved;

at the same time, technical feasibility, commercial realities and related implementation risks cannot be ignored: an "all-electric" energy demand would require a multiple of electricity production capacity and infrastructure all the way to the consumer, which looks hardly possible—regardless of the huge cost this would imply. Hence, the gas industry might be tempted to just sit and wait for such realities to sink in.

However, waiting for such hard truths to come to the surface of political reality will not suffice, the reason being that political intervention into the direction of an "all-electric world" will take away the level playing field and reduce the potential for gas already in the near future. This is aggravated by the financial capacity of the (currently) very strong German economy, which leads political decision takers to believe that it is possible to finance the *Energiewende* and to accept cost levels which may not be acceptable to many other economies. This results in major risks for the role of gas, as the transition into the low-carbon world, which according to many experts would strengthen the role of gas (in any case, compared to other fossil fuels), may take a different course: through an acceleration of introducing (renewable) electricity into all sectors of energy consumption, which, as a consequence, would accelerate the reduction path of gas.

Such context will be explained further below—building upon the current status of the *Energiewende*—and will be followed by an attempt to (1) describe the principles for a successful and affordable transition into the low-carbon world combined with (2) how the gas industry needs to change its communication approach to relevant stakeholders. The current political barrier for gas needs to be broken, in the best interest not just of the gas industry (which alone would not be convincing), but in the best interest of an effective and efficient low-carbon society that remains competitive in the global markets without putting welfare of its people at risk.

In order to keep the story as crisp as possible, the author had to make some compromises regarding the description of certain quite complex developments, which would certainly deserve more attention. Nonetheless, the key messages have hopefully been sufficiently explained.

4.2 German Energiewende: Delivery on Promises to Date

4.2.1 Background/How It All Started

Climate protection has been put seriously on the map of German policy makers at the end of the last century following the Rio climate conference. An ambitious target for reduction of GHG emissions (40% reduction by 2020 compared to 1990) was set, combined with the establishment of the Renewable Energy Act ("EEG") providing substantial financial support for the introduction of wind and solar energy. This was combined with a first attempt to phase out the 21 nuclear power plants in Germany by 2022 and resulted in an agreement to that effect between the government, consisting of Social Democrats and the Green Party, and the relevant energy companies in 2001, the Nuclear Consensus ("Atomkonsens").[8]

For years renewable energy as well as climate protection did not see a major boost, which was to change once the conservative/liberal government had come to power in 2009. One year later, the so-called energy concept was developed: major parts of the nuclear exit strategy were made undone and permit durations extended—which was combined with the establishment of the *Energiewende*: ambitious plans to achieve the GHG reduction targets were put in place and were supposed to provide a political platform for accepting the extension of the nuclear power plans: now with permits until 2036. Interestingly enough, when the energy landscape of the *Energiewende* was sketched out, gas did not feature—which was explained in the political arena at the time stating that gas was an obvious choice for the modern energy world and did not need to be mentioned. There were doubts about that statement already there and then.

What looked like a very clever political move, i.e. to combine growth in renewable power with prolongation of nuclear power, turned out to be highly questionable in wide parts of the German society, where major opposition grew against the prolongation of nuclear power generation. When the Fukushima disaster occurred, the German government seized the opportunity for a massive turnaround: the complete exit out of nuclear by 2022 was decided (including the immediate shutdown of seven nuclear power plants—thereby referring to unacceptable risks of nuclear power plants which had become obvious to the government in the Fukushima disaster). In parallel, the transformation of the energy sector, in particular through growth of renewable energy, was confirmed.

The political dynamite of such decisions was anticipated, and in order to ensure wide acceptance in the German society, the so-called Ethikkommission (Ethics Commission) was established and relevant groups of the society, including various sectors of industries, unions, churches, were requested to develop a plan how the *Energiewende* could be ensured as a sustainable effort of the entire society. When the report of the Ethics Commission was released in May 2011, the 2020 targets for the reduction of GHG emissions in all sectors of energy consumption were emphasised, and gas finally got back on the agenda, in particular as a very important tool also to reduce CO_2 in the power sector.[9] Germany apparently took the reduction of GHG emissions very seriously.

4.2.2 Implementation of "Energiewende" so far

With the experience of 5 years of *Energiewende*, it is worthwhile to assess the decisions taken to transform the energy sector, the drivers for these decisions and their impact. It reveals that a strong and costly emphasis on moving the power generation sector into renewable energy did not achieve a lot on climate protection. It also reveals that, interestingly enough, this is not considered as a real issue, but goes largely unnoticed in the public domain.

4.2.2.1 The Targets (in Brief)

It was obvious that the combination of exiting nuclear and developing an energy landscape which would very much rely on renewable energy would pose major challenges on the German economy if not society as a whole.

At the political level, climate protection played a key role. With the aim to create a largely renewable-based economy by 2050, a reduction path for GHG emissions has been developed: by 2050, GHG emissions shall be reduced by 80–95% compared to 1990, and by 2020, a reduction target of 40%, which had been established by the previous government, was confirmed again. This was combined with, e.g. (1) the target to reduce primary energy consumption by 50% by 2050 (compared to 2008) and (2) a 60% reduction of total energy consumption. In combination, this meant nothing but the complete reset of the German energy sector by mid of this century—see details in Fig. 4.2. What was left open was how to get there (Fig. 4.3).

Whilst the target to reduce GHG emissions by 40% compared to 1990 appears to be very ambitious compared to EU targets of 20%, it needs to be noted that the reduction of German GHG emissions was "boosted" by the overhaul of the East German economy following the German reunification post 1990, which contributed substantially to a reduction of around 20% already by 2000 (see also Fig. 4.4 below).

	2014	2020	2050
Greenhouse gas emissions (compared to 1990)	-27 %	min -40 %	min -80 to 95 %
Increase in share of renewable energy in final energy consumption			
Share in gross final energy consumption	13.5 %	18 %	60 %
Share in gross power consumption	27.4 %	min 35 %	min 80 %
Share in heat consumption	12 %	14 %	
Share in transport sector	5.6 %		
Reduction of energy consumption and increase in energy efficiency			
Primary energy consumption (vs. 2008)	-8.7 %	-20 % →	-50 %
Final energy productivity (vs. 2008)	1.6 % per a	2.1 % per a	
Gross electricity consumption (vs. 2008)	-4.6 %	-10 % →	-25 %
Primary energy demand buildings (vs. 2008)	-14.8 %	→	~ -80 %
Heat demand buildings (vs. 2008)	-12.4 %	-20 %	
Final energy consumption transport (vs. 2005)	1.7 %	-10 % →	-40 %

Fig. 4.2 *Energiewende* goals set by German Government. *Source* author's elaboration on Federal Ministry for Economic Affairs and Energy (2015a, b)

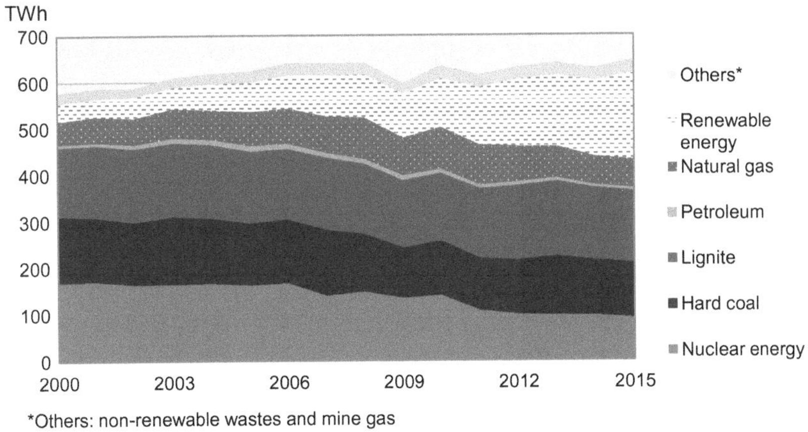

Fig. 4.3 Gross electricity generation in Germany, 2000–2015. *Source* author’s elaboration on AG Energiebilanzen e.V. (2016a, b)

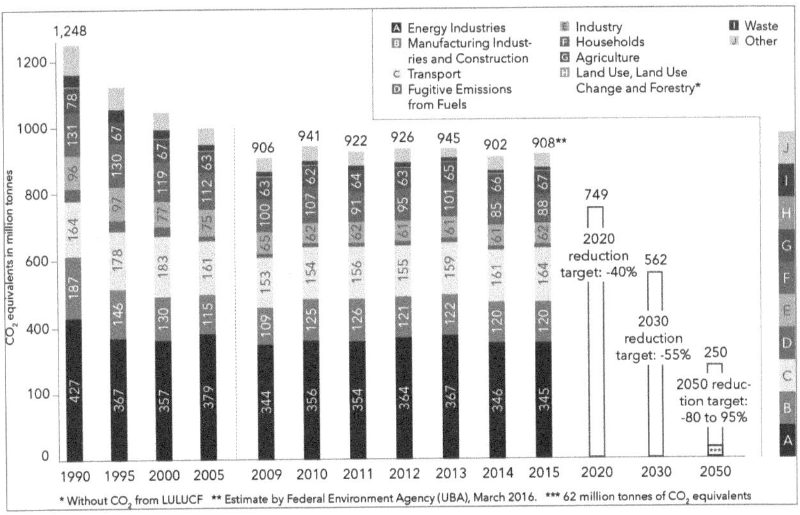

Fig. 4.4 German CHG emissions 1990–2015. *Source* author’s elaboration on Federal Environment Agency/UBA (2016)

Next to the climate targets, which are at the core of the political agenda[10] another pillar for changing the energy landscape is obviously a strong build of renewable energy. In order to accelerate such development, a subsidy scheme was introduced that provided the investors with attractive incentives under the Renewable Energy Act ("EEG"). Finally, energy efficiency was confirmed as a very important tool to reach climate targets.

4.2.2.2 Changes to Energy Landscape and (Limited) Impact on Climate Goals

Since these principal decisions were taken, some fundamental adjustments to the German electricity landscape have taken place, whilst in other energy-related sectors only limited progress was made. A very short overview:

1. Electricity sector
   The process to exit from nuclear energy by 2022 was started, and since then, from the 17 remaining nuclear power plants, eight were shut down in 2011 and one in 2015. The incentive scheme for renewable energy resulted in massive investment in wind and solar. By the end of 2015, renewable energy accounted for 30% of the overall electricity production.
   Interestingly enough, different from many expectations the remaining conventional electricity production came from coal rather than gas—the reason being very simple: coal-fired power generation has been cheaper than gas-fired power generation. Looking at the reasons for the differences in cost, two elements come into play: (hard-) coal prices went down due to global supply/demand developments and the European Emissions Trading Scheme did not provide for CO_2 prices which would incentivise gas-fired power generation over coal. Finally, the German lignite-based power generation (relying on domestic production) remained competitive due to very low marginal cost.

As a result of all this in 2015, the German power generation mix resulted in 18.3% hard coal, 24.0% lignite, whilst gas accounted for only 9.4% (compared to 14.1% in 2010).[11]
This revamp of the German power generation had a major impact on the market conditions. Actually, wholesale (base) prices have gone down by one-third from around 45 €/MWh to around 30 €/MWh between 2010 and 2015—next to increasing renewable energy this was mainly a result of decreasing global coal prices. The prices for households have gone up nonetheless by 15%, largely because the levy for renewable energy ("EEG Umlage"), which was imposed in order to finance the growth in renewable energy, has tripled from 20.5 €/MWh to 61.7 €/MWh between 2010 and 2015 and is now around double the wholesale price.

2. Heating sector
The heating sector has not yet received a lot of political attention. For newly built houses, additional rules were established for improved insulation and adjustments were made for heating systems, which are now required to provide for some renewable elements (e.g. via heat pumps, wood pellets or combining gas with solar). However, in an economy with only around 1% newly built houses per year, the existing buildings provide the real issue for reducing GHG emissions—and here, very limited effort at the political level was visible. There is a mix of reasons: on the one hand, this is resulting from the focus on the electricity sector; on the other hand, though, this is also due to not having an obvious solution, which would provide for material CO_2 reduction at acceptable cost. As a consequence of high cost involved, acceptance of the many millions of house owners and tenants—who also represent the electorate (sic!)—policy makers were hesitant to address this important area of GHG emissions.
3. Transport sector
In the transport sector, climate protection has also turned out to be non-trivial. With growing traffic, even more efficient cars are struggling to make up for the increase of traffic as a whole. Comparing the CO_2 emissions between 2010 and 2015, no major changes can be seen, actually a slight increase. Between 1990 and 2015, CO_2 emissions have not decreased.[12]

In the view of many, there are very exciting developments around electric mobility which get a lot of attention at all fronts. Obviously, up until now they have been immaterial in terms of scale, nor have they made a difference in terms of CO_2 emissions: driving an electric car ultimately just results in a transfer, the emissions from the car to the power plant. This is relevant for the city affected, but not for the climate impact as a whole. Given not just the climate challenges, but also environmental issues in the mobility sector (NOX, particles), it is surprising that so little has happened in terms of seriously tackling this issue also at the political front.

4. Impact on GHG emissions
   Whilst the cost of rebuilding the power generation landscape has now reached a level of more than €25bln/a alone for the renewable energy levy mentioned above, the overall reduction of GHG emissions was basically immaterial.
   Figure 4.4 illustrates that since 2010 hardly any progress has been made regarding the challenging climate protection targets. The reasons for this are easily explained:

- In the electricity sector, increasing coal-fired power generation has made up for reduction of GHG emissions resulting from increasing renewable energy, this aggravated (from a CO_2 perspective) by closing down nuclear power plants.
- As described above, the heating market has not seen any material change as is the case for the transport sector, where efficiency gains are made up by ever increasing traffic.

4.2.2.3 Role of Natural Gas

Whilst natural gas had been promoted as an "ideal partner of the renewable energy", the reality is more complex, to put in mildly:

a) The share of natural gas in the power generation sector has gone down both in absolute and in relative terms. Back in 2010 around

14% of German electricity production was gas fired, whilst in 2015 the share had gone down by 40% to only 9%.[13] Combined with a bad economic outlook, many companies closed down or mothballed their gas-fired power plants, as they expected not to earn their short-term marginal cost at wholesale prices of 30 €/MWh and below.

b) In the heating sector, which accounts for one-third of the German CO_2 emissions, natural gas has been able to keep its important role. In 2015, 49.3% of all buildings relied on gas-fired heating and gas also accounted for approximately 76.5% of new heating systems built into German houses. This compares very much with the numbers of 2010: 49.0% of all houses and 72.9% of new appliances were gas fired.[14]

c) In the transport sector, gas does not play a serious role (yet). Less than 0.5% of German cars are gas (CNG) fired, and also, the recent developments around LNG to be used in trucks and ships have not yet had an impact in Germany.

Where does that leave gas: its relative role has diminished in the electricity sector, which goes against previous plans and expectations, where the electricity generation was considered a major growth area. In the heating sector, the relative role for gas was maintained in a slowly shrinking market, whilst gas played no visible role in the transport sector. This is not a growth story, and today's position is under threat, which will be elaborated upon further below. Expectations that gas could play a major role on the journey towards a low-carbon world have clearly not been met, as can be seen looking at the development of the primary energy consumption in Germany described in Fig. 4.5.

Also against the background of climate protection, natural gas is not meeting its potential:

a) CO_2 emissions in the power sector could be reduced further, if gas would increasingly replace coal—instead, existing modern gas-fired power plants remain idle.

b) In the heating sector, modern gas technology provides a lot of potential to reduce CO_2 emissions in the existing buildings and at considerably lower cost than many efficiency efforts around additional insulation, etc.

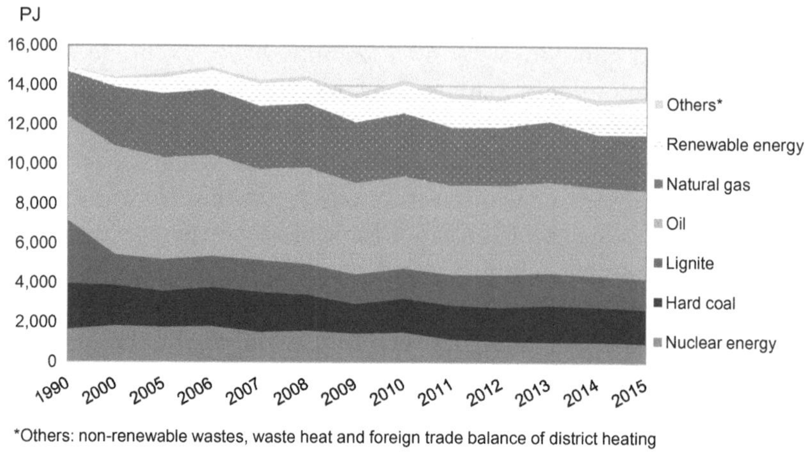

Fig. 4.5 Primary Energy consumption of Germany 1990–2015. *Source* author's elaboration on AG Energiebilanzen e.V. (2016b)

c) In the transport sector, natural gas would not only provide potential to reduce CO_2 (around 25% less than petrol, and this number would go up, if renewable methane is being used). It would also help addressing today's major environmental issues around NOX and particles: Germany is exposed to EU proceedings because the emission limits are not met in more than 20 major cities. Gas could help with technology which is readily available and comes at affordable prices: CNG cars have basically zero particles, and NOX is reduced by more than 80%.

4.2.2.4 Climate Goals: What Went Wrong?

As described above, GHG emissions have not been reduced in any materiality since the beginning of the *Energiewende* in 2010/2011—despite major investments and a lot of political emphasis in Germany, which claims to be a "global role model" for meeting the climate challenges of this century.

The reasons for this failure are obvious: whilst the climate target was maintained, the actions taken had actually only limited impact on

GHG emissions. To cut it short, targets and actions taken did not sufficiently correlate.

1. The political focus was clearly only on the electricity sector and the introduction of renewable energy. At the same time, the increase of coal-fired power generation was "accepted", which comes as a surprise looking at the clear statements regarding climate protection. This policy is not centred around GHG emissions, but around manoeuvering the power generation landscape into renewable energy—which, as the recent history proves, is not necessarily the same, if the power generation set-up is not holistically addressed with a clear target to reduce GHG emissions.
2. The heating and transport sectors, which account for well above 50% of the GHG emissions, have not been in the political focus at all. As a reduction of emissions creates cost in many cases, no progress was being made.

To pinpoint this aspect: whilst climate protection is claimed to be at the heart of the *Energiewende*, it is not at the heart of the sub-targets (and actions!) for its implementation. GHG emissions are only a side aspect at this stage. This is surprising, as (1) the GHG emissions target for 2020 has been set and repeated at the highest political level[15] and (2) climate protection can't wait: if GHG emissions continue unabated, the global budget for the twenty-first century may be exhausted before the middle of the century already.[16] Ranked No. 6 of global GHG emitters, Germany is not fulfilling this self-imposed leading role in climate protection, as it is not sufficiently taking goal-oriented action to achieve the necessary immediate reduction of GHG emissions.[17]

4.2.2.5 Climate Protection and Political Actions—with Missing Links

Interestingly enough, the non-progress on GHG targets goes largely unnoticed. Policy makers and media refer to the major achievements on reduction of GHG emissions since 1990, thereby ignoring that, as described above, these reductions had hardly anything to do with the Energiewende. In the political discussion, it is widely accepted, also within the media sector, that the current efforts do not meet the 2020 GHG targets. This is amazing for two reasons: (1) it is uncontested that climate change requires immediate action,[18] (2) the huge macroeconomic cost related to the current efforts should require tangible results on climate protection. Also, the public support of the Energiewende is still there. But will this continue to be the case?

Trying to understand why this phenomenon, it is the author's view that Germany and her people apparently have convinced themselves that the current move into renewables provides the best solution for the journey into the low-carbon world. Questions around efficiency and effectiveness around climate goals are not being raised or remain ignored. As much as Germany has a history of thorough analysis and implementation, in her current approach to climate protection Germany does not play to that strength. It does not even take her 2020 emissions targets seriously, although they are at the core of the *Energiewende*. Instead, political focus has shifted to setting ambitious targets for 2050 (and beyond) and describing that vision—this is apparently a lot easier to communicate and to gain trust from the electorate.

What is widely missing is the acknowledgement that the political efforts need to be measured against the impact on GHG emissions and related cost. This would be a meaningful performance indicator for effective and efficient climate protection. Such approach, of course, would take away the full focus on renewables and shift the attention also directly towards GHG impact. Renewables would no longer be an end in itself, but just one very important instrument amongst others to support the climate protection. As an example: decarbonising the electricity sector cannot simply rely on growing renewable energy, if coal-fired power generation remains accepted for decades to come. The modern energy landscape of the twenty-first century needs to combine renewable and conventional energy sources in an optimised way, oriented at climate impact, security of supply and cost. Looking at the results of the huge efforts in previous years, this is not the case in Germany.

4.3 Outlook: Is Gas Part of the Problem or Part of the Solution?

More recently, the focus on renewables has reached another level. Further growth of renewable power is envisaged, and with that other challenges appear: apart from building the necessary (grid) infrastructure to manage the massive increasing flows of renewables and to bring the power to the place of consumption, the question has come up how to make best use of the electricity produced: even with better infrastructure, increasing renewable power would result in more and more superfluous electricity under current demand and supply structures. As a result, the so-called Sektorkopplung (coupling of energy sectors) has come into play, and it gains more and more political ground: using electricity also in the heating and transport sectors is considered not just as an emergency outlet in times of redundant renewable electricity, but now being considered a visionary and serious option to also rebuild those sectors—and this is to start now.[19] Examples:

a) In the heating sector, electricity would be used not just in electric heat pumps, but also for district heating through "Power to Heat" or by creating renewable methane via "Power to Gas", with a target to replace non-electric heating system.
b) In the transport sector, electrified cars would take over the role of the combustion engine. Germany has not only a stretched target to achieve one million cars by 2020 (with less than 50,000 in 2016)—compared to around 45 million cars today.

As a result, in such "all-electric world" with very limited need for fossil fuels, natural gas is considered as part of the (fossil) problem and not as part of the solution. Needless to say that oil and coal would share the same fate.

This approach puts a question mark also to the expected development of the respective shares of primary energy by 2050 as reflected in the so-called reference scenario, which was developed for the German Ministry of Economics in 2014 (see Fig. 4.6)[20]. Such scenario referred

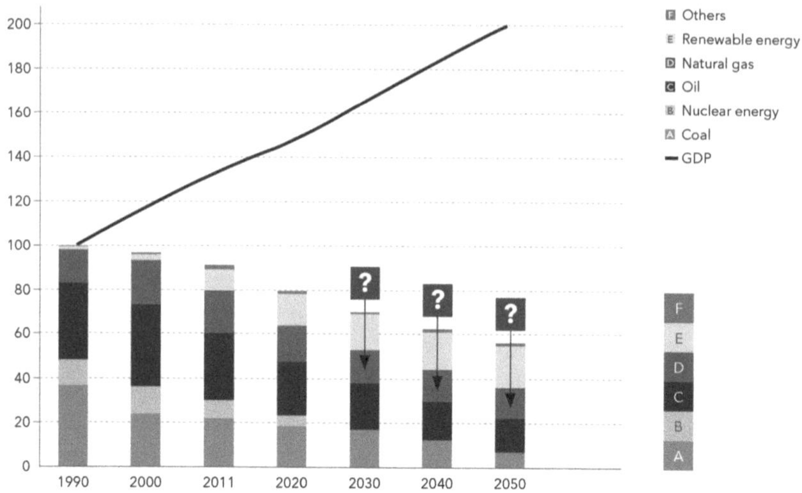

Fig. 4.6 Primary Energy Consumption (Index 1990 = 100). *Source* author's elaboration on Prognos/EWI/GWS (2014a)

to a major decrease of hard coal and lignite whilst gas would gradually go down, but would still substantially contribute to the overall energy demand. The author would challenge whether such scenario is still the basis for the current discussions mentioned above: looking at an accelerated growth of renewable energy, enhanced electrification of heating and mobility sector—and this combined with the extension of permits for coal-fired power plants[21]: there would not be a lot of space for gas left.

4.3.1 Political Drivers for Future Implementation of the Energiewende

As described above, German energy policy is largely focused on establishing renewable energy and the rebuilding of all sectors of energy consumption—this is considered to pave the way towards the aspired reduction of GHG emissions by more than 80% by 2050.[22] However, such approach appears to be simplistic and also dangerous: its focus on the aspired electric endgame runs the risk to block the view to find more efficient ways to decarbonise society. Actually, it looks like a major bet: neither is there a proof of concept, nor a description of how the enormous additional amount of electricity required could be generated (and stored and transported to the place of demand!), nor has the cost been estimated or compared to alternative solutions. Scenarios are being used to justify the desired cause of action—this is not enough.

Nevertheless, the political debate about the future development is centred around this vision. Discussions how to move the heating market and the transport sector into the low-carbon world are increasingly dominated by proposals to support the electrification of these sectors. Reducing GHG emissions through modern and affordable gas (or oil) technology is under pressure, as this is considered as locking in a technology which would not be necessary in 20 years. A recent plan of the German Ministry of the Environment suggested to ban—or as was stated later: to have renewable heating considerably more attractive than fossil heating—any fossil heating system in new buildings as of 2030.[23] With nearly 15 years lead time, this is an amazing judgement on how to shape the heating sector without clarity or proof on future developments and without having assessed the different options.

When it comes to taxation of energy, there is very visibly sentiment to increase taxes on fossil fuels—also to fund the subsidies for renewables. Obviously, this would be a "double-whammy" for gas: (1) deterioration of gas compared to renewables and (2) the subsidy scheme for renewables looks a lot better, which would allow to camouflage the overall cost of the build-up of renewables to the German economy.

And it is not just proposals, political actions have already been taken to the benefit of electric solutions:

a) In the heating sector, electric heat pumps are legally privileged over gas-fired solutions in newly built houses through a beneficial "primary energy factor" which helps heat pumps achieving efficiency requirements, although—compared to gas-fired appliances—heat pumps are less efficient from a CO_2 perspective and more expensive for the consumer.

b) In the transport sector, electrified cars—which will for sure play a relevant role at some stage in the next decade—get an enormous political attention and supporting measures are generated, including the recent subsidy scheme by the German government.

Why is the German *Energiewende* approach so rigid in its focus to accelerate the build-up of renewables—and at the same time (1) not looking for alternative more cost-efficient solutions and (2) turning a blind eye to the limited impact on the reduction of GHG emissions? Answering this must be speculative: it could be a deep and somewhat ideological interest to become independent from resource imports, combined with the strong belief in the ability to deliver an energy landscape that can (largely) rely on electricity. Whether the claim to become a role model for the world is still serious appears to be questionable. In any case, the rigidity is a consequence of not acknowledging the risks related to feasibility, cost, workplaces and welfare coming with the current approach.

4.3.2 Consequences for Natural Gas

For natural gas, which had been the fuel of choice for decades, things have changed: ten years back gas was key in the heating market and seen as the natural follower on nuclear power generation; back in 2010, it was still seen as the natural partner of the renewable sector. Now, there is a serious threat that gas will be increasingly sidelined by the political agenda mentioned above.[24]

As a consequence, already today questions are being raised whether further investment in gas infrastructure should be pursued, be it for new Pipelines into the European market,[25] be it for gas-fired heating systems at the customer end, which as mentioned above is targeted to be banned (or made unattractive) as of 2030.[26] For the gas industry, such discussions create already now negative "noise" and move the entire product into an unfortunate corner, which is considered as a pure necessary evil and not as a contributor to climate protection. In a world where energy use—from an environmental and political perspective—has become increasingly relevant and also emotionally loaded, the gas industry has to watch this space. Securing outlet for gas remains a core challenge in a world where as a result of improving upstream technology gas supplies will not be an issue for the foreseeable future.

4.3.3 Case for Change? The Gas Industry Can't Wait for Realities Sinking in

Coming back to the scenarios mentioned above, international experts of the scenery may recognise that such scenarios are based on bold assumptions, e.g. regarding (1) the ability to scale up renewable power also for the heating and transport sectors whilst ensuring security of supply, which will require major technical progress for storing power or (2) the ability to reduce overall energy consumption through efficiency gains and consumer behaviour. They may also refer to missing cost calculations and the lack of comprehensive assessment of the economic impact of the envisioned transformation.

Being close to economic and technical realities and with some skepticism towards the assumptions used in these scenarios, one might just claim that these scenarios are not realistic and, hence, conclude the realities will sink in and keep the gas industry on track.

This is a line of argumentation, which is based on the advantages of gas as a key player on the road towards an affordable low-carbon world: the integration of renewable energy and natural (or renewable) gas provides the path. It also refers to market forces, which would reward the value of gas. Examples: if in Germany currently three out of four new heating systems are gas fired, customers obviously know what to do. In the power generation sector, a move in prices might push coal out and result in higher gas usage—and downward gas price movements in 2016 seem to indicate that gas is becoming more competitive with coal again. Last but not least, even in the transport sector the huge environmental and climate issues around diesel should create space for gas-fired (CNG and LNG) technologies, which are readily available and scalable at short notice.

However, relying on market forces will not be good enough. Why is this? Relying on market forces implies that market forces are working and not hampered by third-party intervention. The experience of five years of *Energiewende* and the related approach of the political sector have proven the somewhat unambiguous and unconditional preference of supporting renewable energy over other solutions. This may soon be complemented by the preferential treatment of electric solutions over other alternatives—with a vision to establish an energy landscape which is predominantly relying on electricity.

Pursuing this vision without measuring it against other alternatives (on the basis of technical feasibility and/or cost) puts an end to ideas around maintaining a level playing field and to the belief that market forces will provide balanced results. Some examples[27] of today's political debate for future steps confirm this point:

1. Discussion regarding the increase of taxes on fossil fuels in order to finance renewable energy.
2. Preferential treatment of (electric) heat pumps over gas-fired solutions, including a debate to abolish gas-/oil-fired heating altogether or, at least, make them commercially unattractive.
3. Strong support of electrical transport with purchase subsidies being only one aspect.

In such a business environment, obviously winners have been picked through political intervention. And this approach may also be pursued in the future as long as the aforementioned vision provides the platform for an uncontested way forward. For gas, this results in the risk of being marginalised over time—and this is at an accelerated pace which is not justified against the background of the climate challenges at hand.

Politics picking winners in the energy market is not new, and to be clear, an economy and its determining political bodies are entitled to make choices and to steer developments towards desired solutions—this is their role and this is why they get elected (or not). This was, e.g. the case when nuclear energy was established in the 60s/70s and again, when it was decided to phase it out in 2011—and of course, those decisions were under deep discussion on impact, risks and economic cost.

And the same should apply here: the vision of an economy which is predominantly relying on (renewable) electric supplies leaves many questions open at this stage, as was established above. Feasibility, scalability, cost and economic impact on society – all these topics would require a closer look to understand how climate protection can be accomplished in a way that the future energy landscape is effectively reducing CO_2, the cost incurred is under control compared to alternative solutions and security of supply is ensured.

In particular, the core of the *Energiewende*—CO_2 reduction—has not yielded any visible success, and it may be claimed that relying predominantly on (renewable) electricity is in the end just a big bet, which puts unnecessary risk and uncertainty on the entire economy and its people. However, these topics do not get sufficient attention in the political and public debate, which has many elements of a "post-truth" type of discussion, which is not sufficiently driven by facts, but by agenda and ideology.

To be clear: of course, there is a substantial discussion going on also in the political arena with a view to the future of the *Energiewende*. The recent Climate Action Plan 2050 of the Ministry of Environment[28] has triggered quite some opposition both in the Ministry of Economics and in the Chancellery ("Kanzleramt"). Also, ideas to accelerate the phase-out of coal-fired power generation were adjusted under heavy opposition of the power industry and influential unions.[29] Last but not least, concerns about the competitiveness of the German economy have been raised, e.g. by the federal industry association "BDI".[30] Trade unions are concerned about the negative impact on employment, interestingly enough around lignite production[31]; other potential areas may soon become affected as well, e.g. around the automotive sector.

However, the gas industry also cannot just wait for other stakeholders to address the loose ends of the *Energiewende*. Actually, the author has repeatedly received questions, why the gas industry is so hesitant to address the relevant issues around the *Energiewende*—and these challenges are very legitimate.

4.4 Reshaping the Political Debate: (Communicate) Gas as a Pillar of the Low-Carbon World

As described above, the *Energiewende* needs to be refocused, as much as the development of renewable energy is a meaningful tool for the modern energy landscape in the twenty-first century. It is not an end in itself, though climate protection is at the heart of any "*Energiewende*" in the world.

Given the huge consumption of conventional fuels in globally growing energy markets, effective reduction of GHG requires the optimisation of the energy landscape in an integrated play of renewable and conventional energy. This is not just in order to fulfil the big energy demand, but also to maintain the required high levels of security of (electricity) supply and to ensure that the cost for reducing GHG is kept at acceptable low levels.

Natural gas as the most environmentally friendly conventional fuel has a strong place in such development; this is the more so as the existing gas infrastructure also provides the means for growing renewable methane (e.g. via power-to-gas or biogas) into the system. This is not a call for gas to be picked by policy makers as another winner next to renewables, but a firm statement that reshaping the energy sector requires a focus on effective reduction of GHG emissions—and the key performance indicator for this is the GHG reduction and the cost coming with it. Comparing alternatives of GHG reduction, options with lower cost of GHG reduction should be preferred—unbiased towards any technology, neither towards gas nor oil, not biased towards renewables.

Looking at the current public and political debate in Germany, this parameter is undervalued if not ignored in a political and media environment which is almost exclusively focused on renewable energy. Given the strong focus on renewables, this shift in thinking and approaching the climate challenge will require proactive communication. This communication should not just focus on "gas is better than coal/oil", but it needs to address why the transformation of the energy landscape into a truly low-carbon world will be more successful—and may actually be the only feasible option—if the potential of gas is being utilised.

4.4.1 Become Part of the Paris Agreement Process, not a Victim

The Paris Agreement at the end of 2015 provides a landmark agreement for climate protection. Although many critics refer to unclear commitments and potential reopeners for the undersigning countries, it will turn out to be a game changer for climate protection—provided the global economies find ways to reduce GHG in the most efficient way, so-called climate efficiency.

For gas, this means the opportunity to remain on the energy map as part of the solution within the low-carbon energy environment. Building upon the parameter of climate efficiency gas can make a major contribution to the journey towards the low-carbon world. Paris must have made it very clear that a shift towards GHG efficiency is necessary, which puts gas—amongst others—back into the game also in economies like Germany where the focus so far has been on just growing the renewable sector.

Looking at the goals of the Paris Agreement, one might challenge why natural gas would play a role, if GHG emissions should basically be brought close to zero by the end of the century. However exciting it may be to consider the far end of the spectrum, be it 2100, be it 2050, climate protection can't wait. Projections indicate that the 2° target (even more obvious for the 1.5° target agreed in Paris) remains unachievable, if GHG emissions do not get reduced in the near future.[32] As a consequence, betting on 2050 or 2100 is not enough—we need to act now and we need to act decisively on the reduction of GHG emissions.

Using Germany as an example, an effective reduction of GHG emissions has not been accomplished, since the *Energiewende* was launched in 2010/2011.[33] An approach which focuses on climate efficiency would immediately put new questions on the table in the various sectors, i.e. how to effectively reduce GHG emissions within the entire energy sector (power and heat and transport), thereby comparing the various options also with a view to cost. Based on these parameters, gas solutions would clearly feature a lot higher in all sectors and Germany could make progress also on near time reduction of GHG emissions.

4.4.2 Climate Efficiency as Key Parameter for Transforming the German Energy Landscape: A Storyline

The "gas story" does not start with how gas compares to other fossil fuels and why it is preferential—it has to address why gas is and will remain good for society; this is currently undervalued and requires more emphasis. In this context the disruptions resulting from the climate challenges need to be reflected: society is to be understood as the modern energy-efficient society of the 21st century, which is driven by meeting the climate challenges.

In short key messages to include:

1. Germany is the 6th largest CO_2 emitter in the world. The focus of German energy policy on growing renewable sector is respectable; however, it does not adequately address the critical question of reduction of CO_2 emissions.
2. The last five years have proven that despite huge (technical and in particular financial) efforts to grow the renewable sector, the impact on CO_2 emissions was negligible. Sticking to this focus, the targets for 2020 will not be met. Setting more aggressive targets for 2030 or even 2050 does not solve the issue, but remains just visionary.
3. As relevant as renewable energy will become in the future, effective reduction of GHG emissions requires a holistic view on the various sectors of energy consumption. Such view considers not just the "share of renewables", but the effective level of GHG reductions and the related cost (so-called climate efficiency).
4. Climate efficiency as key performance indicator allows to develop alternatives of GHG reductions and enables conscious decisions between these alternatives based, in particular, upon feasibility, sustainability, cost, existing infrastructure and social acceptance.

5. The current focus on expanding renewable energy (for many with a vision to electrify also other sectors of energy consumption) is in principle based on the following elements:
   a. Make the electricity sector renewable
   b. Reduce use of energy as much as possible (efficiency plus consumer behaviour)
   c. Electrify also the heating and mobility sector
6. Considering (renewable) electricity as the single most important solution for all sectors of energy consumption places a major bet on the (1) technical and political[34] scalability of electrification (both production and related infrastructure), (2) ability to drive energy efficiency and consumer behaviour, (3) ability to ensure security of (electricity) supply at any point in time and, last not least, (4) ability and willingness to bear the immense additional financial burden of a massive revamp of energy market (also compared to other options).
7. Picking renewable energy as the winner and neglecting the reduction of GHG emissions as the central target of climate protection also put the support of the people at risk. They have to pay the bill of the additional cost and subsidies and may not accept a lot longer the implications of an *Energiewende*, if the pressing issues around reduction of GHG emissions remain untackled.
8. Once it has become clear that the reduction of GHG emissions could be much more effective in an approach which integrates renewable and conventional energy, today's support of the *Energiewende* through society gets redirected. This is even more so, as other societies around Germany have different approaches to climate protection, which does not go well with the self-established understanding of Germany as a role model for the world.
9. Gas, be it natural gas or increasingly renewable gas, can play a major role in such energy transformation. It delivers on necessary elements: (1) GHG friendly, (2) available, (3) at fair prices, (4) flexible in use, (5) developed infrastructure, also for the use of renewable methane.
10. Such approach no longer compares "good" renewable energy with "bad" fossil fuels, but provides a more differentiated and promising approach to the energy challenges of the twenty-first century, in Germany and elsewhere.

4.4.3 Communication Approach

Effective communication requires to address the stakeholders not just with the necessary arguments, but to also get across that it is beneficial for them to seriously grapple with the line of argumentation.

Looking at the addresses, obviously policy makers are a very important target group. But there are others, who are very relevant: given the impact on society, also other stakeholders come into play, e.g. unions, media, consumer protectionists, other affected pressure groups, NGOs.

4.4.3.1 The Core Issue: Many Relevant Stakeholders Are in a Different Place

Getting these messages across should not be too difficult—one would presume. However, as described above, in the public and political domain the climate challenges get frequently reduced to building renewable energy combined with an exit from fossil fuels. Of course, each additional piece of renewable energy, which replaces a fossil fuel, is good news for the reduction of GHG emissions—and as a fact this has sunk in with the media and with the electorate as well. When it comes to transform the entire energy landscape in a journey towards the low-carbon world, such approach is overly simplistic. The Paris Agreement and the pledges of the supporting countries have made it obvious that the timely reduction of GHG emissions needs to be put into the centre of activities—and the last five years of German *Energiewende* have demonstrated that the focus on renewable growth alone does not necessarily reduce GHG emissions.

It will not be an easy task to get this adjustment to the *Energiewende* across. This is not due to the complexity of target setting in a world of climate protection—actually, this is pretty obvious. The real issue is that the shift of the central performance indicator from "increase renewables" to "climate efficiency" results in a more complex process towards finding the right path towards the low-carbon world. It requires to measure various options how to effectively and efficiently reduce GHG emissions.

In the power market, this would result in a debate around how renewables, gas and coal should coexist. This discussion is not held in an integrated manner at the moment. As an example, one may use the power market: the increase of renewables is not assessed against other options to reduce GHG emissions (considering the overall cost to economy); coal has received a pretty "protective" political approach, which was certainly not driven by climate protection and was not considered in an integrated way between renewables and conventional fuels.[35] The same applies for challenges around security of power supply.

Similar discussions would be necessary in the heating and in the transport sector: the focus would be moved away from, "how can we electrify" to "how can we effectively reduce GHG emissions"—and for the mobility sector also NOX and particles would be part of the assessment. This would put serious question marks to quite a few political decisions where electric solution has received preference over alternatives, which would be at least as effective for the climate.

This line of argumentation is more complex and more difficult to communicate—compared to just focusing on increasing renewable energy.

But such debate would have one major advantage: it would provide more sustainable solutions, as they can be related to have made the best use of economic resources and to direct impact on climate protection. This may become very important in a scenario where the current unconditional support of the *Energiewende* diminishes as a result of limited impact on climate, high cost and/or negative consequences for the economy.

Summarising the above, the gas industry has a major task to break through this barrier. This requires a careful analysis how to address the relevant stakeholders, which the author tries to sketch below.

4.4.3.2 Getting Out of the "Fossil Corner"

A key element to be addressed is that all conventional fossil fuels are considered as bad—and gas is just "less bad". Putting renewables and gas as a pair of opposites needs to be overcome. It is misleading, when it comes to climate protection.

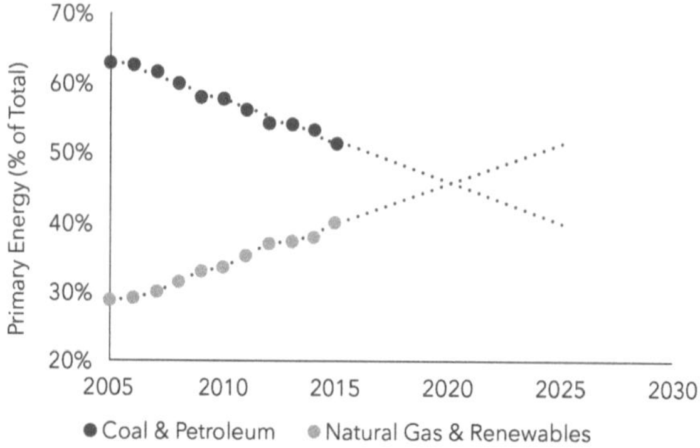

Fig. 4.7 US Primary Energy Trends. *Source* Carroll (2015)

In line with the argumentation in 4.4.2, above, this will require a much higher emphasis on the pressing climate challenges and the options how to ensure the reduction of GHG emissions in an optimal way. The integration of renewables and gas needs to be at the core of describing the progress towards the low-carbon world.

Repeating the shortcomings of the *Energiewende* in its current orchestration would be one way, but this would not be sufficiently solution oriented. Instead, also the positive effects of an integration between renewables and gas in the various sectors need to be described and communicated. It is ultimately a description of an optimised transition period towards the low-carbon world—in line with the Paris Agreement, which provides a consistency which should be welcome by the relevant stakeholders. The graph in Fig. 4.7 illustrates one way how to describe the journey into the low-carbon world. This combination of renewables and gas (as opposed to oil and coal) you would not find in a German government description of the process of climate protection—but you should!

4.4.3.3 Gas Is not Alone

The deficiencies of the current approach regarding the *Energiewende* are not just impacting industries related to conventional fuels—that element might even be welcome for those who do not see value in conventional fuels. However, not using the technical and commercial potential, which gas has for an improved climate protection, affects society as a whole and also very specific parts of the economy. Just two examples:

1. Households are prepared to contribute to climate protection, but they will need to be convinced that they do so at minimal cost—and in many cases, gas-fired heating systems provide solutions for the reduction of GHG emissions at lowest cost; the same holds true if cost for additional insulation is compared to a new heating system. These elements need to be reflected.
2. In the transport sector, the automotive industry is confronted with major challenges around reducing CO_2, NOX and particles; especially in Germany, the big car manufacturers need to carefully consider how to deliver on the environmental and climate requirements. As much as electric cars may become scalable at some stage in the (distant?) future, today's challenges need to be addressed, and the combustion engine will remain key for solving this issue in the coming years. CNG and LNG provide strong solutions to that effect. Needless to say that this is relevant also for the automotive value chain as a whole and related workplaces. Given the size of the German automotive industry with about 800,000 direct employees alone, this demonstrates the exposure of this industry when it comes to the impact of *Energiewende*.

Acknowledging the impact on households and key industries, it is clear that also consumer protectionists and the very influential trade unions have a genuine interest in this topic as well. They need to be engaged in this discussion.

A very good example for this is the 2015 automotive industry dialogue ("Branchendialog"), where the German Ministry of Economics, the German vehicle's manufacturers association (VDA) and "IG Metall" (trade union for heavy industry, engineering and electronics) have agreed on a target to increase tenfold the share of gas-fired cars by 2020.[36]

To become more effective in its communication, the gas industry needs to address the legitimate concerns in combination with and/or in parallel with other relevant players to increase the weight of the arguments.

4.4.3.4 Increasing Public Awareness

The hard facts about the limited success of the *Energiewende* on climate protection that were addressed above under 2 are not new or hidden somewhere. They stem from publicly available data and can be easily retrieved. Also, some journalists have picked this up, but the principle discussion around how to make the *Energiewende* more effective on climate protection has not yet started.

This may be because of the widespread support on the *Energiewende* as a whole. Whilst this support is very important in terms of securing a sustainable support for this big task, it becomes a disadvantage, if the weaknesses in the implementation get somewhat camouflaged by emphasising the successes of increasing renewable energy in the power sector and a bold vision for the far future.

It is very important that the awareness of the hard facts as well as potential other options get communicated clearly, and this not just to politicians, but to other stakeholders as well. As indicated in 4.4.3.3, the economy as well as the citizens is impacted. They should be put in a position to make conscious and well-informed decisions, which includes that there is clarity about the ultimate result (climate protection) and the alternative options available. As a matter of fact, they should demand such information.

4.4.3.5 Providing Solutions, Which Are Good for Society (and Policy Makers!)

The *Energiewende* is a task of the entire society and impacts each citizen. Addressing the (alternative) options for this enormous task makes it necessary to be convincing at the "society level", i.e. it is required to explain why gas is good for society as a whole and what the contribution to the *Energiewende* is. The related content has been described above and summarised under 4.4.2. For further debate related to the various energy sectors, this will of course also entail to provide more detail.

Then, there is another major task: the demand to rethink *Energiewende* such that climate protection gets to the core of the matter also needs to resonate with all the relevant stakeholders, in particular with policy makers. They would need to shift away from the close, if not exclusive, focus on renewable energy and get themselves into the more complex territory how to optimise the integration of renewable energy and conventional fuels. This should be achievable in a combination of (1) the acknowledgement that on limited progress on climate targets will be achieved in the current approach (2) and a political reality which creates more pressure on policy makers to deliver on the promises of the Paris Agreement. Gas, providing constructive target-oriented solutions, should find a way into the core of the political debate.

4.4.3.6 Addressees

The aforementioned gas storyline (ref. 4.4.2.) does not start with gas nor does it end with gas. The development of a climate-oriented society is a task of the entire economy and of all relevant stakeholders and groups. As a consequence, also the communication of the gas storyline is not just a bilateral engagement between gas industry and policy makers. A successful communication of what gas contributes to the climate-oriented energy landscape of the twenty-first century requires engagement with all relevant stakeholders. Apart from other industries, consumer protectionists and trade unions which were mentioned above under. 4.4.3.3, this applies for media as well as think tanks, NGOs, etc. The vast majority of those stakeholders are receptive to hearing more about the intricacies of the energy market transformation and to address the issues searching for climate-efficient solutions—this is the beauty of having common goals. Of course, there may be some difference in views how to get there. But the gas industry should also seek allies for this most important task.

4.5 Summary

1. Germany—the 6th largest emitter of CO_2 worldwide—considers herself a front runner on climate protection and in the implementation of the Paris Agreement of late 2015.
2. Germany's massive investments in renewable energy have not resulted in a visible reduction of GHG emissions since 2010/2011.
3. When increasingly relying on electric solutions also for the heating and mobility sector, the German economy is placing a major bet on its ability to largely electrify the economy, thereby relying on renewables. This—somewhat ideological—bet is based on assumptions around, i.e. (1) scalability of renewables and related infrastructure, (2) ability to ensure security of (electric) supply at any time and (3) ability of electric heating and electric mobility to provide solutions in an effective and economically viable way. These requirements may actually put certain parts of the economy under unnecessary economic stress and may result in a loss of support for the *Energiewende* altogether.
4. In its focus to increase renewable power production instead of reducing GHG emissions, the German *Energiewende* undervalues the potential of gas for the modern low-carbon energy landscape of the twenty-first century.
5. Instead, redirecting the focus towards a climate-efficient transformation of the energy sector would result in an integrated approach, which seeks to combine increasing renewable energy in an optimised way with conventional energy sources and link this to efficiency gains. This would also be in line with the ambitious targets of the Paris Agreement and help achieving a more efficient reduction of GHG emissions already in the near future.
6. The gas industry cannot rely on these factors to sink in. Whilst gas was a fuel of choice and politically supported for many decades, this has changed. The potential of gas for contributing to climate protection needs to be emphasised. Gas must get out of the corner of being fossil and, hence, just less bad than other fossil fuels. In an environment with obvious elements of "post-truth" politics against gas,

natural gas must prove the point of delivering solutions for the low-carbon society.

7. Proactive communication is required to make an impact on the relevant stakeholders. This includes getting other key stakeholders into the debate, e.g. other industries, consumer protection groups, trade unions and the like.
8. Gas has all it takes to remain a valuable asset of not just the German energy landscape in the twenty-first century. But this will not fall into the laps of the gas industry, it will require major efforts together with partners to fight for this role and communicate the benefits with the relevant political and societal stakeholders.

Notes

1. Including, in particular, CO_2 as the primary greenhouse gas.
2. European Commission (2016).
3. See Fig. 4.2.
4. Evangelical Church in Germany (2016).
5. Federal Ministry for the Environment, Nature Conservation, Building and Nuclear Safety (2016), pp. 13, 16.
6. Federal Ministry for the Environment, Nature Conservation, Building and Nuclear Safety (2016), pp. 16, 18.
7. Federal Ministry for the Environment, Nature Conservation, Building and Nuclear Safety (2016), pp. 13, 15.
8. German Bundestag (2012), Dokumente.
9. Ethics Commission for a Safe Energy Supply (2011), pp. 47, 81.
10. Federal Government (2013), p. 50.
11. AG Energiebilanzen e.V. (2016a), see table "Bruttostromerzeugung in Deutschland ab 1990 nach Energieträgern".
12. See also Fig. 4.4 ("Transport emissions"), below.
13. AG Energiebilanzen e.V. (2016b), Evaluation Tables of the Energy Balance for Germany 1990–2015, Primary Energy consumption of Germany, p.8.
14. Federation of German heating Industry, BDH (2016).
15. Federal Government (2013), p. 50.
16. World Resources Institute (2016), see ipcc-infographic.

17. Suhr (2016), see infographic.
18. Federal Ministry for the Environment, Nature Conservation, Building and Nuclear Safety (2016), p. 13.
19. Federal Ministry for the Environment, Nature Conservation, Building and Nuclear Safety (2016), pp. 13, 16.
20. Federal Ministry for Economic Affairs and Energy (2014b), p. 17.
21. See also Sect. 1.2.
22. See Fig. 4.2.
23. Federal Ministry for the Environment, Nature Conservation, Building and Nuclear Safety (2016), pp. 32–33.
24. See Fig. 4.6.
25. Energy Union Choices (2016), pp. 10, 12.
26. See also Sect. 3.1.
27. See also Sect. 3.1.
28. Federal Ministry for the Environment, Nature Conservation, Building and Nuclear Safety (2016).
29. See also Sect. 1.2.
30. Federation of German Industries, BDI (2016), see publication.
31. Westdeutsche Allgemeine Zeitung (2016), see article.
32. See also Sect. 2.2.5.
33. See Fig. 4.4, above.
34. Will the people, many of whom are already now contesting grid expansion, provide the necessary support for further massive expansion of the power grids, which would be required?
35. See also 1.2, above.
36. Federal Ministry for Economic Affairs and Energy (2016), p. 2.

Bibliography

AG Energiebilanzen e.V. 2016a. *Bruttostromerzeugung in Deutschland ab 1990 nach Energieträgern*. Berlin: AG Energiebilanzen e.V. http://www.ag-energiebilanzen.de/4-1-Home.html.

AG Energiebilanzen e.V. 2016b. *Evaluation Tables of the Energy Balance for Germany 1990–2015, Primary Energy consumption of Germany*. Berlin: AG Energiebilanzen e.V. http://www.ag-energiebilanzen.de/4-1-Home.html?_sm_au_=iHV2p3wpvH414WPH.

Carroll, D. 2015. *U.S. Primary Energy Trends*. Oslo: International Gas Union.

Energy Union Choices. 2016. *A Perspective on Infrastructure and Energy Security in the Transition.* Energy Union Choices. http://www.energyunion-choices.eu/wp-content/uploads/2016/07/EUC_Report_Web.pdf.

Ethics Commission For a Safe Energy Supply. 2011. *Deutschlands Energiewende - Ein Gemeinschaftswerk für die Zukunft.* Berlin: Ethics Commission for a Safe Energy Supply. https://www.leopoldina.org/fileadmin/redaktion/Wissenschaft/pdf/Ethik-Kommission_Deutschlands-Energiewendel.pdf.

European Commission.2016. *Paris Agreement.* Brussels: European Commission. http://ec.europa.eu/clima/policies/international/negotiations/paris/index_en.htm.

Evangelical Church in Germany. 2016. *Kirchen und Verbände fordern höhere deutsche Klimaziele.* Berlin: Evangelical Church in Germany. http://www.ekd.de/klimagerechtigkeit/news_2016_04_20_03_klimaziele_kirchen_brot_welt.html.

Federal Environment Agency. 2016. *Treibhausgas-Emissionen in Deutschland.* Berlin: Federal Environment Agency. http://www.umweltbundesamt.de/daten/klimawandel/treibhausgas-emissionen-in-deutschland.

Federal Government. 2013. *Koalitionsvertrag.* Berlin: Federal Government. https://www.bundesregierung.de/Content/DE/_Anlagen/2013/2013-12-17-koalitionsvertrag.pdf?__blob=publicationFile.

Federal Ministry for Economic Affairs and Energy. 2015a. *Energiereferenzprognose.* Berlin: Federal Ministry for Economic Affairs and Energy. http://www.bmwi.de/EN/Service/publications,did=746824.html.

Federal Ministry for Economic Affairs and Energy. 2015b. *Fourth "Energy Transition" Monitoring Report.* Berlin: Federal Ministry for Economic Affairs and Energy. http://www.bmwi.de/EN/Service/publications,did=746824.html.

Federal Ministry for Economic Affairs and Energy. 2016. *Branchendialog mit der Fahrzeugindustrie - Gemeinsame Erklärung der Beteiligten am Spitzengespräch am 1.Dezember 2015.* Berlin: Federal Ministry for Economic Affairs and Energy. https://www.bmwi.de/BMWi/Redaktion/PDF/B/branchendialog-mit-der-fahrzeugindustrie-gemeinsame-erklaerung,property=pdf,bereich=bmwi2012,sprache=de,rwb=true.pdf.

Federal Ministry for the Environment, Nature Conservation, Building and Nuclear Safety. 2016. *Klimaschutzplan 2050.* Berlin: Federal Ministry for the Environment, Nature Conservation, Building and Nuclear Safety. http://www.bmub.bund.de/themen/klima-energie/klimaschutz/klima-klimaschutz-download/artikel/klimaschutzplan-2050/?tx_ttnews%5BbackPid%5D=3915.

Federation of German Heating Industry. 2016. *Marktentwicklung Wärmeerzeuger 2005–2015*. Berlin: Federation of German Heating Industry, BDH. http://www.bdh-koeln.de/fileadmin/user_upload/Daten_Fakten/BDH_Marktentwicklung_2005-2015.pdf.

Federation of German Industries. 2016. *Klimaschutzplan 2050 – Mehr Gründlichkeit, weniger Hast*. Berlin: Federation of German Industries, BDI. http://bdi.eu/artikel/news/klimaschutzplan-2050-mehr-gruendlichkeit-weniger-hast/.

German Bundestag. 2012. *Der Einstieg zum Ausstieg aus der Atomenergie*. Berlin: German Bundestag. https://www.bundestag.de/dokumente/textarchiv/2012/38640342_kw16_kalender_atomaustieg/208324.

IHS Markit. 2016. *Global Energy Demand*. London: IHS Markit.

Prognos/EWI/GWS. 2014a. *Entwicklung der Energiemärkte - Energiereferenzprognose*. Berlin: Prognos/EWI/GWS.

Prognos/EWI/GWS. 2014b. Commissioned by Federal Ministry for Economic Affairs and Energy (2014), p. 17.

Suhr, Frauke. 2016. *Die größten Klimasünder 2015*. Hamburg: Statista. https://de.statista.com/infografik/6050/laender-mit-den-hoechsten-co2-emissionen/.

Westdeutsche Allgemeine Zeitung. 2016. *Verdi und Länder machen mobil gegen Klimaschutzplan*. Essen: Westdeutsche Allgemeine Zeitung. http://www.presseportal.de/pm/55903/3339975.

World Resources Institute. 2016. *Understanding the IPCC Reports*. Washington: World Resources Institute. http://www.wri.org/ipcc-infographics.

Author Biography

Ludwig Moehring started 2010 at WINGAS as managing director and is responsible for sales in Germany and Europe. He has worked in the energy industry since 1992. After working for BEB Erdgas und Erdoel in Hannover, he moved to Shell Gas & Power in London and subsequently The Hague, where at Shell Energy Europe he was responsible for natural gas sales to large resellers and industrial key accounts and the procurement from international producers in Northwest Europe. He is also President of ASUE, an industry body dealing with efficient energy consumption, and on the Supervisory Board of Zukunft ERDGAS.

5

The Role of Gas in the European Electricity System of the Future

Fabio Genoese

5.1 Introduction

In the public debate, natural gas is generally considered an important fuel for future electricity generation, because it is the cleanest of all fossil fuels. The carbon intensity of modern coal-fired power stations easily exceeds one of the modern gas plants by a factor of two. Moreover, gas-fired units are well suited to follow rapid swings in supply and demand due to their flexibility. In the future, these balancing tasks will become more and more important given the intermittent character of the supply of wind and solar power. As a result, gas seems to hold out the promise of being a key pillar of the energy transition and the perfect partner of renewables. As will be shown in this chapter, there is, however, evidence that demand for gas for the purposes of power generation peaked as early as 2010.

F. Genoese (✉)
SciencesPo, Paris, France
e-mail: fabio.genoese@engie.com; fabio.genoese@sciencespo.fr

M. Hafner and S. Tagliapietra (eds.), *The European Gas Markets*,
DOI 10.1007/978-3-319-55801-1_5

5.2 Setting the Scene

The EU power sector represents a major consumer of natural gas, despite a decline in recent years. In 2014, some 94 bcm of natural gas were used to generate electricity,[1] which corresponded to 23% of total EU gas consumption. As shown in Table 5.1, the volumes of gas consumed for power generation saw a steady reduction from 2010 on, and this declining trend has been more strong than the declining trend in total EU gas consumption, which explains why weight of the power sector decreased from almost 30 to 23% in the same period of time.

Mild temperatures and a sluggish economic recovery are generally put forward as reasons for the decline in demand for gas for power generation: Heating demand is lower during mild winters, thus lowering the output of combined heat and power (CHP) plants, which are typically gas-fired. Moreover, total electricity demand in 2015 was still almost 6% below pre-crisis levels. Needless to say, this has negatively affected the output of gas-fired power stations.

However, these two factors alone are not sufficient to explain the recent trend. There are at least three more drivers to look at, namely (1) the competition between coal and gas in power generation and its fundamental drivers, (2) the policy push for renewables and (3) the mechanics of pricing power. The future role of natural gas in the EU power sector will largely depend on how these factors will evolve, which is why a more detailed assessment of these is presented in the next sections, following a brief presentation of the past and current role of gas-fire generation in the EU power sector.

Table 5.1 Total EU natural gas consumption and gas consumed in the EU power sector

	2008	2009	2010	2011	2012	2013	2014
Total (bcm)	519	488	525	477	467	461	409
Power sector (bcm)	145	131	154	138	115	104	94
Power sector (%)	28	27	29	29	25	23	23

Source Author's elaboration based on Eurogas (2016)

5.3 Determining Factors for Demand for Gas for Power Generation

5.3.1 The Current Role of Gas in the EU Electricity Generation Mix

As of 2015, renewables are the number one source of electricity generation in the EU, making up for 29% of gross production (see Fig. 5.1). With a share of 15%, gas ranks fourth in the electricity generation mix, after nuclear and coal,[2] which make up for 27 and 26%, respectively. A few years earlier, natural gas used to have a significantly higher share in total power generation. In 2010, gas and coal accounted for almost the same share, namely 23 and 25%, respectively.

In absolute terms, gas-fired generation decreased by approximately 280 TWh from 2010 to 2015, which is comparable to the size of the Spanish power sector. No other power generation source saw such a strong

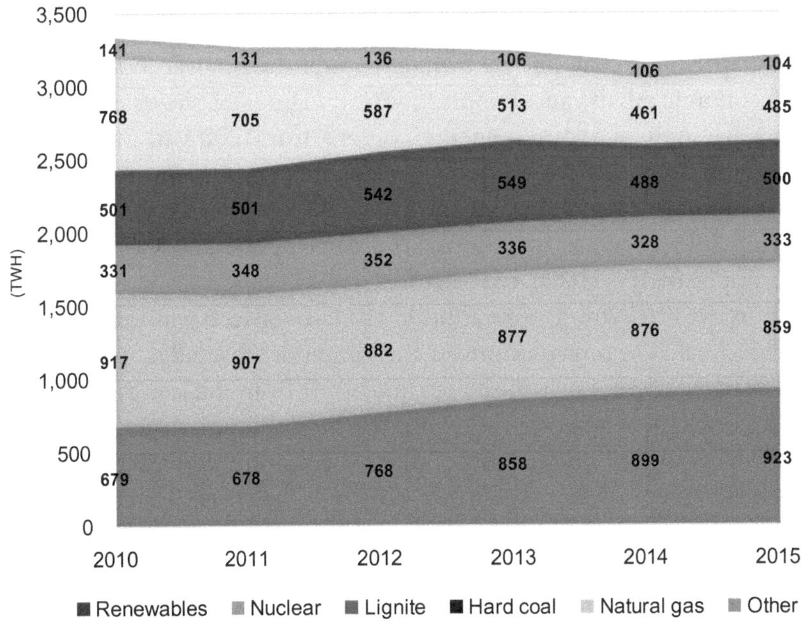

Fig. 5.1 EU-28 electricity generation mix (2010–2015). *Source* Author's elaboration based on Agora Energiewende (2016)

decline, not even nuclear power, which took a severe hit of −60 TWh after the phase-out decision of the German government following the Fukushima disaster. Overall, total power generation saw a decline of 130 TWh, a combined effect of efficiency measures and the economic downturn. The output of renewable energy sources increased by roughly 240 TWh in the same period of time, largely pushed into the market by dedicated policies. Interestingly, coal-fired generation in 2015 was almost unchanged compared to 2010.

5.3.2 The Competition Between Coal and Gas in Power Generation

Fuel prices are the main determining factors for the (variable) electricity production costs of thermal units such as gas- and coal-fired power plants. While coal market is considered to be a global market with comparable prices across the world (net of transportation costs), natural gas prices tend to vary more from region to region. A well-accepted source of reference prices used in EU power sector assessments is the German import price for coal and gas, which is published by the Federal Office for Economic Affairs and Export Control. Figure 5.2 shows gas and coal prices per unit of thermal energy content from 2004 to 2015 as well as the ratio between gas and coal prices. This ratio amounted to 2.0 in 2010 and saw a sharp increase to 2.8 in 2013, with gas and coal prices moving into opposite directions. This development was driven by an oversupply in the global coal market, partly caused by the shift from coal- to gas-fired power generation in the USA, which was made possible by the US shale gas revolution. At the same time, overall demand for gas increased, partly because of the Japanese shift from nuclear power to gas-fired electricity generation following the Fukushima disaster, which saw all of the country's nuclear power plants being shut down in order to be checked. In recent years, the gas-to-coal ratio has been decreasing again, with both commodity prices moving in the same direction: downwards. Yet, with a value of 2.5, the ratio is not as low as it used to be in 2010.

Fuel prices help to understand the competitive disadvantage of gas-fired generation over coal-fired generation. This disadvantage is partly compensated by the fact that gas-fired power plants (more precisely

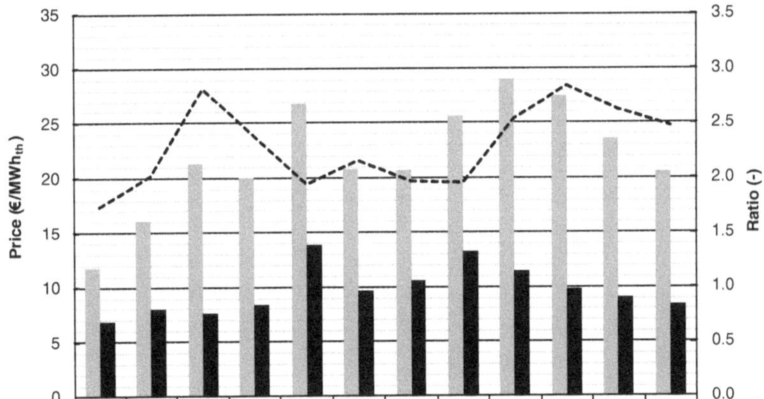

	2004	2005	2006	2007	2008	2009	2010	2011	2012	2013	2014	2015
Gas price	11.8	16.1	21.3	20.0	26.8	20.9	20.6	25.7	29.0	27.6	23.5	20.6
Coal price	6.8	8.0	7.6	8.4	13.8	9.7	10.5	13.1	11.4	9.7	9.0	8.3
Ratio:Gas-to-Coal	1.7	2.0	2.8	2.4	1.9	2.2	2.0	2.0	2.5	2.8	2.6	2.5

Gas price Coal price Ratio:Gas-to-Coal

Fig. 5.2 German import prices for coal and gas (2004–2015). *Source* Author's elaboration based on BAFA (2016)

CCGT plants, i.e. combined cycle gas turbines) have a higher conversion efficiency compared to coal-fired units. This means they require less thermal energy to produce the same amount of electrical energy.

Carbon pricing also shifts the balance more towards gas, due to the higher carbon intensity of coal. In the period from 2010 to 2015, the average EU price for carbon allowances decreased from €22.5 per tonne of CO_2 to €7.7/tonne. Most of this decline can be ascribed to an overallocation of carbon allowances. The original design of the EU Emissions Trading System (ETS) was based on a supply of allowances that was fixed *ex-ante*, irrespective of any changes in demand. Yet, the economic downturn following the 2008/2009 crisis came along with a decrease in industrial output, lowering the demand for carbon allowances and therefore putting a downward pressure on carbon prices.

Combining all these fundamental drivers—fuel prices, efficiency and carbon prices—reveals that *new* gas- and *old* hard-coal-fired power plants had similar variable production costs in 2010, as shown in Table 5.2. In this case, the adjectives "new" and "old" are merely used

Table 5.2 Variable production costs of CCGT and coal from 2010–2015 in €/MWh$_{el}$

	2010	2011	2012	2013	2014	2015
"new" CCGT	39.2	47.1	50.8	47.4	41.2	36.9
"old" hard coal	40.9	46.6	36.8	29.6	29.1	29.1

Source Author's elaboration based on BAFA (2016)

as a proxy to define a difference in conversion efficiency of the two technologies. To be precise, the production costs shown in the table are based on a value of 60% for CCGT, while an efficiency of 38% is assumed for the hard coal plant.

Based on these assumptions, variable production costs were in the range of €40/MWh in 2010 for the two technologies, while an *old* hard coal plant offered a considerably less expensive way of producing electricity in 2015 than a *new* CCGT unit.

These fundamental drivers of coal- versus gas-fired power generation illustrate the decreasing competitiveness of gas vis-à-vis coal. Yet, the massive setback of gas-fired generation in total electricity generation would not have occurred in such a short timeframe without the rise of a new electricity generation source: renewables.

5.3.3 The Policy Push for Renewables

The EU renewable electricity generation has seen a remarkable rise over the last ten years, and this rise was largely driven by the policy target of 20% renewables by 2020, which foresees roughly 35% renewables in the power sector. It is important to note that this sector-specific number is not legally fixed but merely resulting from the national renewable energy action plans (NREAPs) defined by each EU member state in 2009. In fact, the deployment in the power sector was slightly above target in the year 2014, albeit counterbalancing the underperforming transport sector. Figure 5.3 shows the actual and forecasted share of renewables in the EU-28 power sector.

The existence of a legally binding target led to the question of suitable measures to ensure target achievement. To this end, EU member states implemented dedicated policies that pushed renewables into the

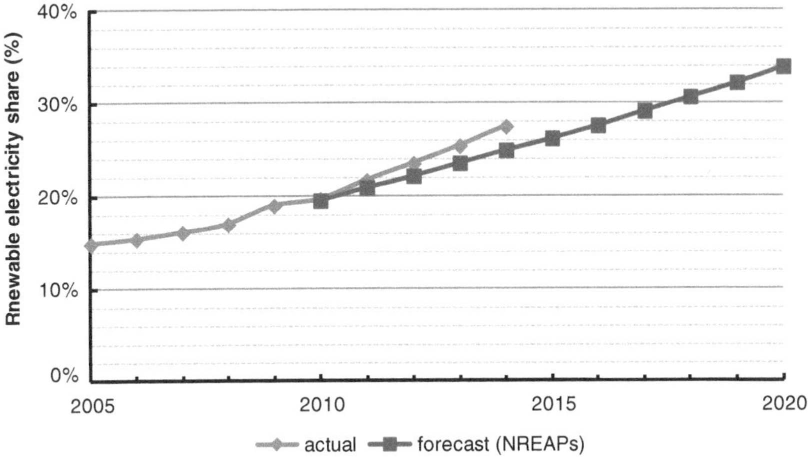

Fig. 5.3 Share of renewables in the EU-28 power sector (2005–2020). *Source* Author's elaboration based on Eurostat (2016)

system. The most commonly used types of support schemes include feed-in tariffs, feed-in premiums or green certificates (see Held et al. 2014). While it is beyond the scope of this book to introduce the exact functioning of these schemes, it is important to note that a dedicated policy instrument inserting new electricity generation in an already well-supplied market effectively decreases demand for all other technologies. In the EU, this came at a time when total demand was anyhow stagnating after the 2008/2009 dip caused by the economic crisis.

The policy push for renewables decreased the size of the market where coal, gas and other conventional sources were competing. Following the price dynamics shown in the previous subsection, this contraction in market size mainly affected gas and not coal, due to the higher variable generation costs of gas. Certainly, the global fuel price dynamics did favour coal here. Some analysts also point out that the price decline in the EU ETS was linked to the policy push for renewables. This is subject to debate. While it is correct that pushing renewables into the market is a form of CO_2 abatement, its impact on the EU ETS largely depends on whether the volumes entering the market were considered when the EU ETS cap was defined and on whether the actual volumes entering the market are in line with the forecasted volumes. The need for consistency between deployment plans for renewables and the emission caps has been pointed out by the European Commission since 2005.[3] Thus, there is consensus that only overachieving pre-defined deployment plans for renewables would put a downward pressure on carbon allowance prices. What complicates the matter is the fact that deployment *targets* for renewables cover three sectors, namely electricity, heating/cooling and transport. Yet, the EU ETS covers power generation and industrial facilities only. Therefore, it is challenging to achieve full consistency between the deployment policies for renewables and the definition of the EU ETS cap.

5.3.4 The Mechanics of Pricing Power

The last point to consider is the mechanics of how power is priced in the EU. The current market design foresees that wholesale electricity markets remunerate the energy delivered, i.e. volumetric payments. As discussed in the previous sections, gas currently loses out compared to coal under current market conditions (i.e. under current coal, gas and carbon prices).

There is an ongoing debate whether the current EU power market design is the most efficient to reach given renewables and climate targets.[4] Based on current projections, it is expected that renewables will satisfy 50% of electricity demand in 2030. The increase will mainly be driven by intermittent sources such as wind and solar (see Fig. 5.4). Such a high share of intermittent generations has implications on how the power system will be run: conventional generation will mostly be relegated to back up tasks to ensure that demand can also be met in times when the wind does not blow and the sun does not shine. As such, 50% of conventional capacity could be on stand-bye for at least 80% of the time in 2030. Today, less than a third of the fleet is operated like this. From an economic perspective, it is more efficient to let this task be performed by technologies with relatively low investment and fixed costs—simply due to the assets' low utilisation. This would generally favour gas over coal plants, which are less expensive to run but

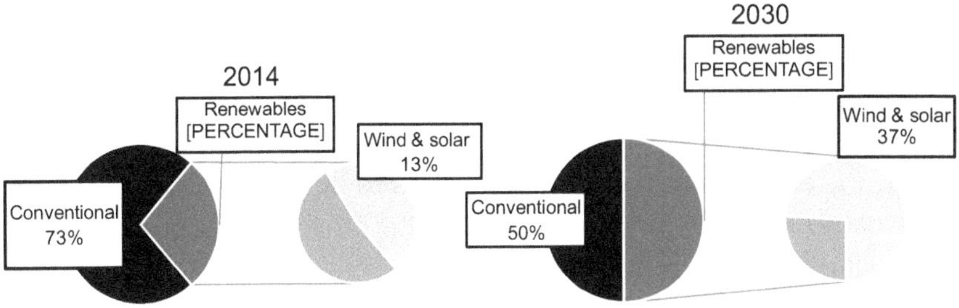

Fig. 5.4 Actual and projected EU power generation mix. *Source* Author's elaboration based on Eurostat (2016) and Resch et al. (2016)

more expensive to build and maintain than gas units. Yet, under the current market design, units can only compete for lowest cost delivery of energy, not for lowest cost availability of capacity. This is aggravated by the fact that the market does not send any long-term price signals for the delivery of energy, which could be interpreted as price signals for the availability of capacity.

5.4 The Future Outlook for Gas in the EU Power System

5.4.1 Fundamental Drivers

Given the recent trends, a growth of gas for power generation can only occur, if coal-fired generation decreases. As discussed in the previous sections, the record-low price of coal is a major reason for the competitive advantage of coal over gas. This price results from an oversupplied coal market at global level—a situation which appears to be here to stay, especially if more countries opt for carbon emission reduction goals, which is to be expected after the Paris 2015 agreement. Theoretically, a further decrease of EU gas prices could also shift the balance towards more gas-fired power generation, although carbon pricing would be a more effective tool, at least in theory. It remains to be seen whether the most recent reforms of the EU ETS will provide more visibility and certainty on the long-term scarcity of carbon and thereby uplift the price of carbon allowances.

Yet, the elephant in the room continues to be renewables, considering that the policy push for renewables is here to stay. This push implies a decreasing market size for all other resources, which will likely lead to a relatively clean power sector with a "dual optimum" mainly based on renewables and coal—unless of course the EU power market design is reformed and geared towards long-term price signals and/or an explicit remuneration of capacity.

5.4.2 Review of Existing Scenarios

A well-known outlook for demand is published by the association of Europe's gas grid operators, ENTSO-G.[5] It is used to estimate future infrastructure requirements. According to their "Green Transition" scenario, gas demand for power generation will increase from last year's 94 billion cubic metres (bcm)[6] to 167–232 bcm in 2035. At present, 2010 marks the record year for gas-to-power consumption with some 154 bcm. The report does not make forecasts about the consumption beyond 2035. Such a forecast horizon is not unusual in the private sector, because business plans do not extend to 2050.

Eurogas, the trade organisation representing the European gas industry, also projects an increase in its "Environmental Scenario" to 230 bcm by 2035,[7] assuming a rebalancing of the power mix towards more renewables and gas. By 2035, the share of gas for power generation would increase to 33%, while the share of coal would decline to 6%. Renewables would cover 44% of electricity demand.

But these projections should be taken with a grain of salt for at least two reasons. First, it should not be taken for granted that gas easily regains market share over coal. A massive fuel switch from coal to gas is not going to happen in the absence of strong carbon pricing or policies targeting the phase-out of coal. Second and probably more important: the market size for conventional power generation will continue

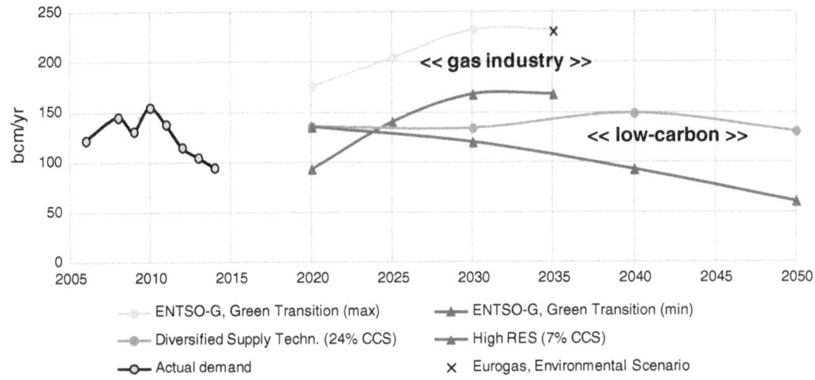

Fig. 5.5 Competing scenarios for gas demand for power generation (bcm/yr). *Source* Author's elaboration based on Genoese et al. (2016b)

to shrink. consumption of all fossil fuels including gas will have to decrease significantly, if the EU's long-term climate policy objectives of reducing GHG emissions by at least 80% below 1990 levels by 2050 are to be met.[8]

As stated in the EU low-carbon roadmap, an almost carbon-neutral[9] power sector forms the central assumption behind *all* decarbonisation scenarios.[10] This implies a declining use of fossil-fuel-fired power generation technologies, including gas-fired power stations. The decline could be partly offset by carbon capture and storage (CCS), a technology that captures carbon dioxide emitted by fossil-fuel-fired plants and hence would allow for a continued use of coal or gas. Yet, even in the EU's low-carbon scenario with the highest share of CCS (24% of EU power generation), gas demand would not rise above 150 bcm and therefore would remain below the 2010 peak (see Fig. 5.5).

Such a scenario would require that CCS reaches technological maturity and becomes economically viable, which is currently not the case. Given the lack of CCS demonstration projects worldwide, one could argue that it is optimistic to assume that CCS could capture a market share of 24%. Other low-carbon options could be diffused more rapidly, displacing CCS. In the EU's low-carbon scenario with less CCS (7% of EU power generation), gas demand would even decrease to 60 bcm by 2050, which is less than the 8-year low reached in 2014.

Hence, considering 2050 objectives, it is reasonable to assume that demand for gas for the purposes of power generation has already peaked in 2010.[11]

5.5 Summary and Conclusions

This analysis suggests that the EU power sector will not act as a driver of growing gas demand, because (1) gas is unlikely to replace all coal-fired generation in the short- to medium-term and (2) the consumption of all fossil fuels including gas will have to decrease in the long run.

Notable game changers in the short- to medium-term include stricter carbon pricing measures or dedicated policies to phase out coal. A change in power market design could also contribute to a fuel switch from coal to gas. For the long run, it is more difficult to identify game changers. The flexible nature of gas-fired technologies alone does not imply that gas will always be deployed next to renewables.

A phase-out of fossil-fuel-fired generation is required to reach the EU's long-term climate policy goals. Yet, these goals should not be equated with a power system running 100% on solar and wind. Hydro power and biomass will continue to account for a significant share of installed capacity. Similar to gas, these technologies are also considered flexible and therefore well suited to follow rapid swings in demand and supply, thereby compensating for the intermittent availability of wind and solar. But it is also true that if we want more renewable electricity and less fossil fuels, we have to find smarter ways of storing and using power. Driven by price signals, the demand side could react to the availability of intermittent sources, e.g. by partially reducing consumption on a cloudy, windless day. Low-cost storage would allow for this adaptation without a loss of comfort. gas turbines are likely to be part of such a system but they would increasingly be used as a measure of last resort (backup)—and thus not be consuming significant volumes of natural gas.

Notes

1. See Eurogas (2016).
2. Unless noted otherwise, the term "coal" is used as an abbreviation for the combined output hard coal and lignite.
3. See "Further guidance on allocation plans for the 2008 to 2012 trading period of the EU Emission Trading Scheme", Communication from the Commission, COM (2005) 703 final, 22 December 2005.
4. See Genoese and Egenhofer (2015) as well as Genoese et al. (2016).
5. See ENTSO-G (2015).
6. See Eurogas (2016).
7. See Eurogas (2013).
8. See "A Roadmap for moving to a competitive low carbon economy in 2050", Communication from the Commission, COM (2011) 112 final, 8 March 2011.
9. GHG emissions of the power sector would have to be reduced by some 95% below 1990 levels.
10. See Commission Staff Working Paper Impact Assessment accompanying the Communication from the Commission, Energy Roadmap 2050, SEC (2011) 1565 Parts 1 and 2, 15 December 2011.
11. Still, compared to last year's low of 94 bcm, a slight recovery of gas demand until 2030 would be consistent with the EU's low-carbon objectives.

Bibliography

Agora Energiewende. 2016. *Energy Transition in the Power Sector in Europe: State of Affairs in 2015*, April.

BAFA. 2016. *Import Border Prices for Coal and Gas from 2004–15*, German Federal Office for Economic Affairs and Export Control, August.

ENTSO-G. 2015. *Ten Year Network Development Plan 2015*, April.

Eurogas. 2013. *Long-term Outlook for Gas to 2035*, October.

Eurogas. 2016. *Statistical Report 2015*, January.

Eurostat. 2016. *Actual and Projected Share of Renewable Energies*, August.

Genoese, Fabio and Christian Egenhofer. 2015. *Reforming the Market Design of EU Electricity Markets*, CEPS Task Force Report, July.

Genoese, Fabio, Eleanor Drabik, and Christian Egenhofer. 2016a. *The EU Power Sector Needs Long-Term Price Signals*, CEPS Special Report No. 135, April.

Genoese, Fabio, Eleanor Drabik, and Christian Egenhofer. 2016b. *Gas Demand for Power Generation Peaked as Early as 2010*, CEPS Commentary, February.

Held, Anne, Mario Ragwitz, Malte Gephart, Erika De Visser and Corinna Klessmann. 2014. *Design Features of Support Schemes for Renewable Electricity*, report within the project "Cooperation between EU MS under the Renewable Energy Directive and interaction with support schemes", January.

Resch, Gustav, Lukas Liebmann, Sebastian Busch, Christoph Zehetner. 2015. *Scenarios on Meeting 27% Renewable Energies by 2030*, report within the project "Dialogue on a RES policy framework for 2030 (Towards2030-dialogue)", December.

Author Biography

Fabio Genoese is a German-Italian national, with over 8 years' experience in energy. Mr Genoese started his career in 2008 as consultant to energy utilities and the German government. From 2013 to 2016, he was in charge of electricity policy & regulation at CEPS, a Brussels-based Think Tank. He is currently Smart Energy Expert at Tractebel, a global business unit of ENGIE, and Visiting Professor at Sciences Po in Paris. Mr Genoese holds a doctoral degree in Energy Economics as well as a Master's degree in Physics.

6

The Prospects for Gas in the European Transportation Sector

Chris Le Fevre

6.1 Introduction

Natural gas has been used as transport fuel in vehicles for many years though primarily in the form of compressed natural gas (CNG). Natural gas has a number of environmental and financial advantages over oil-based fuels in most markets though, with some minor exceptions, it has always struggled to gain significant market share in Europe against the more efficient and established fuels. Interest in the sector has been revived recently by the emergence of LNG as a fuel which has some additional cost and technical advantages over CNG, in particular, as a fuel for heavy road vehicles and ships.

This chapter looks at the financial and environmental case for natural gas as a transport fuel. It considers these advantages in the main transport sectors of passenger cars, road haulage vehicles and marine

C. Le Fevre (✉)
Oxford Institute for Energy Studies, Oxford, UK
e-mail: chris@lefevreconsulting.com

M. Hafner and S. Tagliapietra (eds.), *The European Gas Markets*,
DOI 10.1007/978-3-319-55801-1_6

shipping and the resulting levels of demand that might emerge. The analysis primarily focusses on the situation in EU countries though other nations are mentioned where relevant—for example, Norway is an important player in the marine shipping sector.

6.2 The Main Markets for Natural Gas as a Transport Fuel

Transport is a major consumer of energy accounting for around 28% of global total final consumption of energy in 2013. Oil dominates the sector providing some 93% of transportation consumption, whilst over 64% of oil consumption was in the transportation sector.[1] These figures are broadly mirrored in the EU where transport was 33% of final energy consumption in 2014 (EC 2016b). The total consumption of 394 MTOE was primarily on road transport (73%) with 13% going to aviation and 12% to marine and inland waterways. Oil products dominate and diesel had the largest share (51%) followed by gasoline (20%), kerosene (13%) and fuel oil (9%). Natural gas had a 0.7% share in 2014. (Eurostat 2016) European energy consumption in transport has been falling due to a combination of reduced economic activity and improved energy efficiency.

Eurogas (2016) provides data on natural gas consumption in transport by country. In 2014, transport represented only 0.4% of natural gas sales in the EU though this was an increase of 8.5% on 2013. The countries where gas consumption in transport is significant (i.e. >1 TWh/a) are shown in Table 6.1. This demonstrates that with the notable exception of Italy, gas consumption in transport is still a very small proportion of the total.

The figures in Table 6.1 exclude natural gas used in shipping in the form of LNG. This is still a very small market although Norway is at the forefront. In 2014, LNG usage in shipping in this country was only 124 million m^3 increasing to 133 million m^3 (approximately 1.4 Twh) in 2015.[2]

Table 6.1 Natural gas in transportation in EU, 2014

Country	Total gas consumption (TWh)	Of which transport	
		TWh	%
France	421.3	1.2	0.3
Germany	824.6	2.3	0.3
Italy	655.2	11.1	1.7
Spain	301.4	1.2	0.4
Other	2224.8	2.8	0.1
EU 28	4427.3	18.6	0.4

Source Eurogas (2016)

Whilst there are some discrepancies on numbers between sources, it is clear that natural gas in transportation is still very much a minority fuel—the next section considers what factors might change this.

6.3 The Case for Natural Gas as a Transport Fuel

The case for natural gas is driven by both demand and supply factors. From a demand perspective, the attractions of the fuel are primarily financial and environmental:

- The financial advantage is based on price versus competing fuels. In road transport, the comparison will normally be with diesel/gasoline and the price paid is materially impacted by national taxation levels. In maritime markets, the competing fuel is primarily fuel oil (HFO) though diesel (referred to as marine gas oil, MGO) is also important in some cases. Marine fuel prices are generally untaxed so the price comparison is usually on a commodity basis.
- The environmental advantage relates to emissions. In transport, the main concern is over emissions of carbon dioxide (CO_2), nitrogen oxides (NO_x) and particulates (PM) plus, in maritime transport, sulphur oxides (SO_x). CO_2 and nitrous oxide (N_2O) are greenhouse gases, and their global warming potential (GWP) is typically measured on a CO_2 equivalence basis. $SO_{x,}$ NO_x and PM can cause health

problems, and SO_2 is also responsible for acid rain. The combustion of natural gas generates lower levels of CO_2 emissions and virtually no No_x, SO_x or PM emissions compared with diesel or fuel oil. On the other hand, methane is an important GHG (30 times the global warming potential of CO_2 though the gas remains in the atmosphere for a much shorter period), so any emissions that might be associated with production, transportation, dispensing or incomplete combustion[3] of natural gas are usually included.

On the supply side, the main drivers in Europe are:

- The increased availability of LNG and LNG terminals;
- gas suppliers seeking new markets in the face of stagnant or falling demand in more traditional markets.

These drivers have to be weighed against three critical issues:

- Switching to natural gas will almost certainly entail higher upfront vehicle costs
- The so-called chicken-and-egg syndrome which reflects the unwillingness of manufacturers or buyers to invest in Natural gas vehicles/vessels (NGVs) until there is a widespread network of refuelling facilities, whilst fuel infrastructure providers will be unwilling to make such investments until there is evidence of significant and growing NGV ownership.
- The risk of switching to a new, relatively untried, fuel compared with a known, safe and generally improving current option. This challenge is particularly relevant in an established sector such as transport. The market has existed for a long time and the existing fuel supply chain is efficient and responsive to market needs. A "new" fuel such as natural gas therefore has a huge element of inertia to overcome.

All of the above factors have to be assessed in the context of the perspective of the various key stakeholders in each market. Le Fevre (2014) suggests that in broad terms, there are four groups of stakeholders:

- Vehicle or vessel owners/operators,
- Manufacturers (OEMs) and distributors of vehicles and vessels,
- Fuel suppliers and refuelling infrastructure providers,
- Government and other policy makers.

In addition to the stakeholder actions, the pace and scale of roll-out will be determined by a number of global/regional factors that may be only tangentially linked to the issue of LNG in transportation but could nevertheless be crucial determinants. These include:

- Global energy prices and interfuel competition
- Technological developments
- Levels of economic activity
- Demand for transport

The following sections look at how all of the elements interact in the context of three major market user categories[4]:

- Passenger cars and light commercial vehicles
- Heavy goods vehicles
- Maritime shipping

6.4 Cars and Light Commercial Vehicles (LDVs)

Accurate statistics on natural gas vehicles (NGVs) are difficult to obtain. Figures from the vehicle manufacturers (ICCT 2016) include LPG in their definition whilst up-to-date numbers from the NGVA (a trade association) are not publicly available. Published figures indicate that across the EU, the market share for passenger NGVs is around 0.4% (Ricardo 2016), and this figure is largely due to Italy having by far the largest share of new registrations (14.3%) followed by Sweden with just 1.6%.[5] In addition, Germany has over 90,000 NGVs though this is a relatively small share of the total.[6] The size of NGV fleets is reflected in the number of CNG filling stations[7], the majority of which are located in Italy and Germany.

Despite expectations of growing numbers of NGVs, recent trend suggests at best a levelling off in interest. The share of new NGV registrations has fallen from a high point in 2010 of 3.7% for the EU and 21.6% in Italy. Over the same period, electric vehicle registrations (including hybrids) have grown from 0.6 to 1.9%.

From a stakeholder's perspective, the main attraction for buyers of NGVs is the relatively low price of the fuel compared to petrol or diesel and the potential for improved environmental performance. This latter factor is also attractive to governments.

The price advantage is primarily due to the difference in taxation, and this can vary significantly from one country to another. This saving has to be weighed against the higher cost of an NGV and the relatively limited range. Research by the author (Le Fevre 2014) indicates that a vehicle would have to have an annual usage in excess of 20,000 km or higher in order to achieve a payback within 3 years.

Measuring the environmental advantages of natural gas compared to other road transport fuels is complicated though the most commonly accepted approach for passenger vehicles is to measure well to wheel[8] (WTW) emissions which incorporate the entire life cycle of the fuel from production to combustion including extraction, separation and treatment, transportation, refining and distribution to the tank of the vehicle. The GHG emission performance[9] of NGVs is around 18% better than conventional gasoline cars and 6% worse than diesel vehicles (Ricardo 2016)—though in comparison with the latter, NGVs have virtually no NO_x or PM emissions.

Research in the UK suggests that car buyers base their choices on three key parameters of fuel efficiency/running costs, size/practicality and vehicle price. Fuel efficiency is important from both a financial and an environmental perspective though environmental factors per se do not appear to play a major role in influencing the purchase decision, and buyers typically viewed lower emissions as a "bonus".[10] In this context, NGVs do not offer a particularly compelling alternative to other options.

The same could be said for policy makers seeking a low-carbon transport solution since, unlike electric vehicles, NGVs do not present a pathway to a zero-carbon transport system. However, using biomethane materially improves the environmental performance of NGVs and whilst it is unlikely that supplies would be sufficient to meet the requirements of a large scale switch to CNG, this could represent an important niche for the future in some markets.

The less-than compelling case for NGVs has been reflected by the relatively restricted range of CNG models available from manufacturers who appear to be investing much of their effort into pure electric and hybrid vehicles.

The present evidence suggests that strong growth of NGVs in this sector is unlikely. Nevertheless, the development of biomethane coupled with continued state support for CNG cars in markets such as Italy and Germany means that the sector will retain some significance and could expand in the future. The huge size of the passenger car market means that even a small growth in market share could have a relatively large impact on gas demand and so should not be ignored.

6.5 Large Road Vehicles

Heavy-duty vehicles (HDVs) are typically defined as those in excess of 16 tonnes gross laden weight. NGVA statistics from 2014 show that there were around 5,000 natural gas fuelled HDVs using either CNG or LNG. This represents a very small share of the total population of such vehicles though the scope offered by LNG has reawakened interest in the sector in recent years.

Around 5% of total EU CO_2 emissions are estimated to come from HDVs[11] but reducing emissions from these vehicles is more challenging than passenger cars. Whilst electric vehicles can play a role for urban delivery vehicles and operations such as waste collection, the technology is less suited to long haul and regional usage. These journeys can account for up to 70% of HDV carbon emissions so alternative fuelling options could have a major impact. Natural gas can deliver CO_2 savings (particularly where biomethane is used) and there are also benefits arising from reduction in air pollutants such as NO_x and PM.

As explained above, stakeholder positions are key to assessing the overall market potential. The key decision maker in this sector is the

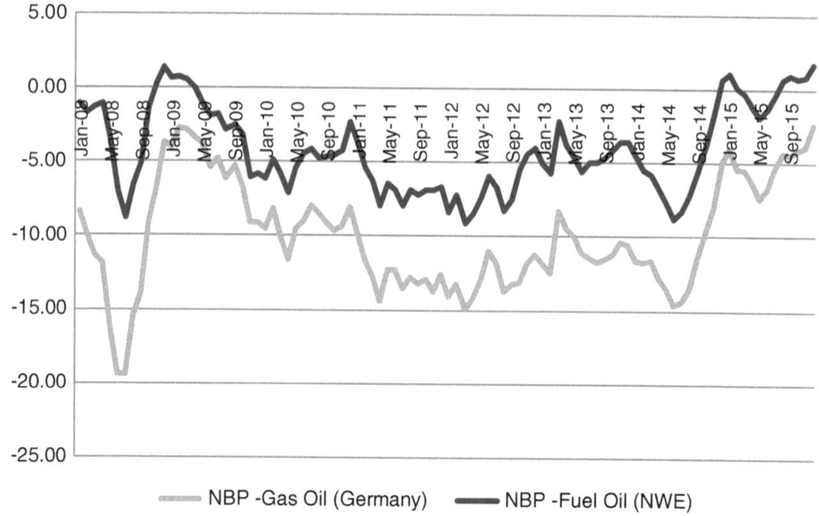

Fig. 6.1 European marine fuel price differentials with NBP gas prices in US$/mmbtu (negative differential means gas is cheaper). *Source* Argus. (NWE is Northwest Europe)

vehicle owner/operator, which in many European countries means operators of large fleets. These can be haulage providers, parcel distributors or large retail organisations with extensive distribution chains.[12] The decision to opt for NGVs is typically taken at the time of vehicle renewal and the financial case is crucial, although environmental considerations are becoming increasingly important.

The financial case is a trade-off between the discounted price of gas versus the higher capital cost for an NGV. Though the premium paid for NGV trucks is reducing, they still cost 20,000–€40,000 more for an equivalent vehicle in Europe (Kantor 2014). The commodity price of gas in Europe is typically well below the price of diesel (see Fig. 6.1) though the actual differential will depend heavily on taxation at the point of sale. Taxation rates for diesel vary widely across the EU and the picture can be further distorted by the availability of rebates to road hauliers in certain countries. The sales-weighted average tax rate (after rebates and excluding VAT) for truck users has been calculated at around €0.44 per litre (Transport & Environment, 2015)—approximately 40% of the average price for diesel in 2014. Levels can vary between countries, and the UK is a particular outlier with a diesel tax rate of over €0.70 per litre. Taxation rates for natural gas in transportation are much lower (typically a third or less of the diesel equivalent[13]) though the cost of refuelling is likely to be higher.

These savings must be weighed against lower efficiencies, the increased vehicle cost and slightly higher operating costs. The payback will depend crucially on distances travelled and for vehicles operating distances of around 100,000 km; the payback would be between 3 years and 4 years in Europe (Le Fevre 2014) with the UK and Italy achieving the fastest payback due to the greatest differentials in taxation.

Refuelling infrastructure and vehicle availability are also important concerns for operators. Whilst a wider range of trucks are becoming available, there is nothing like the extensive maintenance and repair network that is available for diesel vehicles. The refuelling network is developing—particularly along the so-called blue corridors described below—whilst some large operators are investing in their own facilities. Nevertheless, until vehicles can establish a significant track record with regard to reliability and the refuelling network is expanded, many operators will be reluctant to change from a well tried and tested option of diesel—particularly if there is little pressure from their customers or government.

Truck and engine manufacturers as well as infrastructure providers are likely to reflect this reluctance and resist over-committing to new investments until their customer needs are clearer. This so-called chicken-and-egg syndrome will act as a brake on rapid roll-out in this sector in most countries.

For government stakeholders, the key issue relates to environmental performance. Establishing an effective comparative environmental metric for freight vehicles is difficult and the focus has either been on engine rather than vehicle performance in a test environment or controlled field studies.[14]

Evidence to date shows a typical tank-to-wheel (TTW) CO_2 saving of 12 to 16% (65% for biomethane) and reduced NO_x and PM emissions by 85.6 and 97.1%, respectively.[15] However, when taking the full supply chain emissions and the extent to which unburnt LNG

Table 6.2 Comparison of CNG v LNG for truck operators

Consideration	LNG	CNG
Range/utilisation	Preferred where maximum range is important and utilisation high	Preferred for back to base operations with low mileage
Vehicle weight	Preferred for heavy weight vehicles	Preferred for light/medium weight vehicles
Refuel time	Preferred where time to refuel needs to be minimised	Preferred where there is plenty of time to refuel—e.g. overnight
Tank space	Preferred where space is limited	May be preferred if there is space for many tanks

Source Le Fevre (2014)

is contained in the exhaust (methane slip), comparative performance is worse. A study from Ricardo (2016) indicated a 5 to 13% increase in GHG emissions though as Wurster et al. (2014) point out technical improvements should mean that LNG will be better than diesel by 2030 as efficiencies improve. Furthermore, these metrics ignore the significant reductions in NO_x and PM that NGV HDVs could deliver with the consequent health benefits.

One other question facing operators is the choice between LNG and CNG. LNG has a number of advantages over CNG—in particular, it can deliver more power per unit of volume and so range is much greater—though CNG is often more readily available. The main considerations are summarised in Table 6.2.

The large commercial vehicle market has a number of characteristics that make it an attractive prospect for NGVs. The financial and environmental benefits are such that they can be an attractive proposition for buyers and operators. For policy makers, they provide an answer to growing concerns over urban air pollution they emit virtually no NO_x or particulate matter. Manufacturers are starting to produce a wider range of NGV options though the market is still at a very early stage.

The prospects are therefore reasonably promising though growth is unlikely to be particularly rapid. The key obstacles will be uncertainty over fiscal treatment of gas versus other transport fuels and the availability of reliable vehicles and refuelling infrastructure. The role of the EU is tackling these issues is considered in Sect. 6.7.

6.6 Marine Bunkers

In the marine sector, global shipping is almost wholly reliant on oil. The IEA (2016) estimate global marine bunker fuel consumption is in the region of 4 million barrels of oil/day (mb/d), which equates to 4.4% of total oil product consumption. Approximately 80% is in the form of low-price, low-quality fuel oil with much of the remainder being diesel—often termed marine gas oil (MGO). Cargo ships account for some 90% of global consumption. Other vessels such as passenger ships, fishing vessels, tugs and naval craft consume the rest.

Gas in the form of LNG is a feasible marine fuel; it has been used as the prime source of propulsion in LNG tankers for many years and is becoming established in other ships in parts of Europe.[16] The prime driver is emission control legislation known as MARPOL VI though there are also some potential price advantages. The other important factor is a relative scale of LNG demand for ships versus road vehicles making it an attractive new sector for fuel marketers and bunkerage ports. Ships can be converted though LNG is more likely to be chosen for new-build vessels.[17] Refuelling can be achieved from either onshore bunkers, LNG barges or LNG trucks.

The key stakeholder for assessing overall market potential is the International Maritime Organization (IMO), a specialised agency of the UN whose responsibilities include the prevention of marine pollution. Annex VI of the MARPOL convention[18] established emission standards for oceangoing ships for SO_x, NO_x and particulate matter both globally and in defined emission control areas (ECAs). The standards include a global limit of 3.5% sulphur in fuel established in August 2012 and a 0.1% limit in ECAs from January 2015. In addition, a global 0.5% limit is planned to apply by January 2020 though this may be deferred to 2025 depending on the outcome of a fuel availability study that is due to report in 2018. The NO_x limits are based on engine size and date of installation.

The ECAs in Europe presently comprise the Baltic, North Sea and the English Channel, and the MARPOL requirements were enshrined in EU legislation in 2012.[19]

The other important stakeholders are the vessel owners and operators who must decide how to meet the MARPOL requirements. There are three options:

- Convert the vessel to LNG—the fuel has no sulphur emissions and virtually zero particulate matter. However, conversion is expensive and LNG may not be readily available.
- Install emission-scrubbing technology to remove the sulphur from the fuel oil which is also expensive and may not be approved by some authorities.
- Switch to more expensive MGO, this can be done relatively easily and is particularly attractive if vessels are operating in ECAs for only part of the time.

A comparison of these options in terms of emissions is shown in Table 6.3.

Thomson et al. (2015) demonstrate that for a range of simulations, LNG has a materially improved performance for both GHGs and other emissions. Aagesen (2013) suggests that in the longer term a significant proportion of vessel owners (particularly container line and cruise ship operators) are considering LNG for both financial and environmental reasons.

Table 6.3 Comparison of emissions by marine fuelling option

Fuelling option	SO_2 Tonnes/y		NO_x Tonnes/y		PM Tonnes/y		CO_2 eq '000 Tonnes/y	
Vessel type	1	2	1	2	1	2	1	2
Heavy fuel oil plus exhaust scrub-bers	1.8	5.5	268	821	3.9	11.8	14.8	45.4
Marine Gas Oil	3.4	10.5	300	921	8.2	25.0	13.0	40.0
LNG	0.2	0.6	48	146	0.2	0.5	12.6	38.4

Sources: DMA 2012, *Ricardo* 2016
Note Vessel type 1 = coastal tanker/bulk carrier, vessel type 2 = large RoRo

The financial issue is critical for operators since fuel costs which can range between 58 and 78% of a vessel's operating expenses (DNV 2012). The commodity price of natural gas has generally been below that of fuel oil and gas oil in Europe though the gap has fallen since 2013. This is illustrated in Fig. 6.1 which shows the differential between the traded gas price at the UK's NBP minus the equivalent price of gas oil and fuel oil. The figure shows that natural gas has generally been cheaper than both oil products though the differential has narrowed as a result of the relatively sharper fall in oil prices.

In practice, the actual price paid by LNG marine fuel users will depend on factors such as point of delivery and other contractual terms. Various pricing arrangements are beginning to emerge. These include:

- "Hub plus" pricing—this is where the LNG price is linked to a gas trading hub such as TTF or NBP and a predetermined markup is added,
- "Oil product minus" pricing—where there is a guaranteed margin against a competing fuel such as FO or MGO.

Hub prices in Europe may display marked seasonality which is not so apparent in oil indexation, and oil product prices are better understood and more widely available than gas prices in some markets. For these reasons, buyers may prefer an oil minus arrangement rather than a gas Hub plus approach. There is also the possibility of a new LNG bunkering index once liquidity has reached a satisfactory point though this would be strongly influenced by overall LNG pricing dynamics which could be a drawback for some buyers.

The benefits of reduced LNG fuel costs have to be considered against the higher capital charges for a new or converted LNG-fuelled vessel. These relate primarily to the higher costs of a LNG-fuelled engine and of the storage and delivery system. Ricardo (2016) estimates that new-build LNG-fuelled vessel premiums range from €4.3 million for an inshore tanker/bulk carrier to €16.7 million for a large RoRo vessel.

Semolinos et al. (2013) have analysed the economics of new-build LNG shipping versus the two alternatives of HFO plus scrubbers and burning MGO:

- For LNG to be more attractive than HFO plus scrubbers, a discount at or below $2/MMBtu is required for most vessel types (tankers, container ships, ferries and most cargo vessels). Very large bulk carriers with high fuel economy require a discount of nearly $5/MMBtu.
- For LNG to be more attractive than the MGO option, a discount of between $2 and $4/MMBtu is required for most vessel types and very large bulk carriers require a discount of around $6/MMBtu.

This analysis suggests that LNG has strong attractions on financial grounds in the European markets though the advantage over fuel oil has narrowed somewhat.

A further important driver in the marine market will be the readiness or otherwise of ports to develop LNG bunkering facilities. Europe would appear to be well placed in this regard. The three largest bunkering ports—Rotterdam, Antwerp and Gibralter—are all either supplying LNG fuel or are planning to do so. A survey by Lloyds Register (2014) reported that 76% of European ports expected to have LNG available by 2020. GIE[20] report that there are 26 LNG bunker loading facilities in Europe with a further 5 under construction.

There is still a great deal of uncertainty over the eventual level of take-up of LNG as a fuel in the maritime market. The main barriers to growth are uncertainty over the timing of future MARPOL restrictions on FO, concerns over the competitiveness of LNG versus competing fuels and doubts over the availability of cost effective bunkering facilities.

The provision of state support towards the additional costs of shipping and refuelling could be crucial in the early stages. In this regard, Norway has lead the way. In May 2015, it was reported that 81% of all LNG-fuelled vessels were sailing in Norwegian waters[21] and most bunkering facilities are located there.

6.7 The Role of the EU

The European Union is likely to play an important role in determining the scale and scope of the adoption of natural gas as a transport fuel. EU policy with regard to fuel in transport is driven by two main objectives.[22]

- A desire to reduce dependence on imported oil for transportation where it accounted for an import bill of €1 billion per day in 2011 and a deficit in trade balances of some 2.5% of EU GDP,[23] and
- The reduction of vehicle emissions which account for around 20% of the EU's total CO_2 emissions and are increasing,[24] in order to achieve the gradual decarbonisation of transport.

These objectives have spawned a range of initiatives that have lent some support to the adoption of natural gas as a transportation fuel though many of these have been in the context of a broader drive towards reducing overall emissions. The most important initiatives impacting on natural gas are:

- The Fuel Quality Directive,[25] which sets tougher standards regarding pollutants in fuels and which, together with the Renewable Energy Directive, targets a 10% share of energy from renewable sources in transport in community energy consumption by 2020.[26]
- The Alternative Fuel Infrastructure Directive (European Commission 2014) which sets out timelines for the development of marine- and road-based refuelling facilities for alternative fuels including natural gas, hydrogen and electricity. This requires common technical standards by 2016, CNG/LNG refuelling stations every 400 km on key networks and at "sufficient" seaports by 2025. The LNG Blue Corridors project is one initiative established to meet these arrangements. They report that at the end of 2014, there were around 1,300 LNG-fuelled trucks in Europe of which nearly half were in the UK with the Netherlands and Spain accounting for most of the rest.[27]

In a working document published in July, the Commission (2016b) notes that most of the focus has been on cars and light commercial vehicles. Initiatives to reduce emissions from HDVs did not emerge until 2014 and these were primarily directed towards monitoring and reporting.

6.8 Growth Projections for LNG as a Transport Fuel

The increasing interest in natural gas in transportation has stimulated a number of forecasts of future demand. The market, however, is still very small and so minor differences in growth assumptions can result in a wide range of outcomes over a 20-year horizon. Furthermore, regional forecasts are complicated by uncertainty over marine fuel demand which is often assessed on a global basis.

Nevertheless, most forecasts agree that for marine demand, the pace of MARPOL roll-out and the future level of gas versus oil price differentials are likely to be the key drivers. For road vehicles, the pricing issue is the most important determinant (which requires a view on future levels of taxation as well as commodity price differentials) together with the extent to which government support might be available to assist with the capital cost of refuelling facilities or new vehicle purchase.

A study for the EU (EU 2016a) indicates that LNG could provide about 3% of fuels used in heavy goods vehicles by 2030 (and 6–8% by 2050) and about 4% of fuels used in inland navigation (7% by 2050). Uptake of LNG would also take place in international shipping, especially in the short sea shipping segment, providing about 4% of overall bunker fuels by 2030 and 10% by 2050. CNG in buses could also increase. Biomethane is also expected to have an increased market share of between 0.2% and 1.5% by 2050 though this is modest compared to liquid biofuels. Overall demand for gas could reach 2–6% of total energy demand in transport by 2050 equivalent to between 8 and 24 bcm of natural gas

A study by Cedigaz (2014) suggests base case global demand for LNG in marine transport could exceed 100 bcm by 2035, and it would be reasonable to assume at least 20% of this would be for the European market. The same study forecasts base case demand in Europe for LNG for road transport in excess of 20 bcm which suggests a total LNG demand of 40 bcm by 2035.

It is clear that there is a wide range of possible outcomes even on a 10-year horizon. LNG could become a significant source of demand by 2030 though it would appear that some of the higher range forecasts are unlikely to materialise at present rates of build-up.

6.9 Conclusions

Natural gas as a transport fuel clearly has a role to play in Europe as it provides both financial and environmental advantages over other fuels. The prospects are enhanced by the combination of demand for cheaper, more environmentally friendly fuels coinciding with a period of potential gas over-supply and increased pricing flexibility.

The advantages of natural gas are, however, not sufficient to give it an exclusive edge over emerging alternatives in all sectors of the transportation fuel market. The prospects in the passenger car market do not look particularly promising as the trend seems increasingly towards electric vehicles, though if government support continues, it will still feature in some countries. The strongest prospects are in the marine and road freight markets, and here there is a strong likelihood that gas (including biomethane) will feature as a transportation fuel in the coming 20 years. Developments in the maritime sector are likely to be most important as this will provide a platform of significant scale to allow road-based usage to develop subsequently. This is likely to be primarily via LNG though CNG will also feature in land-based applications.

The key determinants are likely to be whether gas prices remain competitive both with existing fuels and new alternatives and the extent to which government support specifically incentivises natural gas and biomethane.

The early stage nature of the market means that there are likely to be periods of stop–start investments in infrastructure. During periods of under-investment, users will be reluctant to commit whilst over-investments will lead to under-utilisation and failure to make promised returns. Timescales will be extended by the fact that most decisions to switch to gas will take place at the point of vehicle/vessel renewal. This underlines the fact that it is still too early to form a definitive conclusion on overall prospects, but if conditions remain favourable, the acceleration in take-up could become quite rapid.

Notes

1. IEA (2015).
2. https://www.ssb.no/en/energi-og-industri/statistikker/energibalanse/aar-forelopige/2016-05-20?fane=tabell&sort=nummer&tabell=266659.
3. Incomplete combustion with methane venting to atmosphere is referred to as methane slip.
4. Bus, rail and inland waterways also have potential for natural gas usage though relative volumes are small.
5. ICCT 2016. Includes LPG vehicles.
6. http://www.dena.de/en/topics/energy-efficient-mobility.html.
7. LNG is not a suitable fuel for small vehicles.
8. This combines the well to tank (WTT) and tank to wheel (TTW) measures.
9. Measuring the comparative environmental performance of vehicles is a highly complex issue. For more information on the subject, see Le Fevre (2014), Ricardo (2016) and Edwards et al. (2013).
10. Lane and Banks (2010).
11. http://newsroom.unfccc.int/.
12. Large retailers may outsource distribution but require their contractors to use clean fuels. For examples of users, see http://www.boconline.co.uk/en/clean-technology/liquefied-natural-gas/lng-transport-fuel/boc-customers/index.html.
13. See http://ec.europa.eu/taxation_customs/resources/documents/taxation/excise_duties/energy_products/rates/excise_duties-part_ii_energy_products_en.pdf.
14. See https://www.gov.uk/government/uploads/system/uploads/attachment_data/file/448049/low-carbon-truck-trial-2.pdf for an example.
15. PWC (2013). https://www.gov.uk/government/uploads/system/uploads/attachment_data/file/448049/low-carbon-truck-trial-2.pdf (2015).
16. DNV (2015) report there were 63 LNG-fuelled vessels (excluding LNG carriers) in operation as at May 2015 with a further 76 on order.

92% of the existing vessels are in Europe and the vast majority are in Norway.
17. Some new vessels are being built to be adaptable to LNG in the future. See DNV (2015).
18. IMO (2005).
19. Directive 2012/33/EU.
20. http://www.gie.eu/index.php/maps-data/lng-map.
21. DNV-GL 2015—excludes LNG tankers.
22. See http://ec.europa.eu/transport/themes/urban/cpt/.
23. European Commission (2013)(COM 17) page 2.
24. See http://ec.europa.eu/clima/policies/transport/index_en.htm.
25. European Commission (2009a).
26. European Commission (2009b).
27. http://www.unece.org/fileadmin/DAM/energy/se/pp/geg/geg2_jan2015/ai9_2_Pilskog.pdf.

Bibliography

Aagesen. 2013. LNG-fuelled deep sea shipping—The outlook for LNG bunker and LNG-fuelled newbuild demand up to 2025. Lloyds Register Group.

Cedigaz. 2014. LNG in Transportation – A report for Cedigaz by Le Fevre C., Madden M. and White N.

DMA. 2012. North European LNG Infrastructure Project. Danish Maritime Authority.

DNV. 2012. Shipping 2020, a report on technology uptake for the maritime shipping industry for DNV Maritime Oil and Gas. DNV.

DNV-GL. 2015. In Focus—LNG as Ship Fuel, LNG Report No. 01 2015. DNV-GL.

Edwards, R., Larive, J-F., Rickeard, D., and W. Weindorf. 2013. Well to tank report Version 4.0, JEC Joint Research Centre-EUCAR-CONCAWE collaboration, EUR 26028 EN-2013.

Eurogas. 2016. Statistical Report. Eurogas.

European Commission. 2009a. The Fuel Quality Directive DIRECTIVE 2009/30/EC.

European Commission. 2009b. The Renewable Energy Directive DIRECTIVE 2009/28/EC.

European Commission. 2013. European alternative fuels strategy, Brussels, 24.1.2013 COM(2013) 17 final.
European Commission. 2014. Alternative Fuel Infrastructure Directive DIRECTIVE 2014/94/EU.
European Commission. 2016a. The EU Reference Scenario 2016. Energy Transport and GHG July 2016. Trends to 2050.
European Commission. 2016b. A European Strategy for Low-Emission Mobility. COM 2016.
Eurostat. 2016. Consumption of energy statistics. Brussels.
IEA. 2015. World Energy Outlook, Nov 2015.
IEA. 2016. Medium Term Oil Market Outlook Report, Market Analysis and Forecasts to 2021.
ICCT. 2016. European Vehicle Market Statistics, Pocketbook 2015/2016, The International Council on Clean transportation.
IMO. 2005. *MARPOL*. International Maritime Organisation: Annex VI prevention of air pollution from ships.
Kantor. 2014. Regulatory implications of new developments in the gas supply chain, Report by Kantor Management Consultants for Agency for the Cooperation of European Regulators. ACER.
Lane and Banks. 2010. Low Carbon Vehicle Partnership Car Buyer Survey: Improved environmental information for consumers, Research conducted by Ecolane & Sustain, June 2010.
Le Fevre. 2014. The prospects for natural gas as a transportation fuel in Europe. Oxford Institute for Energy Studies. Working Paper NG 84.
Lloyds Register. 2014. LNG bunkering infrastructure survey 2014—The outlook of Ports on provision of LNG bunkering facilities. Lloyds Register Marine.
PWC. 2013. LNG as a fuel – why it's the next big thing for the transport sector, http://www.icisconference.com/uploads/assets/2013_PwC_LNG_FOR_THE_TRANSPORT_SECTOR_v1.pdf.
Ricardo. 2016. The role of natural gas and biomethane in the transport sector, Report for Transport & Environment, Brussels.
Semolinos, et al. 2013. LNG as Marine fuel—Challenges to be overcome, presentation at LNG 17, Houston Apr17, 2013.
Thomson, H., Corbett, J., and Winebrake, J. 2015. Natural gas as a marine fuel. Energy Policy, 87(2015), 153–167.
Transport & Environment. 2015. Europe's tax deals for diesel.
Wurster, R. et al. 2014. LNG as an alternative fuel for the operation of ships and heavy-duty vehicles, LBST, March 2014.

Author Biography

Chris Le Fevre has worked in the energy sector for 35 years. He is presently a Senior Visiting Research Fellow at the Oxford Institute for Energy Studies and an independent energy consultant specialising in strategic, commercial, regulatory and operational issues in the natural gas sector. He has held a variety of positions to executive director level in Transco (now National Grid Gas), British Gas and Shell and served on the Board of the Northern Ireland Utility Regulator.

7

Global LNG Market Outlook and Repercussions for Europe

Anne-Sophie Corbeau

7.1 Introduction

Europe[1] plays a unique role in global liquefied natural gas (LNG) markets. It provides a market of last resort when supply is plentiful, and a market from which supplies can be diverted when other markets are tight. But LNG imports are also influenced by the balance between Europe's gas demand, production and pipeline imports. Europe is the largest gas importer in the world and relies on both LNG and pipeline gas to meet its import needs, which represent almost half of European gas demand. Of the two, LNG accounts for only around 10% of Europe's gas consumption, leaving pipeline gas in a dominant position. The only other market where such an interaction between pipeline and LNG supplies occurs is China. The respective volumes of pipeline gas and LNG that are imported ultimately depend on total import needs, contractual commitments, the respective price of the different supply sources, supply developments in Europe's pipeline suppliers and the state

A.-S. Corbeau (✉)
KAPSARC, Riyadh, Saudi Arabia
e-mail: Anne-Sophie.Corbeau@kapsarc.org

M. Hafner and S. Tagliapietra (eds.), *The European Gas Markets*,
DOI 10.1007/978-3-319-55801-1_7

of global LNG markets. Geopolitics also plays an important role. The European Commission (EC) has been vocal about the need to diversify away from Russian pipeline gas and to consider LNG as a possible alternative. Against this backdrop, Europe has plentiful and underutilized regasification capacities, liquid and well-developed gas trading hubs in Northwest Europe and a relatively well-developed pipeline infrastructure that brings a certain degree of flexibility to the process of transporting gas across Europe.

Looking forward, Europe will continue to be influenced by this duality of external and internal factors, resulting in a wide range for estimates of LNG imports. Such extreme variations have already taken place. LNG imports halved from 2010 to 2014: While LNG was diverted to high-priced Asian importers, pipeline gas filled in the gap. Similarly, flexibility in pipeline contracts allowed more LNG to reach Europe when there were no other markets for it. Until now, these substantial volume variations have been manageable.

The extent to which LNG will be available—or not—for European gas markets could have far-reaching consequences for both pipeline gas and LNG suppliers, from market and geopolitical perspectives. Just because substantial regasification capacity exists, it does not guarantee that LNG supply will go there. LNG can still be diverted/re-exported to other markets if the price offered there is more attractive. By contrast, the EC wishes to diversify away from Russian pipeline gas through alternatives from North Africa, the Caspian region and, obviously, through LNG. But this political will may conflict with market realities. First, higher prices to attract LNG back to Europe would likely have a negative impact on demand, while gas has already been struggling against renewables and coal-fired plants. Suppliers, including Norway and Algeria, need certainty about future imports before investing in new infrastructure. Many European LNG suppliers with LNG contracts are at the same time portfolio players optimizing between destinations. This, along with the rise in short-term LNG contracts and spot cargoes, could exacerbate potential variations in Europe's future LNG imports. Tight LNG markets will put Europe in competition with other markets, while oversupplied LNG markets—as is currently the case—mean that large volumes of surplus LNG could be aiming at Europe as the only market able to absorb such quantities. Depending on the quantities at stake, this could put pressure on pipeline gas suppliers, notably Russia. Russia has ample spare gas production capacity and is a low-cost supplier. In this context, Russia may put the resilience of LNG suppliers to the test by defending its market share.

7.2 The Role of Europe in Global LNG Markets (2000–2016)

As of late 2016, 12 European countries are importing LNG: Belgium, Greece, Italy, France, Lithuania, the Netherlands, Poland, Portugal, Spain, Sweden, Turkey and the United Kingdom (UK). The only terminals under construction are two in the Canary Islands and one in Malta. European regasification capacity amounts to 161 mtpa (220 bcm/y) as of 2016, against imports at 38 mtpa in 2015. The average utilization rate of this regasification capacity is therefore low, compared with the world average (32%). Looking forward, those terminals most likely to be built will be for supply diversification purposes, notably in Southeastern Europe and in the Baltics, areas which depend strongly on one supplier: Russia.

The years from 2000 to 2016 can be divided into three distinct periods, with totally opposite implications for Europe. These help to demonstrate how European gas markets interact with global LNG markets according to global LNG supply/demand dynamics and comparative price levels. Such developments may also assist in understanding the future:

The 2000–2011 period was a period of rapid extension for the LNG trade, which grew by a multiple of 2.4. LNG supply grew as new suppliers such as Oman, Egypt, Russia and Peru emerged while Qatar substantially increased its LNG export capacity and became the largest LNG exporter (78 mtpa). LNG demand increased and diversified away from Asia's traditional mature LNG buyers (Japan, Korea and Taiwan). As European gas demand grew rapidly, driven by the expansion of gas in the power generation sector, its LNG imports increased by 40–65 mtpa.

Its share in LNG trade remained constant over 2000–2008 at around 25% before jumping to 28–29% over 2009–2010 (See Fig. 7.1). China and India started importing LNG in the mid-2000s, joined later by Thailand, Latin American and Middle Eastern countries (GIIGNL 2016). Finally, US LNG imports collapsed after 2007 due to the rise of US shale gas.

The 2011–2014 period tells a totally different story: LNG trade witnessed a pause as some LNG export plants faced difficulties. This was not compensated for by the arrival of new LNG trains, while Angola LNG's exports were halted less than 1 year after the plant started. Oil prices at around \$100/bbl until mid-2014 drove oil-linked Asian gas prices to levels above \$15/MMBtu. The global gas market experienced a general tightness due to the strong pull on LNG to meet incremental Asian demand in Japan (+18 mtpa), China and India, as well as in new Southeast Asian LNG importers. The share of Asia rebounded from 61% in 2010 to 75% in 2014. The real change over that period was the collapse of Europe's LNG imports from 65 mtpa in 2010 to 32 mtpa in 2014, the lowest level in over a decade, lowering Europe's share in global LNG trade from 29 to 14%. LNG volumes were diverted away from Europe due to the higher demand and better prices in Asia. This resulted in the interesting phenomenon of LNG re-exports,

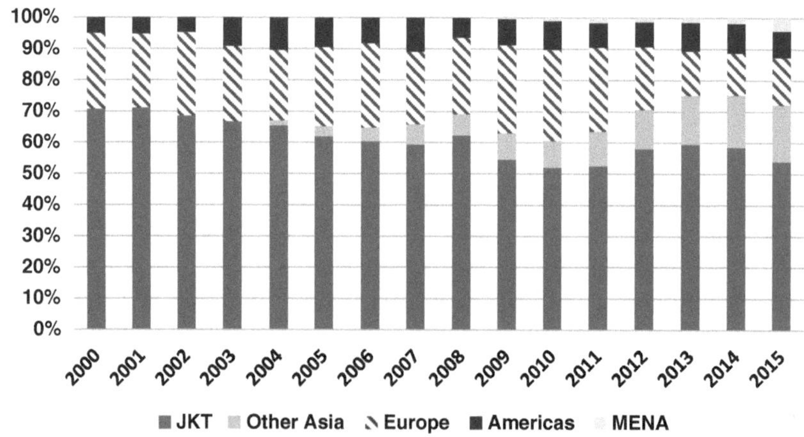

Fig. 7.1 Regional shares in LNG imports. *Source* GIIGNL (2016)

notably from Spain and Belgium, toward the lucrative Asian markets. Simultaneously, European gas demand collapsed, due to the effects of lower economic growth, strong growth of renewables and the improved competitiveness of coal-fired plants.

The period since late 2014 has some similarities with that of 2000–2011. Markets have turned upside down again as a result of declining oil and gas prices in Asia, a convergence between Asian and European gas prices and the start of the wave of new LNG supply. Six new LNG projects have started up in Australia, Indonesia and the USA. But, this is just the start of an unprecedented boom in LNG supply: Most LNG capacity additions, notably from Australia and the USA, will come during 2016–2018. Meanwhile, new countries started importing LNG—Pakistan, Jordan, Egypt and Poland. Imports from the MENA region reached almost 10 mtpa in 2015, while Asian LNG imports dropped—by 2%—for the first time since 2009. The implications for Europe were limited: Its share in the LNG trade rebounded only slightly to 15% as its LNG imports gained 5 mtpa. This was due to the evolution of the European supply/demand balance. Demand rebounded slightly, more because of the return to normal weather conditions than due to any improvement in gas competitiveness. Production dropped as a result of the caps put in place on the Dutch Groningen field. Overall, pipeline gas exporters seem to have benefited more from the rise in import needs than the LNG sector.

The dynamics of individual countries also play a significant role in the evolution of European LNG imports (see Fig. 7.2). The UK perfectly reflects the changes in global market dynamics as most LNG is delivered at around NBP prices. Main UK supplier Qatar diverted large quantities to Asia after the Fukushima nuclear incident, though Qatari cargoes came back in 2015 as Asian LNG demand weakened and regional spot prices started to converge. The story is somewhat similar for Belgium and to a certain extent also for the Netherlands. Spain has suffered from declining gas demand since 2008, notably in the power sector due to the exponential growth of renewables. LNG has been more impacted proportionally by that than Algerian pipeline gas imports. Turkey presents a different case as a growing market—demand gained 10 bcm over 2010–2014—and has limited supply alternatives so

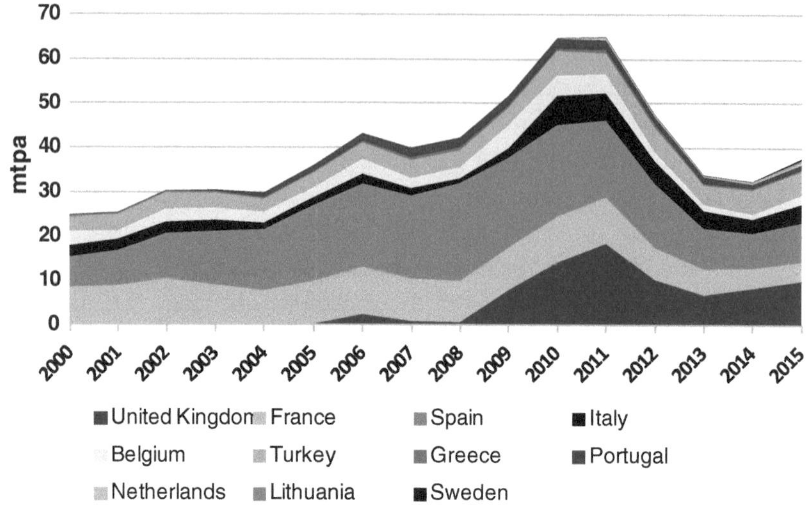

Fig. 7.2 Europe's LNG imports, 2000–2015. *Source* GIIGNL (2016)

that LNG imports have increased regularly. Lithuania started importing in 2014 to diversify away from and put pressure on Russia.

Prices played a determining role in the LNG dynamics between Europe and other regions. Up till 2008, prices in the USA, Europe and Japan were relatively aligned. After 2009, prices diverged due to the combination of three factors:

- The rise in US shale gas production kept gas prices at or below \$4/MMBtu.
- In Europe, renegotiations in long-term contracts with a wider inclusion of spot indexation put downward pressure on European gas prices, which stayed below \$10/MMBtu over 2012–2014.
- In Asia, by contrast, LNG import prices increased substantially after 2011 and occasionally peaked at around \$18/MMBtu, attracting LNG away from Europe.

However, since 2015, European and Asian prices have converged, and LNG re-exports have declined substantially.

7.3 Future LNG Supply

The past few years have been beset by difficulties with existing facilities running short of gas (Egypt), facing war (Yemen and Libya) and continuing technical issues (Angola and Norway). Meanwhile, LNG supplies fell in many LNG exporting countries—Oman, Indonesia, Brunei and Trinidad and Tobago. Since 2014 and the start-up of the Papua New Guinea plant, LNG supply picked up again. Despite the commissioning of two Australian LNG plants—Queensland Curtis and Gladstone—and Donggi Senoro in Indonesia, LNG supply increased by only 6 mpta in 2015 due to the issues cited above. In early 2016, two Australian LNG plants, APLNG and Gorgon, and Sabine Pass in the USA started operating.

7.3.1 An Unprecedented LNG Supply Wave up to 2020

The large expansion of LNG supply taking place over 2015–2020 will bring 150 mtpa of new capacity, on top of that from Angola LNG (5.2 mtpa) (see Table 7.1). This is totally unprecedented in terms of scale; consequently, global LNG markets are expected to be totally transformed. Australia and the USA provide the largest additions—63 and 64 mtpa, respectively. Australia will become the largest LNG exporter, followed by Qatar and the USA. Interestingly, several floating LNG (FLNG) plants are under construction, marking the beginning of a new trend: Kanowit in Malaysia is likely to be the first one operational as the vessel arrived on site in June 2016. Prelude FLNG and Cameroon FLNG are both under construction and planned for 2017–2018. Meanwhile, the small FLNG unit initially earmarked for Colombia could be available for a small field development, but the location is undetermined at the time of writing.

It is worth noting that many trains recently commissioned have been delayed due to the oversupply on global LNG markets as well as to low spot gas and oil prices. This pattern may be repeated over the next 4 years, with LNG plants currently under construction also being delayed. The development of LNG supply up to 2020 will be largely

Table 7.1 LNG plants schedule—2015–2020

Project	Country	Capacity (mtpa)	Online date
Operating as of June 2016			
Queensland Curtis T1	Australia	4.25	Q1 2015
Queensland Curtis T2	Australia	4.25	Q3 2015
Donggi Senoro	Indonesia	2	Q3 2015
Gladstone T1	Australia	3.9	Q4 2015
APLNG T1	Australia	4.5	Q1 2016
Sabine Pass T1	USA	4.5	Q1 2016
Gorgon T1	Australia	5.2	Q1 2016
Gladstone T2	Australia	3.9	Q2 2016
APLNG T2	Australia	4.5	Q4 2016
Gorgon T2	Australia	5.2	Q4 2016
Under construction			
Sabine Pass T2	USA	4.5	Q4 2016
Kanowit FLNG	Malaysia	1.2	2017
Malaysia LNG	Malaysia	3.6	2017
Gorgon T3	Australia	5.2	2017
Ichthys T1–T2	Australia	8.4	2017–2018
Sabine Pass T3–T5	USA	13.5	2017–2019
Sengkang	Indonesia	2.0	2017
Wheatstone T1–T2	Australia	8.9	2017–2018
Cameron T1–T3	USA	12.0	2018
Cameroon FLNG	Cameroon	1.2	2018
Cove Point	USA	4.6	2018
Prelude FLNG	Australia	3.6	2018
Freeport T1–T3	USA	13.2	2018–2019
Yamal T1–T3	Russia	16.5	2018–2020
Corpus Christi T1–T2	USA	9.0	2019–2020
Tangguh T3	Indonesia	3.8	2020
Woodfibre LNG	Canada	2.1	2020
Restart			
Angola LNG	Angola	5.2	Q3 2016

Source Companies' Web sites, KAPSARC research

determined by these additions as well as by the evolution of LNG supplies from existing plants. All the liquefaction plants under construction are likely to be completed by 2020 unless technical difficulties arise, as has been the case with Angola LNG and Gorgon T1. The real question mark is over the start date and the pace of the ramp-up of these LNG plants. Many existing LNG plants face supply issues or higher domestic demand and have been operating below capacity. The question is whether and to what extent such a trend will continue. While it is unlikely that those in Egypt will resume exports at normal levels or that Libya LNG will restart by 2020, potentially Yemen LNG could restart.

We can consider two scenarios for future LNG supply up to 2020 (see Fig. 7.3):

- The high case assumes a relatively timely start for liquefaction plants under construction, based on the information available as this report is being written, in tandem with LNG supply from selected LNG export plants declining. This means that LNG supply will come faster to the global gas markets and reach higher levels, potentially deepening the oversupply phenomenon and lowering LNG spot prices even further.
- The low case assumes both delays for the plants under construction and declines from existing LNG plants. Consequently, LNG output will increase by over 100 mtpa over 5 years, but more slowly than in our first case.

This means that LNG supply can be expected to rise to between 353 and 380 mtpa by 2020.

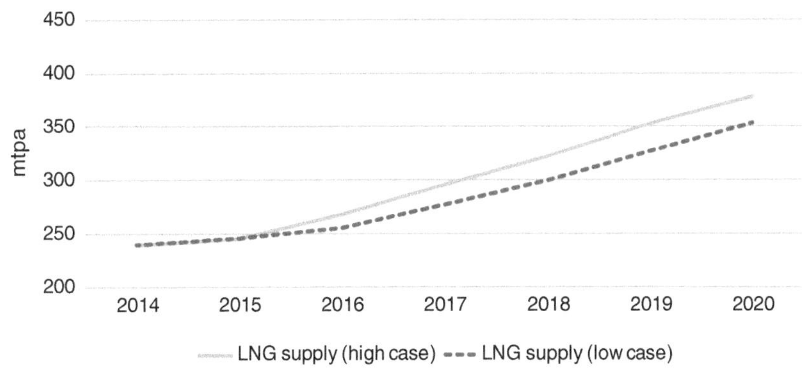

Fig. 7.3 Scenarios for global LNG supply (2014–2020). *Note* Both LNG supply scenarios are the result of bottom-up analysis, looking individually at each LNG exporting country and at the LNG plants currently under construction. *Source* KAPSARC research

7.3.2 Beyond 2020

Looking beyond 2020, the evolution of future LNG supply is characterized by uncertainty. The LNG industry is currently undergoing a familiar boom-and-bust supply cycle. The early stages of this boom have already had obvious consequences on global LNG markets: LNG spot prices collapsed to levels as low as $4/MMBtu in Asia and Europe in mid-2016. However, the biggest increase in LNG supplies, which will take place over 2016–2018, is still ahead of us. The compounded effects of low prices and the surge in capacity are setting the stage for a potential dearth of financial investment decisions (FIDs). Many project FIDs originally intended for 2015 and 2016 have been postponed indefinitely, notably in North America—for example, LNG Canada, Lake Charles LNG, Douglas Channel LNG and Prince Rupert LNG.

Around 1000 mtpa of LNG export projects are planned globally. A large portion of this capacity is concentrated in North America—315 mtpa in the USA and 352 mtpa in Canada—while there are around 90 mtpa in Australia, 70 mtpa in Eastern Africa and 75 mtpa in Russia. US projects have been quite popular due to the brownfield nature of most of them, but the gap between US and Asian gas prices is now insufficient to justify committing to investment. Greenfield Canadian projects suffer from higher costs due to the pipeline and LNG infrastructure that requires to be developed and likely objections from the First Nations, Aboriginal Canadians. Russian LNG projects face sanctions and a difficult access to finance. Australian projects need to bring costs down, while many project developers are still going ahead and bringing trains under construction on line when completed. In Eastern Africa, Mozambique is clearly ahead of Tanzania and Eni's Coral FLNG project seems well advanced, but both countries suffer from uncertain regulatory frameworks and lack of qualified workforce and infrastructure (Corbeau and Ledesma 2016).

Other projects could move ahead in other countries. For example, BP took its FID on Train 3 of the Tangguh LNG project in Indonesia in mid-2016, and there has been a lot of interest in further developing Papua New Guinea's LNG capacity. Meanwhile, many companies have different LNG projects in several countries and can advance those which are the most competitive. In particular, ExxonMobil has several projects strategically placed in various countries and is now expected to acquire a stake in Eni's FLNG (Reuters 2016).

Regardless of which LNG project is considered, all face the same issues: costs and demand uncertainties. In the current price environment, project sponsors must cut costs drastically: While US projects under construction are relatively in line with typical cost escalation figures in the oil and gas industry, the cost of recent Australian projects such as Gorgon LNG has reached record levels. Very few projects would be sanctioned in the current oil and gas price environment, with oil at $40–50/bbl and spot gas prices at $4–6/MMBtu in Europe and Asia. Investors have pledged to work on cost reduction strategies including using multiple, smaller plants and standardized designs, while a few are considering FLNG. However, costs are likely to remain high for greenfield LNG projects located in remote places.

Projects' success will therefore depend as much on their own technical characteristics as on their location. Brownfield expansions are likely to be the front runners: This is why companies are looking at expanding Tangguh in Indonesia, PNG or various US LNG projects, such as Corpus Christi Train 3, Cameron T4 and Freeport T4. Among the potential brownfield expansions, Qatar stands out: It is the world's lowest cost supplier and it benefits from a key geographical position between Asia and Europe. The moratorium on further development of the North Field imposed in 2005 was lifted in early 2017 paving the way for more LNG. Additional exports could also come from the utilization of existing LNG export plants by other countries, a move whereby one country would use the underutilized plants of the second. This could take place in Venezuela/Trinidad and Tobago, Israel/Egypt and Iran/Oman, but the geopolitical aspect of all these projects is decisive. Liquids-rich projects stand a better chance since liquids provide an additional source of revenue at times of robust oil prices. Finally, FLNG is still unproven at this stage, but could in some cases be less expensive than onshore project options. There might be a progressive move toward smaller projects, including FLNG, with overall lower capital costs.

Countries with a stable fiscal and regulatory framework, available infrastructure, reasonable local content requirements, a qualified workforce and political stability will be the front runners. Those that are too demanding in terms of fiscal take, with an unstable, complex or unfavorable regulatory environment, will lag behind. This may leave some gas resources undeveloped despite their considerable potential. The lengthy fiscal discussions in Canada have certainly resulted in the Western projects' missing a key window of opportunity. Some LNG projects may still move ahead due to the pure political backing, once certain concerns have been met. But among all the projects considered, expansions of projects that can be achieved in a timely manner and without significant cost overruns stand the best chances. The ability of LNG producers to keep costs under control and make LNG affordable is key for the future development of LNG demand.

Long-term contracts are still seen as necessary, but there is also a trend from buyers toward contracts of shorter duration—such as 10 years—which better fit the important demand variations they face. Increasing quantities are also in the hands of portfolio players, which have the possibility to arbitrage between markets. Looking at the LNG contracts signed over the past 3 years, very few long-term or short-term contracts have Europe as a destination, even though some involve a European player contracting LNG for its portfolio, since this does not guarantee that the LNG will end up in Europe. Beyond the price level, the question of the type of indexation in the long-term contract will be crucial: oil-indexed or spot prices. Buyers have been trying to move away from the traditional oil indexation by using alternatives such as US Henry Hub or European spot prices. But Asia still suffers from the absence of an Asian trading Hub reflecting the region's own supply/demand balance.

The magnitude of the bust to come will be determined by two factors—how long the pause in FIDs lasts and when/if global markets are expected to rebalance. Our view is that markets will rebalance at some point between 2020 and 2025. Given that LNG projects need around 4–5 years of construction on top of a couple of years for negotiations, we will need several projects to commit to proceeding in late 2016/early 2017, if markets tighten as soon as 2020–2021. But many project sponsors are likely to wait until oil and gas prices recover to levels sufficient to enable their projects to be sanctioned, at the same time as having some confidence that gas markets will rebalance within 4 or 5 years. However, some buyers may be ready to help projects that they consider strategic due to security of supply reasons. This implies that these companies would own a stake and bring cash from their balance sheets, and from the financial institutions of their own countries, such as export credit agencies (ECAs).

Future LNG supply reflects expectations of future demand: The perceived appetite for LNG from Asian markets along with the high prices they were paying in the early 2010s precipitated the flurry of FIDs from 2009 to 2015. Similarly, the same uncertainty about future gas, and therefore LNG, demand together with low gas prices is resulting in the current pause in FIDs. Due to the long lead time required to build an LNG plant, the LNG industry is likely to continue to move in cycles over the next few decades, while LNG supply will become increasingly more flexible and the role of portfolio players will increase as well. This brings considerable uncertainty to the supply outlook post-2020, emphasizing the role of Europe as a tool to manage oversupply or tightness, as the only market other than China which can use both LNG and pipeline supply. Because of its liquid trading hubs and important underutilized LNG import capacity with third-party access, Europe is more likely to provide some demand-side flexibility. However, there might be limits to how little LNG supply Europe needs or how much it can absorb, not only due to technical constraints, but also due to the ability and willingness of Europe's other suppliers to let LNG act as a shock absorber and potentially benefit from that situation. The limits of this flexibility could thus be tested by geopolitical developments, notably the relationship with Russia, which intends to keep a key role in Europe's gas supply.

7.4 Understanding LNG Demand Dynamics in Other Regions

World LNG trade reached 245 mtpa in 2015, with 72% coming from Asia (177 mtpa), 15% from Europe (38 mtpa), 8% from the Americas (21 mtpa) and 4% from the MENA region (10 mtpa).

Looking forward, Asia is expected to remain the largest importer of LNG, while developing markets in Latin America, the Middle East and Africa could grow beyond their current status as niche markets. As described in the overview of 2000–2016, Europe's LNG imports will be influenced by the appetite of other importers and by regional price levels, as well as by the global LNG supply available.

7.4.1 The Mature Markets: Japan, Korea and Taiwan

The group of historical LNG importers—Japan, Korea and Taiwan—imported 140 mtpa in 2014, but only 133 mtpa in 2015 on the back of nuclear restarts and because gas is more expensive than coal. The main factors impacting their future LNG demand are:

- The impact of economic growth and population decline on future energy demand,
- The evolution of nuclear energy policies and
- The role of coal and renewables in the power generation sector.

The three countries have different approaches to nuclear energy. Japan had progressively to shut down all its nuclear power plants after the accident at the Fukushima plant. While some have restarted, they face strong opposition from the population. Korea still expects nuclear to play an important role in its power generation mix, and several units are under construction. Meanwhile, nuclear power has been facing opposition in Taiwan for years and the existing units there are likely to be decommissioned. Meanwhile, the three countries see coal as playing an important role. Many new coal-fired plants are being built or are planned and coal-fired generation has been performing quite well there recently. None of these countries has significant gas production.

Japan's future LNG demand depends largely on the future role of nuclear in the power mix, given that the power generation sector is a key consumer of natural gas. The last Japanese Ministry of Economy, Trade and Industry (METI) energy strategy from 2015 foresaw around 20–22% for nuclear by 2030, equivalent to around 200 TWh, compared to 280 TWh in 2010. On this basis, METI announced that LNG demand would drop to 62 mtpa by 2030, 30% below its peak in 2014. However, its next strategy is likely to be more conservative for nuclear, given the experience with restarting nuclear units so far. Coal is expected to keep an important role in the power mix at around 26%, according to the last energy strategy, but it remains to be seen how this will be compatible with the recommendations of the 2015 United Nations Climate Change Conference (COP21). There is likely to be a further push on renewables, even though the target of 22–24% by 2030 is quite aggressive, and on energy efficiency. In its latest World Energy Outlook, the International Energy Agency (IEA) forecast gas demand would stay around 100–104 bcm over the long term (around 76 mtpa). Scenarios from Asia-Pacific Economic Cooperation (APEC) predict a different path for gas demand, which is seen as dropping in the medium term due to nuclear coming back and then increasing from 2020 onward to around 75 mtpa (APEC 2016).

In Korea, the outlook for future LNG demand used to be bright, but this has changed fundamentally since 2014. In 2014, METI's energy strategy forecast gas demand would increase to 46 mtoe in 2011, with 36 mtpa of imported LNG, to 70 mtoe in 2030 and 73 mtoe in 2035. However, after a period of high gas prices and the economic slowdown of China, these forecasts were massively changed in 2015. METI now predicts LNG demand as 33.96 mtpa in 2022 and 34.65 mtpa in 2029, against 33.4 mtpa in 2015. According to these figures, the expected growth in the residential and industrial sectors will not be sufficient to offset the drop in the power sector. The medium-term outlook for gas demand is relatively bleak, given that 4.2 GW of nuclear capacity is under construction and scheduled to come online over 2017–2019 (World Nuclear Association 2016), and around 8 GW of additional coal-fired capacity is also expected online before 2020. At the same time, the cost of gas-fired generation has been two to three times higher than that of coal or nuclear, resulting in a much lower utilization of gas-fired plants (Accenture 2016). There is also an increased emphasis on renewables. But there are still some forecasts of LNG demand increasing. APEC's business as usual scenarios predict a 20% rise from the peak level of 2013 to around 49 mtpa, on the back of a limited growth of coal over the whole period, as well as nuclear generation gaining 50% over the coming 15 years (APEC 2016). Alternative scenarios all foresee an increase in gas demand.

In Taiwan, the future is brighter. The existing nuclear plants, representing a net capacity of 4.9 GW, are likely to be decommissioned by 2025 as they are only licensed to operate up to 40 years. In November 2011, a few months after Fukushima, a new national energy policy overturned the country's 20-year lifetime extension. Nonetheless, Taipower still asked for an extension for Chinshan (the outcome is still unknown), the first power plant whose license is due to expire. Meanwhile, the construction of Lungmen, the fourth nuclear power plant, has been postponed (World Nuclear Association 2016). The government is planning to expand renewable capacity to around 12.5 GW by 2025 and 17.25 GW by 2030, against 4.7 GW in 2015. There were seven 800 MW coal-fired units under construction as of December 2015, but this capacity replaces older coal-fired plants. This leaves some growth margin for gas. APEC scenarios foresee gas demand growing by between 18 and 62% compared with 2013 levels, reaching between 15 and 21 mtpa by 2030 (APEC 2016).

7.4.2 China and India

China and India are both considered as fast growing markets in terms of absolute demand growth, but their incremental demand is very uncertain. The evolution of their LNG imports depends upon:

- The outlook for gas demand, in particular the competitiveness of gas against coal,
- The development of alternative energy sources (renewables and nuclear),
- The interaction with domestic gas production, including unconventional gas in China and
- The availability and scale of pipeline gas supplies.

Chinese gas demand growth slowed down in 2015 due to the economic slowdown, reaching around 197 bcm. However, it seems to be picking up again in early 2016 as a result of the drop in prices decided by the National Development and Reform Commission (NDRC) in late 2015. Economic activity has a large impact on Chinese energy demand, but the price of natural gas against alternative sources is also crucial. The residential sector is a large consumer of natural gas and there has been a continuous switch from LPG to gas. Industrial gas demand is particularly impacted by the availability and the price of natural gas. In the power sector, gas is still not competitive against coal. The cost of electricity produced by gas-fired plants is around twice more than coal-fired plants (CNPC ETRI 2015), while the government has put in place ambitious targets for renewables. There is, however, a push to move from coal to gas in the power, industrial and heating sectors. CNPC Research Institute of Economics and Technology (CNPC ETRI) estimates that this potential represents around 115 bcm of additional gas demand in the medium term, with 55 bcm in the industry, 40 bcm in the power sector and 20 bcm in heating. Meanwhile, a key area of potential demand growth is transportation. China has around 2 million NGVs on the roads as of mid-2015, including 250,000 heavy-duty vehicles using LNG. A major issue there is the competitiveness of gas against diesel. Forecasts of Chinese gas demand were substantially downgraded from 400 bcm by 2020 to between 269 and 330 bcm, with 300 bcm as the mid case. For 2030, the forecast ranges from 380 to 540 bcm, with a mid case at 455 bcm (CNPC ETRI 2015). The impact of COP21 discussions on natural gas is difficult to assess at this stage.

Chinese gas production includes conventional gas, tight gas, coalbed methane (CBM), shale gas and synthetic gas (syngas). The 2020 targets of 30 bcm for shale gas, 30 bcm of coalbed methane and 50 bcm of syngas look very challenging and there are no targets for the period beyond 2020. Previous targets for unconventional gas were missed in the past: Shale gas production reached 4.5 bcm in 2015 against a target of 6.5 bcm. China aims at producing 144 bcm in 2016 (Bloomberg 2016). In addition, lower oil prices can be expected to have an impact on future upstream investments. gas production is likely to respond to demand developments, and the Chinese government will continue to prioritize indigenous production over imports (Corbeau and Yermakov 2016).

The last factor impacting China's LNG imports is pipeline imports, which reached 33 bcm in 2015 (Hellenic Shipping News 2016). China imports pipeline gas from Central Asia through the Central Asian Gas pipeline (CAGP) (55 bcm/y), as well as through the Myanmar–China Gas pipeline (12 bcm/y). The fourth leg of the CAGP is under construction and planned for 2020, adding 30 bcm/y. More importantly, the Power of Siberia gas pipeline (38 bcm/y) from Russia is under construction, but there are uncertainties as to the completion date. Another Russian gas pipeline has been planned that would reach the Western side of China. Given the oversupply in the Chinese gas market, there may be very little incentive to push forward the completion of either pipeline. However, due to the scale of these pipelines, there will be a major impact on future LNG imports. Recent CNPC estimates show imports between 190 and 270 bcm by 2030 (Natural Gas Asia 2016). The high range of these estimates calls for substantial LNG imports: Based on the capacity planned and under construction, pipeline imports could reach up to 140 bcm, which would leave a gap of 130 bcm—or almost 100 mtpa—to be filled by LNG, but only of 70 bcm, assuming only one Russian gas pipeline, in the lower case.

As for China, there is a wide range for future Indian LNG imports due to the uncertainties over demand and production. Gas production has been declining since from 49 to 29 bcm as production from the KG-D6 gas development has fallen. Unlike China, however, there are no pipeline imports, even though India has been looking at importing pipeline gas from either Turkmenistan or Iran. Potential demand is relatively high, as reflected by official government forecasts suggesting potential demand could reach 272 bcm by 2029–2030, from about 51 bcm in 2015, but supply would reach only 173 bcm, leaving 100 bcm unmet. In this scenario, India's Petroleum and Natural gas Regulatory Board (PNGRB) sees LNG imports rise to 78 bcm (57 mtpa) (PNGRB 2013). It is assumed that most planned terminals would materialize, but also that India will start receiving pipeline gas in the 2020s, which at this stage still looks challenging. LNG import levels are forecast as much higher by PNGRB than anticipated by the IEA (55 bcm) by 2030 (IEA 2015). A major factor determining LNG imports, and ultimately demand, is the price of LNG and how this affects its competitiveness, notably in the power generation sector. Even at around $6/MMBtu, which is the high range of Asian spot prices as of mid-2016, gas-fired plants are not competitive against coal-fired plants running on imported coal (IEA 2015). Meanwhile, domestic coal, or the expansion of renewables, is preferred to imported gas (KAPSARC 2016). Petronet has renegotiated its key contract with Qatar, with prices dropping toward $6/MMBtu during spring 2016 as a result. Meanwhile, there have been delays in expanding gas infrastructure, both LNG import terminals and pipelines. For example, the use of the Kochi LNG terminal is limited at 2% because the transport infrastructure is not there. There is nevertheless an interest in expanding and developing LNG import infrastructure with around 60 mtpa of LNG import capacity planned.

7.4.3 Southeast Asian Markets

The Southeast Asian markets, Thailand, Malaysia, Indonesia and Singapore, imported around 8.5 mtpa in 2015, on top of Pakistan which started importing in 2015. Other countries such as Bangladesh, Vietnam, Myanmar and the Philippines are expected also to become LNG importers. The future demand for LNG in these countries will depend on:

- The evolution of domestic production (besides Singapore),
- The interaction between coal and gas in Indonesia, Malaysia, the Philippines, Vietnam and Bangladesh,
- Interaction with renewables' development (Thailand) and
- Future pipeline imports (notably in Singapore where LNG is expected to replace pipeline gas).

The main reason why Southeast Asian countries are turning to LNG imports is the existing or anticipated decline in domestic gas production. The only exception to that trend is Singapore, which is diversifying away from declining imports from Malaysia and Indonesia, and Myanmar, even though Myanmar is still expected to remain an exporter to China. The trend relative to production is particularly worrisome in the two largest Southeast Asian gas producers and LNG and pipeline exporters, Indonesia and Malaysia. Due to the insular nature of the country and resources being located far from demand centers, Indonesia is planning to develop small-scale LNG in order to give access to power to remote regions. Pakistan and Bangladesh both have important unmet demand as their production level is flattening and their fields are already mature. Pakistan started importing in 2015 and in 2016 launched a tender for a second floating storage and regasification unit (FSRU) to be operational by 2018. In 2016, Bangladesh agreed to a 3.5 mtpa FSRU, located at Moheshkhali, to be provided by Excelerate. The plant is likely to be operational by 2018–2019. Thailand is also likely to see its production peaking before the end of the decade, implying that it has to find more gas supplies to meet increasing gas demand.

One common theme across the region is that gas is struggling to compete with coal. Most governments anticipate coal-fired generation will increase over the coming decades. It remains to be seen how this would be compatible with COP21, but in the absence of a carbon price, gas-fired plants are not competitive. In Indonesia, the new Electricity Supply Business Plan for 2016–2025 from state-owned PT PLN foresees that the bulk of future power capacity additions will be coal-fired. The outlook for electricity in Peninsular Malaysia anticipates that coal-fired plants will represent 64% of total capacity by 2020, against 45% in 2014. As much as 5 GW will be installed, so that coal consumption will jump by 75%. Thailand is also trying to diversify its energy mix away from natural gas, which represented 64% of the power mix in 2014. It plans to increase renewable capacity by 10 GW by 2036, so that renewables could represent 10–15% of the power mix, against gas, accounting for 30–40%, coal for 20–25% and imported hydro for 15–20%. This does not mean that gas demand would decline, since power demand is expected to double over the same timeframe (Ministry of Energy 2015).

7.4.4 Latin America and North America

LNG demand in North America—the USA, Canada and Mexico—was 7 mtpa in 2015. The only Canadian LNG import terminal is located on the same side of the continent as the prolific Marcellus shale gas play, while pipeline capacity from the USA to Mexico has been continuously expanded and more pipeline capacity is being built. Meanwhile, Mexico has large shale gas resources that the recent changes in upstream regulation could help to develop. It is very likely that the region will no longer import LNG at all.

Latin American gas markets have changed considerably since 2000, featuring a rapid growth in gas demand, with a drop in production for two large producers—Argentina and Venezuela—leading to a failure to integrate the region through intraregional pipeline trade. Five countries currently import—Argentina, Brazil and Chile, as well as the Dominican Republic and Puerto Rico—while another two, Uruguay and Colombia, have terminals under construction. Latin America LNG demand depends on:

- The evolution of natural gas production, notably in Brazil and Argentina (which has important shale gas resources),
- The completion of additional LNG regasification terminals in Uruguay, Colombia, Brazil and Argentina
- The development of LNG-to-power solutions in Central America and the Caribbean as a way to replace oil-fired generation and on a yearly basis, the availability of hydro.

Latin America has attracted the largest number of FSRU projects so far, a trend which can be expected to continue. While the first terminals were supported by state companies or large industrial conglomerates with good credit ratings, the next generation is involving smaller—often local—companies. For most countries, it is a question of expanding LNG demand beyond the current imports or completing plants under construction. The exception is Central America and the Caribbean region, where there are currently many projects in countries that are not yet LNG importers—Panama, Honduras, El Salvador and Jamaica. These are struggling to move ahead due to the lack of established customers and infrastructure. So far only two, the Dominican Republic and Puerto Rico, have moved forward, in the early 2000s. There is strong demand from the power generation sector, as gas is the fuel of choice to complement renewables and replace oil products. As in Africa, many projects in the Caribbean are LNG-to-power projects. The expected increase in LNG demand will be driven by the rise in energy demand, with the higher range of LNG imports resulting from a slow development of local gas resources. The flexibility of LNG supply is crucial in this region, as LNG often backs up hydro, which has proven particularly variable over the past few years. Argentina and Brazil therefore heavily rely on spot cargoes and short-term contracts to source their LNG cargoes.

7.4.5 Middle East and Africa

LNG demand is bound to increase strongly in the Middle East and African areas. Kuwait and Dubai have been importing since 2009–2010, later joined by Israel. Jordan and Egypt installed FSRUs in 2015. They had already imported 10 mtpa in 2015, more than twice the volumes in 2014. The main uncertainties are:

- Growth of natural gas demand, especially in the context of increasing domestic gas prices,
- Evolution of domestic gas production in the context of the recent decline in oil prices,
- The relative cost of LNG compared to indigenous gas in the Middle East and regional piped gas in Africa,
- Future of existing LNG export plants in Oman and the UAE and
- Successful development of LNG-to-power solutions in Africa.

In the Middle East, demand for natural gas has been growing at around 6% over 2000–2015, spurred by low gas prices and the availability of domestic production. Gas is essentially consumed in this region in the power and industrial sectors. However, gas reserves are not homogenously spread and Qatar and Iran hold the bulk of them. Gas production has failed to grow as fast as demand except in those two countries; the next generation of gas fields is often non-associated gas, tight gas or sour gas, which are more expensive to develop than the previous generation of associated gas fields. This means that most Middle Eastern countries are gas short and have to import either pipeline gas—like

Oman and the UAE—or LNG. Developing gas pipelines can take decades, so that most countries are now turning to LNG to meet additional demand. Bahrain has advanced plans to import LNG by 2019, while Saudi Arabia recently announced the possibility of importing LNG. Additional facilities are planned in the UAE and Kuwait. Middle Eastern countries tend to have very low gas prices, usually below $4/MMBtu, even though some reforms have been put in train to increase these levels. This is still substantially lower than spot LNG prices, and it is still uncertain at this stage whether countries will turn to LNG or possibly to other alternatives in the power sector. For example, the UAE is building nuclear power plants and has ambitious targets to develop renewables.

In Africa, as many as nine countries, including Egypt, are looking at LNG imports as of 2016, with some, such as Ghana, Morocco and South Africa, being further advanced than the others. Ghana already has an FSRU in place, which could be operational by 2017. Morocco's new facility may be operational only by 2020–2021. The gap between potential demand and production in Egypt is such that even with the Zohr gas field coming online later this decade, there is still ample room for increased LNG demand in the next few years. The evolution of LNG imports after 2020—should these continue—will be determined by the appetite in the long term for gas in the region, the evolution of domestic production and the cost of LNG. The development of LNG-to-power solutions in other African countries is still very much a new thing and its evolution is uncertain at this stage. In addition, some countries are looking at LNG as a medium-term solution to develop gas use before domestic gas resources are developed.

7.5 How Much LNG Is Left for Europe?

The amount of LNG available for Europe is therefore theoretically the amount of LNG supply which has not been absorbed by the other markets. By drawing on official government and national oil company projections, APEC scenarios, the IEA, the Asian, Latin American and Middle East and Africa subchapters in the book LNG market in Transition: the great reconfiguration (Corbeau and Ledesma 2016) and the author's estimates, we have established ranges for LNG demand for the period 2020–2030. This illustrates the potential for Europe to act as a large shock absorber for global LNG developments.

The low ranges of LNG supply to Europe would not fundamentally change the existing situation. Pipeline gas would remain a key component of Europe's gas supply, while LNG would represent between 15 and 20% of Europe's total demand, consistent with a scenario where import needs would gradually increase, while being met by different sources of supply. This does not show a massive diversion of LNG supplies away from Europe. The high ranges of LNG supply looking for a home in Europe would certainly create some stress in the current equilibrium, given the scale of the volumes at stake. This would certainly fit the EC's agenda of LNG supply diversification, but would also require an expansion of the existing capacity well beyond what already exists. It is also questionable whether many FIDs would be taken beyond 2020 if there were such quantities of surplus LNG, given that these FIDs would likely be at very low prices, endangering both existing and future LNG supply projects. Such high volumes would also trigger a response from Europe's traditional pipeline suppliers, from Russia to Norway, Algeria and the Caspian region.

7.5.1 LNG's Interaction with Europe's Supply/Demand Balance

While global LNG markets determine how much LNG is available for Europe, these quantities will also depend on Europe's own supply/demand balance. This will be determined by the evolution of natural gas demand, indigenous gas production and the interaction with pipeline supplies, including Russia, North Africa and the Caspian region. In this context, Norway's gas output can be considered as domestic production, with the exception of the LNG export volumes from Snøhvit.

The different components of Europe's future supply and demand mix are considered elsewhere in this book. Chapter 3 looks at possible scenarios for European gas demand, Chap. 9 at Russia, Chaps. 10 and 11 at Dutch and Norwegian gas production, Chaps. 12–14 at the different pipeline options, ranging from North Africa, the Caspian region, the Eastern Mediterranean region and Iran.

7.5.2 Interactions with Demand

As for most regions, the price at which LNG will be available in Europe will have great significance for its future gas demand. Natural gas has had a hard time competing with coal and renewables in the power sector. While the growth of renewables is mostly determined by policy decisions, the shares of coal and gas in the power sector tend to depend on their respective competitiveness. The recent increase in coal prices since June 2016 has largely improved the competitiveness of gas-fired plants against coal-fired plants (CME group 2016). With coal prices around $60/t, gas-fired plants running on LNG at $4/MMBtu—with a CO_2 price of €10/ton—are competitive even against the most efficient coal-fired plants. Nevertheless, should gas prices rise above $6/MMBtu, gas-fired plants will no longer be competitive. The situation would be slightly better in the UK, as it has established a carbon price floor (£18/ton) which is added on top of the emissions trading system (ETS). Switching potential also varies according to the individual power capacity conditions in each European power market.

7.5.3 The Role of LNG as a Supply Source

While LNG has not taken the same key role in Europe as it has in some Asian countries, it is nevertheless considered as an important source of supply for diversification reasons. In February 2016, the EC published a new LNG and storage strategy, highlighting the benefits of LNG as a diversified source of supply and a way to improve energy security. The Commission nevertheless noted that while Western European countries had access to LNG through numerous import terminals, the situation is different in Southeast Europe, Central-Eastern Europe and the Baltics, even though Lithuania now has its own LNG import terminal. These regions are also more dependent on Russia, while the EC would like to reduce Europe's reliance on this particular supplier.

This strategy does not offer any indication as to how more LNG could be attracted to Europe, other than by making sure that enough infrastructure exists to transport gas across the region, completing the internal market and working closely with international partners to make sure that LNG can be traded freely on global markets. It also recognizes that the cost of LNG will depend on a range of factors and could be high when markets are tight.

7.5.4 The Interactions with Russia

LNG, in particular US LNG, has often been presented as a way to lessen dependence on Europe's main pipeline gas supplier: Russia (Bordoff and Houser 2014). This will depend on how much LNG will be left targeting Europe to threaten Russia's market share. The low-demand volumes outside of Europe would leave more than 100 or even 200 mtpa available for Europe, even though such volumes would depend on how much more supply is brought to the market beyond 2020. The low-demand cases would certainly deter further investments (see Fig. 7.4). Beyond surplus LNG volumes, one key factor will be the pricing environment. It has already substantially changed over the past 2 years, and spot prices in Europe and Asia have remained around $4–6/MMBtu in 2016. These prices are much lower than what was anticipated when the projects were sanctioned. Many LNG projects would struggle to recover their full costs in such an environment, but they will continue operating as long as they can justify their variable costs, in the hope of an eventual price rebound. There are risks of shut-in production among high-cost producers, and risks that new LNG projects may not be able to recover their costs, setting the stage for the next boom-bust cycle in LNG supply—and for wide fluctuations in gas prices (Corbeau and Yermakov 2016).

US LNG projects have a different business model, where the pricing and volume risks are borne by the offtaker of the tolling capacity of LNG terminals, which is often an aggregator. As long as US LNG offtakers can cover their variable costs—cost of gas, transport and regasification—they will continue to lift their LNG. But should prices fall to very low levels, these offtakers could choose to lose the liquefaction fee and not lift LNG. In any case, the low level of spot prices, not covering the full cost including the liquefaction cost, results in losses for the offtakers, and after several years of oversupplied markets, they may call for a renegotiation of their long-term offtake contracts (KAPSARC 2016).

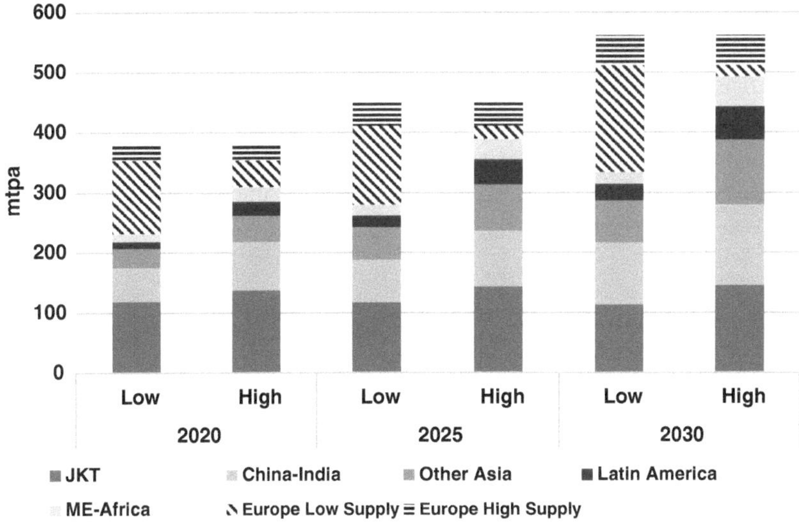

Fig. 7.4 Global LNG supplies and volumes left for Europe. *Source* For total LNG supply, author's estimates from various sources

Meanwhile, Russia has ample spare gas production capacity and is a low-cost supplier. In the case where large volumes of surplus LNG are heading toward Europe, Russia may choose to protect its market share in Europe which, until the early 2020s, is and remains its main export market. Russia may choose to adapt to the threat of large volumes of LNG being supplied to Europe by granting discounts or altering contract price formulas, or potentially by dumping pipeline gas at trading hubs. But this would be a considerable departure from Russia's traditional behavior. Should this not be sufficient to keep LNG at bay, Russia could choose to undercut LNG, notably US LNG, by selling gas below its variable cost. On the LNG side, Qatar is also a low-cost supplier and can arbitrage between markets. It can be expected to react to other suppliers' marketing strategies. The potential outcome of any confrontation between Russian pipeline gas and LNG in Europe will be different in a high (\$15/MMBtu+) than a medium gas price environment (\$8–10/MMBtu+). The price levels at which any price war between LNG and pipeline gas in Europe takes place will influence the perceptions of the seriousness and duration of the challenge for the incumbent low-cost supplier. Lower oil prices reduce the cost of delivered Russian pipeline gas to Europe and make it more competitive, but greatly impact Russia's revenues. Higher oil prices in a context of LNG oversupply and low US gas prices mean that Russian gas may very well be more expensive than LNG delivered to Europe at spot or Henry hub plus prices (Corbeau and Yermakov 2016).

7.5.5 Getting LNG to the Right Place

Europe's current LNG import infrastructure (161 mtpa) seems enough to accommodate large LNG import volumes, but not the large volumes in the Europe High Supply case (Fig. 7.4). Two technical constraints are relevant here. The first is that LNG terminals are rarely operated at 100%. In practice, the average utilization rate is closer to 70%. Additionally, underutilized LNG import capacity may not be located in markets which can absorb large volumes. Europe's intraregional transport infrastructure has been designed to move pipeline gas from east to west and from north to south which implies that there is limited access to LNG in Central and Southeast Europe, due to the lack of LNG capacity and interconnectivity (Fig. 7.5).

One-third of Europe's underutilized LNG import capacity (40 mtpa) is located in Iberia, which is poorly interconnected with the rest of Europe. The region's demand in 2014 was only 31 bcm. In addition, both Spain and Portugal import pipeline gas from Algeria, taking 16 bcm in 2015. The second-largest amount of underutilized capacity is in the UK, which also imports pipeline gas from Norway, the Netherlands and Continental Europe and still has significant domestic production. Should more LNG arrive at the UK LNG terminals, this would likely exceed demand and be sent to Continental Europe via the Interconnector pipeline (IUK). Such an influx of LNG into Northwest Europe could potentially create some congestion issues in Zeebrugge, the point where IUK gas enters the European system. Large quantities of LNG imports into Western Europe would also produce an eastwards shift in intra-European gas flows. But there could be internal constraints

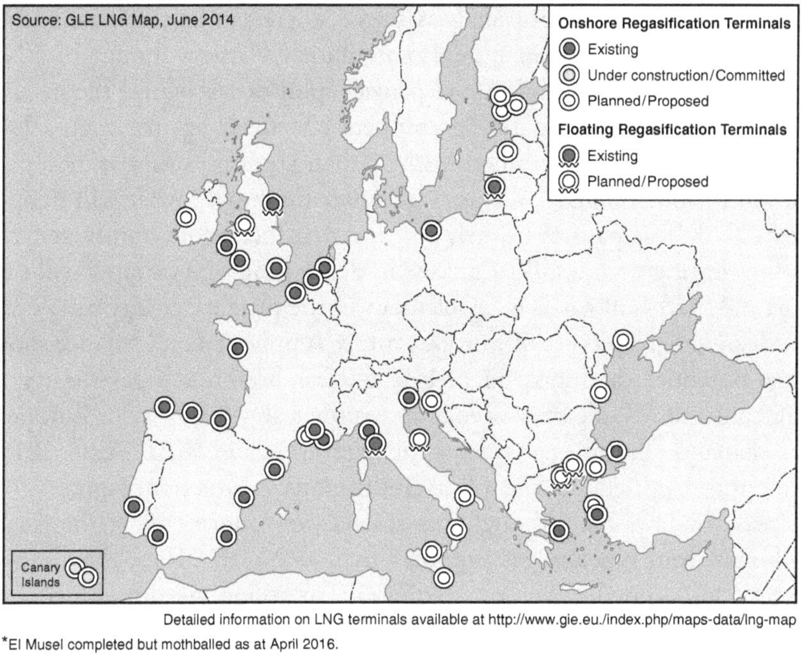

Fig. 7.5 Europe's LNG import terminals. *Source* OIES/KAPSARC (2016)

on how much LNG could go that way, even though the interconnections from west to east have been improved since the disruption experienced in 2009 (Corbeau and Yermakov 2016). One solution to absorb more LNG would be to develop a few additional LNG terminals in regions which currently have suboptimal access to it or improve further the internal transport network in Europe.

7.6 Conclusions

Europe's future LNG import requirements will be determined by the various market and political influences pushing in different directions: The state of global LNG markets, and notably the relative levels of LNG import prices, will be crucial to determine how much LNG will be left for Europe. As our analysis shows, there can be significant LNG demand outside Europe, but also potentially large volumes of LNG looking for a home. While LNG project sponsors recognize the need to make LNG more affordable to compete effectively against coal, a low-price environment combined with demand uncertainties is not conducive to investments, putting a question mark over what will happen post-2020. This pattern is likely to trigger further boom-and-bust cycles in the coming decades. But, success in developing LNG supply will support the success of gas as a key element in the primary energy mix.

Meanwhile, a picture of Europe's future supply/demand balance shows large potential variations. First, it is unclear how much gas the region will consume: While most scenarios assume a slow recovery of European gas demand, the greener ones anticipate a decline (KAPSARC 2015). Still, the expected decline of its indigenous production points to an increase in imports, although the scale is very uncertain at this stage. Volumes from pipeline gas suppliers such as North Africa, the Caspian region, Iran and the Eastern Mediterranean region are quite uncertain, as the later chapters of this book will describe. In this framework, Russia will remain the main supplier to Europe, with the possibility to expand its supplies at prices competitive to LNG, even though Russia prefers to sell at oil-indexed prices.

Over the past years, Europe has often played the role of a shock absorber, resulting in significant variations to its LNG imports. Europe is relatively well suited to such a role due to its large regasification capacity, offering third-party access, the presence of alternative pipeline supplies with flexibility embedded in the long-term contracts and liquid trading hubs. There will likely be limits on how much LNG can be absorbed, particularly if large volumes of surplus LNG are stranded at low prices. Europe's capacity to absorb these volumes effectively may be put to the test, not only due to internal transport constraints, but also because suppliers such as Russia—and Norway, Algeria and those in the Caspian region—are unlikely to see LNG invading their main export market and potentially reducing its market share without any reaction. Large volumes of stranded cheap LNG may also have unexpected consequences for the period beyond 2020, where new LNG supply will be required.

Very little LNG has been contracted by European players for the European market recently, or when it has this has been by aggregators that have the possibility to arbitrage between markets and send the gas somewhere else. An increasing role for Europe as a recipient of surplus volumes would further increase the role of spot and short-term LNG trade.

Against that backdrop, supporting LNG is currently viewed favorably European politicians due to its diversity and flexibility. But the decisive role in increasing LNG supply will be played by the global pricing dynamics that will be discussed in Chaps. 8 and 15.

Notes

1. In this chapter, Europe is defined as EU28, Norway, Switzerland, Turkey, Albania, Bosnia and Herzegovina, Macedonia and Serbia.

Bibliography

Accenture. 2016. Gas Grows Up. Part 1: Developing New Sources of LNG Demand.

APERC. 2016. APEC Energy Demand and Supply Outlook, 6th ed. Tokyo: Asia Pacific Energy Research Center.

Bloomberg. 2016. China's Shale Gas Reserves Jump Fivefold as Output Lags Targe, April 6.

Bordoff, J., and T. Houser. 2014. *American Gas to the Rescue?* New York: Columbia Center on Global Energy Policy.

CME Group. 2016. Coal (API2) CIF ARA (ARGUS-McCloskey) Futures Quotes.

CNPC ETRI. 2015. Prospects of Gas Market in China, December 9.

Corbeau, A.-S., and D. Ledesma. 2016. *LNG Markets in Transition: The Great Reconfiguration*. Riyadh: KAPSARC.

Corbeau, A.-S., and V. Yermakov. 2016. *Will There Be A Price War Between US LNG and Russian Pipeline Gas*. Riyadh: KAPSARC.

GIIGNL. 2016. *The LNG Industry: GIIGNL Annual Report*. Neuilly sur Seine: GIIGNL.

Hellenic Shipping News. 2016. Chinese Gas Imports: Slowing Growth But High Targets, February 1.

IEA. 2015. *India energy outlook*. Paris: World Energy Outlook Special Report.
KAPSARC. 2015. *Natural Gas: Entering the New Dark Age*. Riyadh: KAPSARC.
KAPSARC. 2016a. *How Will Natural Gas Adapt to the New Price Environment*. Riyadh: KAPSARC.
KAPSARC. 2016b. *The Impact of Low Oil and Gas Prices on Gas Markets: A Retrospective 2014–15*. Riyadh: KAPSARC.
Ministry of Energy. 2015. Thailand Power Development Plan 2015–2036. Bangkok.
Natural Gas Asia. 2016. China's Gas Imports May Reach 190–270 bcm by 2030. CNPC, August 4.
OIES/KAPSARC. 2016. LNG LNG Markets in Transition: The Great Reconfiguration. Oxford: Oxford University Press.
PNGRB. 2013. Vison 2030: Natural Gas Infrastructure in India. New Delhi.
Reuters. 2016. Eni Reaches Deal with Exxon on Mozambique Gas Project-Sources, August 5.
World Nuclear Association. 2016. Nuclear Power in South Korea, July 27.
World Nuclear Association. 2016. Nuclear Power in Taiwan, September 26.

Author Biography

Anne-Sophie Corbeau is a Research Fellow II at KAPSARC. She has over 15 years of experience in the energy industry, notably on global gas markets. She recently co-edited the KAPSARC/OIES book: LNG markets in transition: the great reconfiguration (2016). Previously, she worked for the International Energy Agency and IHS CERA. At the IEA, Ms. Corbeau was responsible for managing the research on global gas markets. She was the main author of the 'Medium Term Gas Market Report' and (co-)authored several publications on China, India, trading hubs and LNG markets. She also worked in the fuel cell and hydrogen area.

8

Global LNG Pricing Dynamics and Impact on Europe

Thierry Bros

8.1 Five Revolutions Are Reshaping the Energy Landscape

8.1.1 Three Revolutions on the Supply Side: US Shale Gas, US Shale Oil and Worldwide Renewable

The US shale gas revolution[1] is only the first (and most documented) of three revolutions that happened since the beginning of this century on the supply side. The world has changed thanks to the US shale revolutions (gas first and then oil)[2] and a global quest for renewable. Those revolutions took over a decade but will shape the twenty-first century. Australia followed producing unconventional gas and is now also exporting it. It should take some time for unconventional oil

T. Bros (✉)
SciencesPo, Paris, France
e-mail: thierry.bros@sciencespo.fr

M. Hafner and S. Tagliapietra (eds.), *The European Gas Markets*,
DOI 10.1007/978-3-319-55801-1_8

and gas production to materialise in other places where the resource is available[3] (Argentina, Canada, China, Mexico, Russia, South Africa, etc.), but the US shale revolutions should be exported to a few other countries.

Renewable policies were first designed in Europe (mostly based on subsidies as renewable then was much more expensive than any other form of energy) from 2001[4] to reduce CO2 emissions, but China is now also investing heavily as it needs clean energy to reduce air pollution in its big cities. The three biggest renewable producers in 2015 were EU, US and China. In 2015, China was the second producer of solar energy behind the EU[5], but it has become the biggest single solar producer (before Germany, the biggest contributor to EU solar energy). Thanks to technological improvements, renewable cost has gone down massively and can now compete with traditional electricity production (fossil fuels or nuclear). It can spread all over the world as it is no longer a fancy idea for only rich countries with an ecological mindset[6]. In the 2005–2015 period, on a worldwide basis, renewable production grew by an astonishing 16%pa when total primary consumption grew only by 2%pa. And 2015 was the first year on record in which additions of renewable power generation capacity were higher than those for thermal capacity.

Even if gas reserves (where 50% of the reserves is shared by only three states: Iran, Russia and Qatar) are more concentrated than oil reserves (where 50% of the reserves is shared by four states: Venezuela, Saudi Arabia, Canada and Iran) on a worldwide basis, liquefied natural gas (LNG) allows, as Churchill pointed out, diversity of supply. Thanks to LNG that represents 10% of the global gas supply, any consuming state can increase diversity of supply and hence its energy security. Australia and the USA are exporting their unconventional gas (under LNG) and oil in the global market, increasing de facto diversification of supply for all consumers. For example, Lithuania and Poland, where Russia provides 100% of their gas supply, have both invested in an LNG regas terminal (respectively in 2014 and 2015) to be able to access waterborne LNG to mitigate the Russian risk that was high on the political agenda for those two European states.

8.1.2 On the Demand Side: Efficiency Gains, an Ongoing Revolution

More energy has always been needed to sustain economic growth. Energy was provided first by slaves in Ancient Egypt, Greece or the Roman Empire, then by coal for steam machines during the Industrial Revolution in Europe and finally by oil during the twentieth century on a worldwide level. But for the first time ever, since 2006 (before the financial crisis), the European economy has been able to grow with less energy. This decoupling between economic and energy demands is now a European reality.

Not only have we seen during the last decade, like in the 70s, a demand destruction in front of higher prices (especially in Europe), but most policies are aimed at improving energy efficiency in the long run to avoid burning fossil fuels that has a negative impact on climate change. Europe is now definitively past its peak energy demand and could be followed by other developed states in the coming decade. This new trend is going to continue in Europe with the 2030 EU climate and energy road map[7]. On top, the outcome of the Paris COP 21 in December 2015 is putting further pressure at the world level on all fossil fuels (and in particular coal) that emit CO2. The secular correlation between economic growth and energy was disrupted in Europe in 2006, and this is going to spread all over the world (from developed to developing countries) in the coming decades. Even China is more energy efficient (it uses less energy per unit of GDP) in the past few years.

8.1.3 Energy Storage: The Next Revolution

We are witnessing an energy system where not only supply is widely available and demand is bound to peak, but also some major technological breakthroughs in energy storage should materialise soon. This could completely alter the energy landscape where major companies were dealing with massive inflexible infrastructures (coal extraction, hydrocarbon production, nuclear plants, etc.). The future could look like a decentralised smart system where end-users select the kind of local energy they have (wind, hydro, solar), are able to store it and use it when needed. The intermittency of renewables that was a major obstacle in a centralised electricity transmission system should be solved with new batteries and new storage solutions (power-to-gas, molten salts, etc.). This should allow the share of renewable to continue to grow fast.

It is interesting to note that two of those revolutions were started in the USA (shale gas and shale oil) and two in Europe (renewable quest and energy efficiency). The ongoing fight to achieve the cheapest and most efficient energy storage is global with high prize at stake as this could be the silver bullet to achieve a completely green energy supply. Companies and states are investing heavily to solve this problem, and already, new products like home batteries[8] are appearing on the market. Like renewable in the early 2000s, batteries will be very costly to start with, but cost should go down thanks to Research & Development. Finally, the manufacturing process should reduce the cost of batteries that would then be disseminated in all houses. With financial markets turning their back on coal that faces strong policy headwinds (for climate change risks) and limiting their exposure to oil,[9] vast sums of money are available for those new technologies. The penetration of this dual technology (renewable and storage) could be as fast as mobile phones that leapfrogged landline phones, especially in developing countries. This next revolution is just around the corner and will disrupt completely the energy landscape.

8.2 Global LNG Pricing Shifting Away from a Quasi-Pipe Business

In the early days, before those revolutions, LNG was viewed as a quasi-pipe business with long-term oil-indexed contracts with destination clauses. Dedicated suppliers and consumers had little options to get out of those deals. Two major changes happened:

- At the turn of the millennium, BG Group (now part of Shell) introduced a new business model based on optionality. LNG was able to flow to the region that had the highest prices and markets to mitigate alone the Fukushima disaster. The re-routing of cargoes was a very lucrative business, until the spreads between markets were reduced to the cost of shipping. It is also important to note that this re-routing that accounted for 7% was done thanks only to market principles (no state intervention).
- Cheniere introduced from 2010 for US LNG a new formula not linked to oil any longer but to the US spot price (Henry Hub). The liquefaction plant became de facto a service provider, not a commodity producer.

Those two major changes have and will continue to have profound impacts on the LNG world. In 2016, Japan's Fair Trade Commission (FTC) launched an investigation to see whether the contract clauses restricting the resale of LNG cargoes impede free competition. In case the FTC of the world's largest LNG importer finds the destination clauses are in violation of the competition laws, the existing LNG contracts would be open for renegotiation. Renegotiations and arbitrations have been a major theme for European (mainly pipe) gas contracts in the last 10 years and have allowed the European gas market to now be mostly spot driven. If Japan starts to renegotiate its LNG contracts, the LNG world price formation that is still, according to International Gas Union's (IGU) wholesale gas survey, 69% oil-indexed in 2015 could like the European gas market face tremendous changes in the next decade.

LNG that used to be a small part of the gas market and that was priced like gas on an oil indexation should evolve into a more fungible market like oil but traded on an LNG spot basis! Flexible LNG will not lead to one single worldwide gas price (as seen in oil) as the cost of transport is material but should link all regional prices. This means that the risks and challenges in this industry will need to be completely reassessed.

8.3 Gas Pricing in Europe Was in the Hands of a Duopoly

8.3.1 "Oil-Derived" Spot Gas Prices in Europe Down but not as Much as Oil Prices

Since the RWE-Gazprom arbitrage back in 2013, we estimate that oil indexation represented less than 50% of wholesale gas in Europe. And this was then a tipping point with only one way forward: more spot indexation. The relatively uncompetitive European gas market is now mostly spot driven. According to IGU, oil indexation represented 78% of total wholesale gas pricing in 2005 but only 30% in 2015, while gas-on-gas competition moved up from 15 to 64%. This move away from oil indexation helped Europe to reduce its total gas bill.

All European buyers now want full spot indexation as can be seen with the latest round of arbitrages/renegotiations. But this also means that Norway and Russia, which control more than 50% of total supply for Europe, now have greater market power than ever. Thanks to this new market power, there is a floor for gas prices in Europe at c. 4$/Mbtu, close to the estimated current full cost of producing and shipping Norwegian and Russian pipe gas to Europe (3.5$/Mbtu) Fig. 8.1.

Over those ten years, Gazprom fiercely defended oil indexation before finally changing its stance and selling some gas at auction in September 2015 (at a price higher than its long-term oil-pegged contract prices). A second gas auction was carried out in March 2016 as after the opening of the regas terminal in Lithuania, the gas landscape in the Baltic states is changing. Russia has now different options to sell gas: via legacy long-term contract, via auctions, via Gazprom Marketing & Trading and via Wingas, a fully owned European utility.

The Norway–Russia duopoly had three options for managing gas prices in Europe: (1) to achieve a high price (above the cost of US LNG or of new pipe gas) as in 2012–2013 by tightening supply. This option has not been pursued since early 2014 as it prompted final investment decisions to bring additional "new" gas into Europe that is now on its way; (2) to remain in a tunnel between an EU floor and the incentive for new gas by

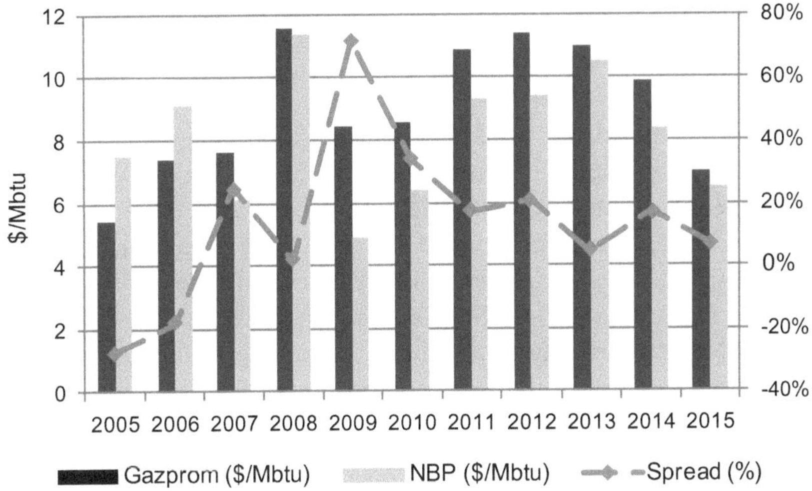

Fig. 8.1 Gazprom realised gas price in Europe vs NBP month ahead. *Source* SG Cross Asset Research/Commodities, Gazprom, Datastream

swinging supply to adjust to demand. This option so far is the best one for long-term rent maximisation; and (3) to engage in a price war by using some of its spare production capacity to shut in US LNG production. This could be an option when/if too much US LNG will be operational.

8.3.2 2015: Poor Supply Growth but More LNG to Europe Thanks to Less Reloads

Worldwide LNG supply in 2015 witnessed another poor growth (+2%). But with NBP and spot LNG in Asia on par, we saw an increase in net LNG berthing in Europe. Europe was, is and will continue to be the "dumping" ground for excess LNG as Asia has contracted enough gas.

The re-export volumes have gone down from 18% of gross imports in 2014 to 9% in 2015. This % should continue to go down in the coming years as re-export is not the best option to arbitrage (in a liquid market, the best way is to send the cargo straight to the location where the margin is the highest). Could this extra LNG pose a threat to the Norway–Russia duopoly?

8.4 Going Forward, the Speed of the LNG Supply Growth...

Europe faces "solidarity cracks" when trying to implement a common gas strategy that should be part of the EU's Energy Union. With Russia being the major gas provider in many Member States, tense EU–Russia relations do not favour gas, even if it is the cleanest fossil fuel. The best example of this "solidarity crack" is the division between the pros and the cons regarding the Nord Stream 2 project Fig. 8.2.

US LNG exports may impact both the pricing in the European gas market, where Norway and Russia control more than 50% of total supply, and the perception of gas. The arrival of this new supply marks the beginning of a new phase of competition. On top of this, LNG from the re-commissioned Angola and new Australian, US and Russian LNGs are set to hit the market in the coming years.

Finally, the abundance of LNG shipping capacity provides a greater connection between all the continents than ever before Fig. 8.3.

By 2020, the USA should have 63 mtpa of liquefaction capacity available, which could translate into an export level of 50 mtpa in 2020 if we assume a load factor of 80%.

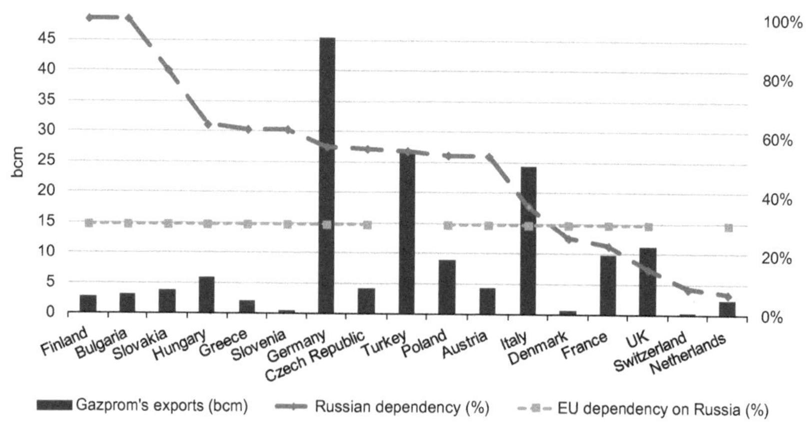

Fig. 8.2 European countries gas dependency on Gazprom

With the commissioning in February 2016 of Cheniere's Sabine Pass train 1, we believe that (1) LNG supply will start to grow by more than 6% pa from 2016e to 2020e Fig. 8.4.

Massive new LNG supply will materialise when the biggest LNG market (Japan) sees demand fall. Now, it is the worst possible timing for this new US LNG as it has no dedicated market. Therefore, Europe

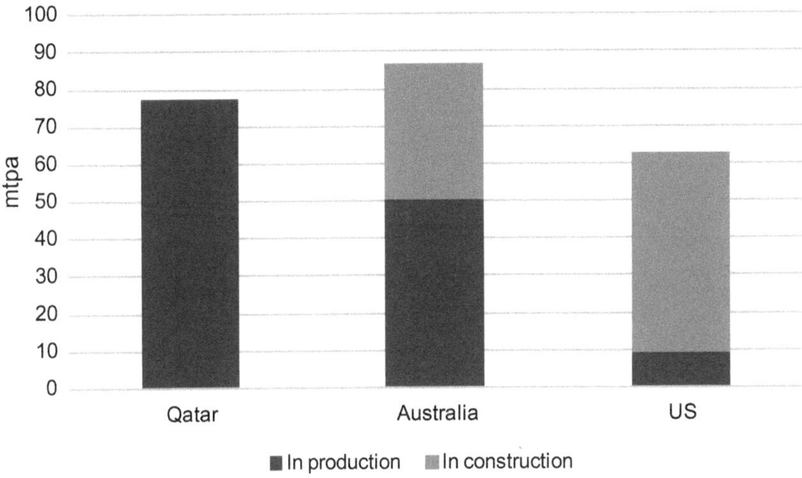

Fig. 8.3 Three major LNG suppliers in 2020e

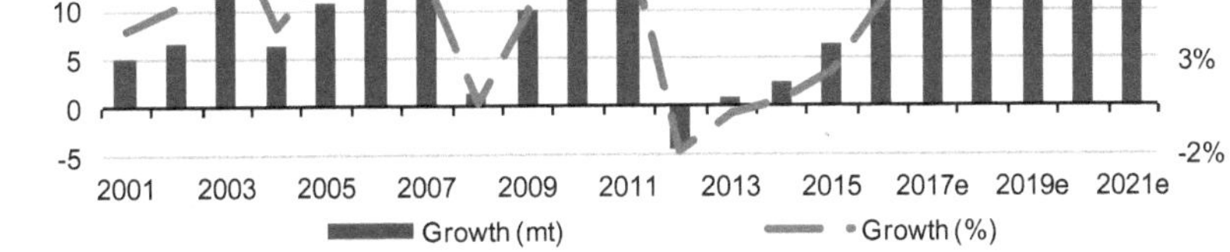

Fig. 8.4 LNG supply growth (Y-o-Y). *Source* SG Cross Asset Research/ Commodities, GIIGNL

will likely be the "dumping" ground for this LNG. Competition always intensifies at the worst possible time Fig. 8.5.

After a 3%pa growth in 2010–2015, we forecast a global supply surge of 8%pa in 2015–2020e. Asia growth will slow down from 6%pa in 2010–2015 to 5%pa in 2015–2020, while Europe after a drop of 10%pa in 2010–2015 will see a resumption of growth (+11%pa). The growth in Latin America and Middle East is also slowing going forward Fig. 8.6.

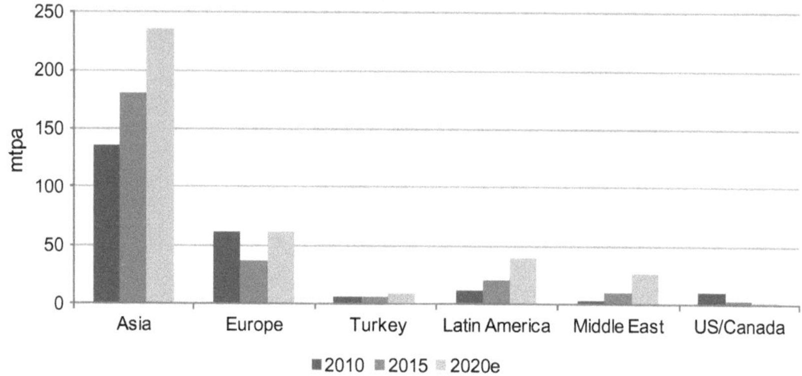

Fig. 8.5 2010–2020e LNG deliveries. *Source* SG Cross Asset Research/ Commodity, HIS for historical

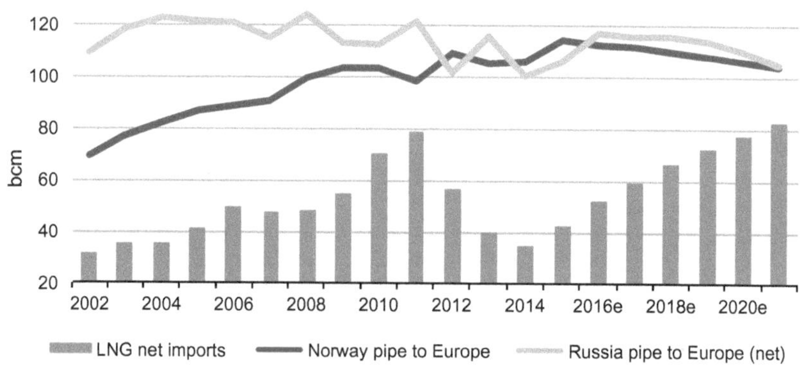

Fig. 8.6 LNG is competing with Russian and Norwegian pipe gas in Europe. *Source* SG Cross Asset Research/Commodities, IEA for historical data

8.5 ... and the Rate at Which It Reaches Europe...

The high level of Russian and Norwegian gas in Q1 16 can be explained by the following: (1) oil-indexed contract prices being lower than spot prices; (2) some of this gas being re-exported to Ukraine; and (3) a desire to reduce the need for US LNG in the coming months. Traditional pipe suppliers are trying to flood the European market before the arrival of any US LNG.

8.6 ... Will Dictate How Russia Reacts

With the steep capex cuts made in Norway since 2014, we expect gas Norwegian production to already have peaked and to slowly decline in the coming years. Norway has, therefore, very little flexibility left in its ability to swing production to balance demand.

Russia has two remaining options for managing gas prices in Europe:

1. To try to keep prices around $4/Mbtu by swinging supply to adjust to demand.
2. To engage in a price war to stop future US LNG production (around $3/Mbtu) by using some of its spare production capacity (Gazprom alone had 150bcm/y of unused production capacity in 2015).

Gazprom has both market power and a lower supply cost (the rouble devaluation resulting from US sanctions on Russia has had the effect of making Russian gas even more cost competitive).

In 2015, after the opening of the Lithuania regas terminal, Gazprom responded by discounting the price of its gas by 23%. So, cutting prices by 1$/Mbtu (24%) would be an interesting way for Russia to make a point. If it manages to do so, it could show its strength and sell additional volumes in Europe Table 8.1 and Fig. 8.7.

Table 8.1 Gazprom's options on a FY basis

	Vol (bcm)	Price ($/Mbtu)	Revenue ($bn)
Swing in supply to try to mitigate falling prices	116	4.1	16.8
Price war to shut in US LNG	122	3.1	13.4

Source SG Cross Asset Research

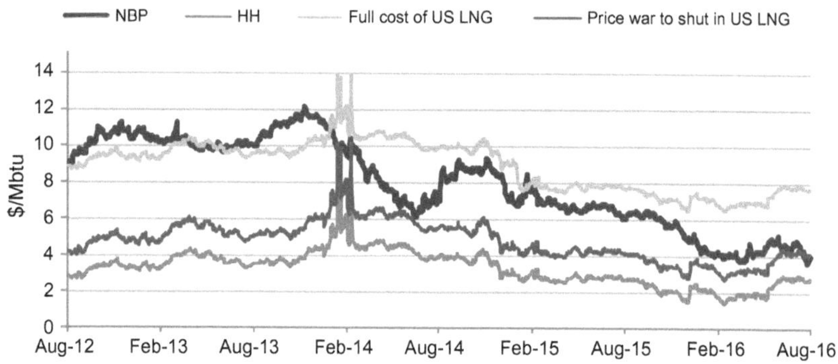

Fig. 8.7 How likely is a price war? *Source* SG Cross Asset Research/ Commodities, Datastream

To assess the full cost of US LNG in Europe, we added the liquefaction tolling (3$/Mbtu), the shipping cost (1.5$/Mbtu), the regas (0.5$/Mbtu) and the historical margin of trading LNG (1$/Mbtu), i.e. a proxy of HH + 6$/Mbtu. With lower shipping rates for LNG, we have reduced the cost of US LNG from HH + 6$/Mbtu to HH + 5$/Mbtu since 2015.

With the full cost below or at the level of NBP, it made sense back in 2011–2013 to FID liquefaction trains in the USA. But as seen on this graph, since 2014, curves have moved and the theoretical shipping of US LNG to Europe will entail a loss. But will US LNG come to Europe? In April 2016, the first shipment of American LNG from Cheniere's Sabine Pass reached Europe. We only expect few distressed cargos to reach Europe in 2016e.

A price war would cost Gazprom $3.4bn in revenue (on a FY basis) but would stop US LNG being produced. A price war would also impact the Energy Union strategy as it would reveal the real cost of diversification of supply. To implement this hypothesis, Gazprom would need to decide to use some of its spare production capacity to push more volumes on the spot/hub markets in Europe on top of its contracted gas.

Sporadically, a price war could take place in the 2017–2020e period as in front of no growth in European gas demand, pipe and LNG supply are available with a level of spare capacity never reached before. But risks remain mostly on the Ukrainian transit side.

8.7 Increase Security of Supply Thanks to LNG

LNG could not only provide lower prices, marginally reducing our dependency on Gazprom, but most importantly improve the perception of gas in civil society and at government level. With the opening of new regas terminals, even the Eastern part of Europe could be interested in using more gas to achieve a faster cost-effective energy transition.

EU–Russia–Ukraine is an unstable *ménage à trois*. According to the Ukrainian Energy and Coal Minister, Ukraine wants to hike the tariff for shipping Russian gas and to introduce a ship-or-pay clause. Gazprom is unlikely to agree to this, if the cost is above the alternative option (Nord Stream). Hence, 2020 will be a notable year as Russia is unlikely to renew the Ukrainian transit contract. As less supply would be available from Russia, US LNG would provide diversification and extra security of supply.

Notes

1. By combining two technologies (fracking and horizontal drilling), US producers have been able, since 2005, to unlock shale gas reserves that before could not be produced on a commercial basis.
2. The shale oil revolution tilted the pricing power away from OPEC as the USA was becoming the biggest worldwide oil (and gas) producer.
3. https://www.eia.gov/analysis/studies/worldshalegas/.

4. Directive 2001/77/EC of the European Parliament and of the Council of 27 September 2001 on the promotion of electricity from renewable energy sources in the internal electricity market.
5. BP Statistical Review—June 2016.
6. In 2015, in Denmark, renewable accounted for 25% of the total primary energy consumption vs. 2% in China (BP Statistical Review—June 2016).
7. http://ec.europa.eu/energy/en/topics/energy-strategy/2030-energy-strategy2030.
8. For example https://www.teslamotors.com/powerwall or http://www.bollore.com/en-us/activities/electricity-storage-and-solutions.
9. Some International Oil Companies like Shell or Total are claiming that they are now more gas orientated than oil.

Bibliography

After the US shale gas revolution—Thierry Bros—Editions Technip.

BP Statistical Review of World Energy, June 2016, https://www.bp.com/content/dam/bp/pdf/energy-economics/statistical-review-2016/bp-statistical-review-of-world-energy-2016-full-report.pdf.

Directive 2001/77/EC of the European Parliament and of the Council of 27 September 2001 on the promotion of electricity produced from renewable energy sources in the internal electricity market. *Official Journal* L 283, 27/10/2001 P. 0033–0040.

European Commission, 2030 Energy Strategy, http://ec.europa.eu/energy/en/topics/energy-strategy/2030-energy-strategy.

Groupe International des Importateurs de Gaz Naturel LIQUÉFIÉ (G.I.I.G.N.L.), The LNG industry in 2015 (Neuilly-sur-Seine: 2015), http://www.giignl.org/sites/default/files/PUBLIC_AREA/Publications/giignl_2015_annual_report.pdf.

The LNG industry in 2015—GIIGNL.

Thierry Bros After the US Shale Gas Revolution (Paris: 2012) Editions Technip.

U.S. Energy Information Administration, Analysis & Projections, World Shale Resource Assessments, http://www.eia.gov/analysis/studies/worldshalegas/.

Author Biography

Thierry Bros is a senior research fellow of The Oxford Institute for Energy Studies, a member of the EU-Russia Gas Advisory Council, an advisor to the World Energy Council - Global Gas Centre and a visiting professor at SciencesPo. In 2016, he founded thierrybros.com to provide independent research, advice, training & expertise on energy markets after 25 years in the energy field, from the policy side to trading floors. Thierry is highly regarded by the energy community, notably accredited as the best European gas analyst for five years in a row (2013-2017).

9

The New Russian Gas Export Strategy After the Ukraine Crisis

Tatiana Mitrova

9.1 Setting the Context

9.1.1 Russian Macroeconomic and Political Situation After the Ukraine Crisis

Over the last couple of years, as Mitrova 2016 pointed out, Russia has found itself in a completely new environment—a "perfect storm" of economic, market, domestic political, and foreign policy-related upheavals.[1] These fundamental shifts occurred in an unpredictable manner due to the coincidence of several external and internal factors:

- increased global supply of hydrocarbons (including the US shale revolution) with aggressive competition from other traditional and new suppliers entering the market (e.g., the USA, Iran, Iraq, Australia,

T. Mitrova (✉)
Energy Research Institute of the Russian Academy of Sciences,
Moscow, Russia
e-mail: mitrovat@me.com

M. Hafner and S. Tagliapietra (eds.), *The European Gas Markets*,
DOI 10.1007/978-3-319-55801-1_9

East Africa, and Brazil) and the resulting low oil and gas price environment;
- global economic weakness and hydrocarbon demand stagnation in Russia's main export markets;
- geopolitical tensions with the West, including USA and EU technological and financial sanctions introduced against Russia as a reaction to the annexation of Crimea and hostilities in eastern Ukraine;
- structural domestic economic crises driven by inefficiency in the regulatory framework, high resource dependency, and a lack of stimulus for entrepreneurship and industrial diversification;
- stagnant domestic energy demand, driven by the economic slowdown;
- increasing problems with the access to financing, since the domestic financial market is weak and underdeveloped, and foreign capital flows are limited by economic sanctions and a poor investment climate.

Russia had to adapt to the much lower oil and gas prices and export revenues, which was extremely painful for the country's economy, for which hydrocarbons were providing more than 50% of the federal budget revenues in 2011–2014 and dropped down to just 34% in 2016.[2] It has become obvious that the oil and gas sector will not be able to generate its formerly high GDP and budget revenue growth rates, challenging the whole economic model of the country's development, which has evolved over the last decade of growing oil prices.

In 2015–2016, low oil prices ravaged all of Russia's key economic indicators. Demand for durable goods shrank by almost half, imports plummeted 35%, trade turnover in rubles fell almost 12%, and foreign investment—which had fallen to almost zero in 2014—was nonexistent in 2015. Inflation increased to at least 15%.[3] So, after the 6–8% GDP growth rates observed in 2004–2008, in 2015 annual GDP contracted by 3.9%, driven by a contraction in domestic demand weighted down by falling real wages, higher cost of capital, and weakened consumer and investor confidence. Economic growth is expected to resume; however, the economic recovery in 2016 seems to be muted and medium-term prospects are weak.

These dramatic changes have serious consequences not only for the Russian state budget and the macroeconomic stability, but for the gas industry and gas export strategy as well.

9.1.2 Transformations of the Russian Gas Balance in the Recent Years

Russia has always been one of the key players at the global gas market. It is among the largest gas resource holders, producers, and exporters, with nearly 17% the global gas reserves, 19% of the global gas production, and 24% of the global cross-border gas trade.[4] It is a dominant supplier both for Europe and for the Commonwealth of Independent States (CIS) and holds great influence over these regional markets.

Domestically, gas is not only the backbone of the Russian energy sector, but also one of the most powerful policy tools. While the oil sector is the revenue provider for the Russian budget, the gas sector has a much broader political agenda, including social issues, influence over the regions, low energy prices for the domestic industries, providing financing for the "projects of state importance" and for the powerful vested interest groups.[5]

After a period of extensive growth and real "Golden Age of Gas" in the 2000s, currently the Russian gas industry is facing numerous challenges. Virtually, all of its external and internal conditions had radically changed for the worse. Fundamental shift in all major components of the country's gas balance (domestic demand, exports, production, and imports) creates huge uncertainty, aggravated by the sanctions and ongoing geopolitical confrontation. Since gas accounts for the major share of the country's primary energy consumption and power generation, the cost of a mistake is extremely high in Russia, forcing the government to be very cautious in this decision-making (and inevitably increasing the uncertainties associated with the future development of the Russian gas market).

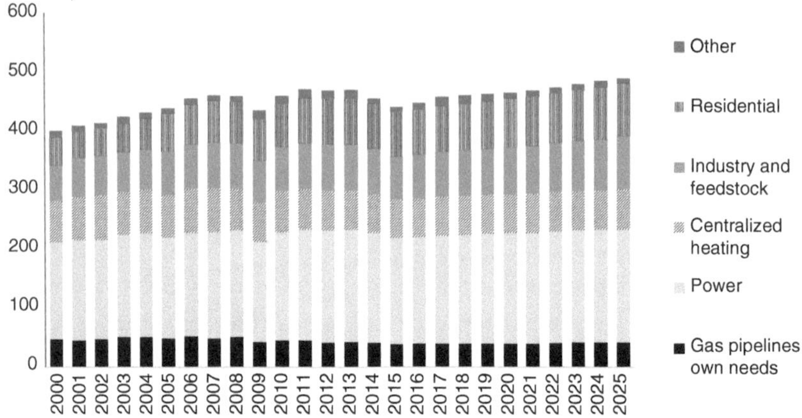

Fig. 9.1 Russian gas demand dynamics by sector in 1990–2013, bcm. *Source* Author's elaboration on Rosstat, ERI RAS, Energy Ministry of the Russian Federation

9.1.3 Domestic Gas Demand Stagnation

Historically, Russian energy balance is dominated by gas, which is providing for 53% of the total primary energy consumption in the country and for 50% of the Russian electricity mix. Domestic consumption of natural gas, which demonstrated steady growth outpacing that of GDP before 2008, had almost ceased to grow in the recent years ridden by ramification of the crisis. Moreover, it is falling for a third straight year. In the power sector, consumption fell by around 10 bcm and the outlook remains weak as new efficient gas units alongside additional base-load nuclear capacity are set to come on line amid slow growth in electricity demand. Moreover, plans to reform the heat market are advancing. This would result in higher tariffs and prompt modernization investments, possibly from 2018, further reducing gas demand. Overall, a poor economic growth outlook exacerbated by the financial stress of low prices is set to weigh on consumption growth[6] (Fig. 9.1).

This demand stagnation is creating serious problems for the Russian gas producers, especially assuming the recent changes in the domestic gas price regulation. In 2006, the decision was made in favor of outpacing growth of the domestic regulated gas prices to effect a phased

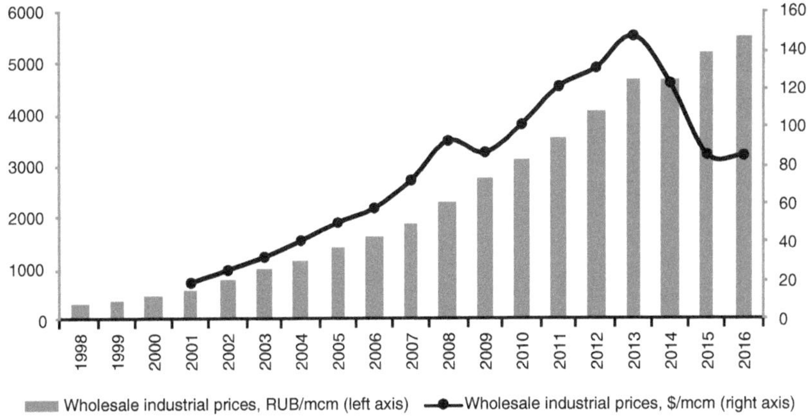

Fig. 9.2 Russian average weighted wholesale gas prices for industrial consumers in 1998–2016. *Source* Author's elaboration on Rosstat, Federal Tariff Service of the Russian Federation

transition to the European netback levels (calculated as export price minus export duty, transportation, and other costs relating to storage and sale)—that is, equal profitability of supplying gas to the domestic market and for exports, ensuring gas price growth on the annual level of 15–25% until it reaches the netback level by 2011 (per a 2006 estimate). With the rise of the oil price, this date was further postponed until 2015–2018. But in 2013, as negative processes such as deceleration of GDP growth, industrial production, and fixed investments became very strong, the Russian government finally decided to freeze gas prices, simply indexing them with the rate of inflation. As a result, the initial 2011 target date to reach netback parity was postponed to 2030–2035 (especially following ruble depreciation in December 2014, when prices expressed in dollars were halved back to the level observed in 2008—Fig. 9.2). Basically, Russia became locked in the framework of low state-regulated domestic gas prices. Such indexation, of course, will eliminate a significant share of gas producers' revenues from the domestic market and will force them in the longer term to reduce their investment program. Moreover, poor economic situation is leading to increasing non-payments, further undermining gas industry revenues

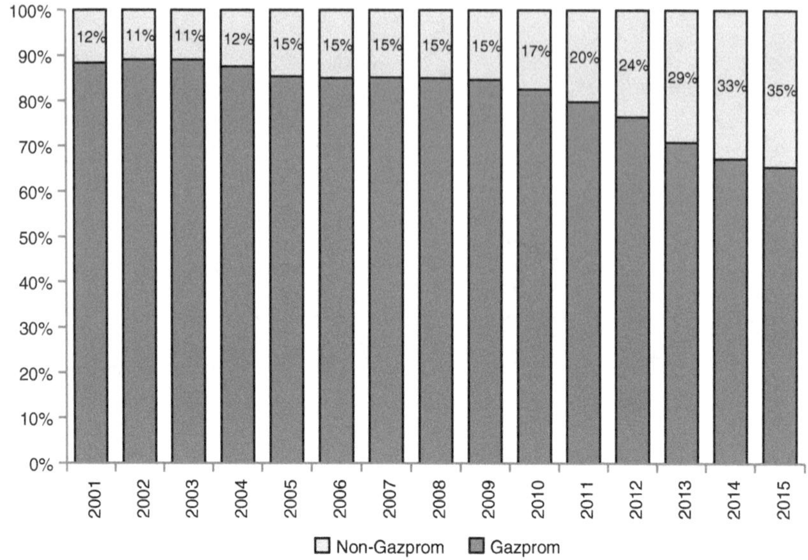

Fig. 9.3 Domestic gas supply structure in 2001–2015. *Source* Author's elaboration on Rosstat, ERI RAS ad Gazprom

on the domestic market and increasing attractiveness of export for the gas producers.

For Gazprom, this situation is even more painful, as it is facing stiff competition in its core domestic market, losing further share to competitors Novatek and Rosneft, who are benefiting from growing sales to the more lucrative industrial segment. Gazprom's loss of market share to its major competitors is driven by the fact that these competitors can sell gas to industrial customers below regulated prices, while Gazprom is not allowed to provide any discounts. As a result, between 2011 and 2015, Gazprom's sales to the domestic market decreased by about 40 bcm, and its market share went from 83 to 65% over the period (Fig. 9.3).

The loss of ground in the domestic market became more relevant for Gazprom in 2015. Due to the fall in dollar-denominated export prices, the differential between the export netback parity level and the average regulated domestic wholesale gas price has narrowed, making the loss of domestic volumes more significant to the overall company's revenue.

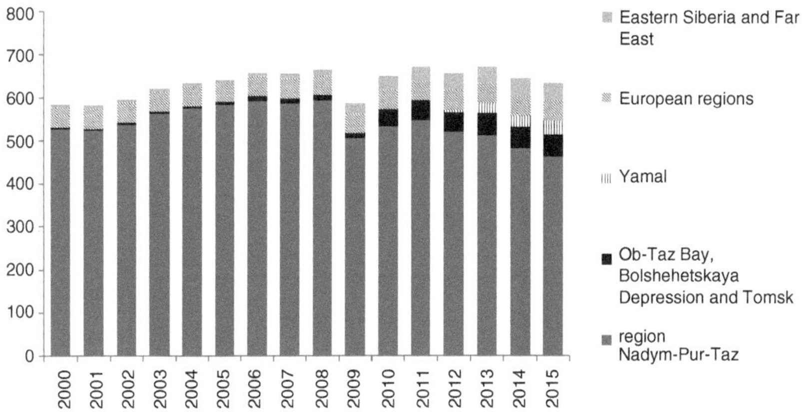

Fig. 9.4 Russian gas production profile and regional structure in 2000–2015, bcm. *Source* Author's elaboration on Rosstat, Energy Ministry of the Russian Federation, CDU TEK

9.1.4 Russian Gas Production: Expanding Gas Glut

Russia was dynamically raising its gas production in the "golden decade" of 1998–2008, mainly due to the Soviet legacy and new fields in the Western Siberia (Nadym-Pur-Taz), but in 2008, production has dropped significantly (Fig. 9.4) and still has not completely recovered due to the demand slowdown in the domestic and European gas markets and lower supplies to the CIS.

This decline is solely demand-driven: On the supply side, Russia was heavily investing since 2008 and now possesses huge spare production capacities of 150–170 bcm per annum, which are supposed to increase up to 250–265 bcm per annum by 2020 as a result of the past investments, including those into oil production, which will deliver additional associated petroleum gas (Table 9.1).

The negative impact of the unfavorable market conjuncture on Gazprom's output was even larger, and pressure on the company is increasing, as strong competition domestically came on top of challenging market conditions. Novatek and Rosneft both recorded steep production growth, reaching 70 bcm (+10%) for Novatek and 62 bcm (+17%) for Rosneft, while Gazprom had to play a role of the

Table 9.1 Russian gas and unutilized production capacities by company, bcm

Company	Production in 2015(bcm)	Unutilized potential and capacities additions under development by 2020(bcm)
Gazprom	406	~155
Novatek	52	~48
Rosneft	42	~48
VIOCs (APG)	46	~15
TOTAL	635,5	266

Source Author's elaboration (2016)

swing producer: Its production reached a new historical low of about 418 bcm (−6%YoY)[7] and its share of total Russian production declined to 66% in 2015 from 83% in 2008 (Fig. 9.5). Non-Gazprom's production is booming against stagnation at Gazprom; there is increasing output by all producers of more profitable wet gas (instead of traditional dry gas). As a result, Gazprom has to constrain its own supplies, while independents actively expanded into end users market.

For many years, the question of gas market reform was a taboo topic in Russia. But recently discussions on market liberalization started, mainly driven by Rosneft and Novatek. After obtaining LNG export liberalization in 2013, they continue their lobbying activity in order to have better access to underground storage and more transparent gas tariff setting, and Rosneft might even acquire the right to export gas via the Power of Siberia pipeline to China. Currently, despite all the efforts from Novatek and Rosneft's side, the government is not developing any regulatory framework to unbundle Gazprom, which means—bearing in mind the long period of time that would be needed to implement such a regulation—that this question is not on the agenda at least for the next few years. The government's reaction is very cautious: It is frightened by the prospect of a transitional period when something might go wrong. These fears are understandable, taking into account the huge economic and political role of gas and its unique role as an internal and external policy tool. Although Gazprom is being increasingly challenged by its competitors to unbundle its gas transmission system, the company will likely maintain certain prerogatives such as

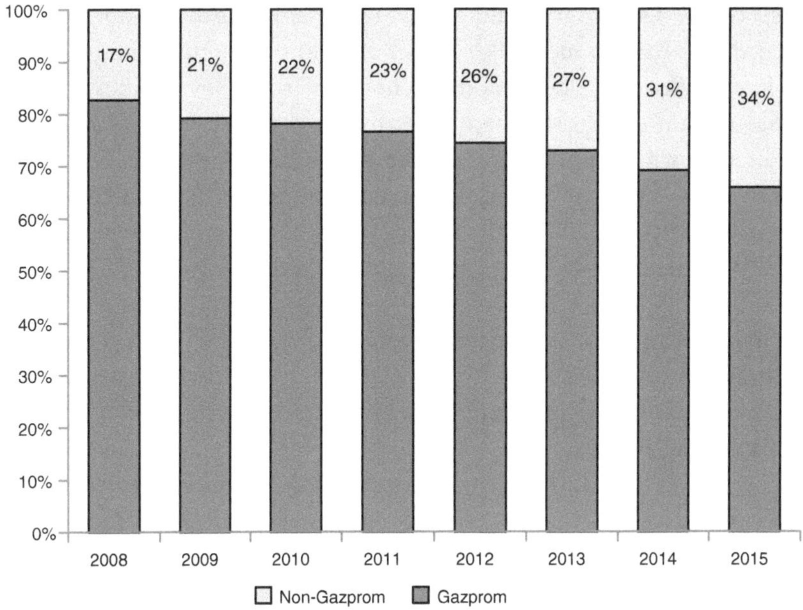

Fig. 9.5 Structure of gas production in Russia by company in 2008 and in 2015. *Source* Author's elaboration on Rosstat, ERI RAS

controlling pipelines and the bulk of gas exports given its obligations to gasify the country, supply gas to the residential sector and distant regions, pay higher taxes, and make strategic pipeline investments. Nevertheless, competitive pressure on Gazprom is increasing. As a result, Gazprom will need to define an appropriate strategy if it wants to stabilize production, by either fighting back competition in the more lucrative domestic wholesale and industrial segment (which would require obtaining the right to sell certain volumes of gas below the regulated price but raises concerns, from a state budget perspective, over the risk of price dumping), or compensating with additional exports.

Looking at the future Russian gas production outlook, we can see that the gas sector undoubtedly has capacities for sustainable production growth: The resource base is huge and is sufficient to meet domestic and export demands. Theoretically, given investment availability and sufficient demand, Russia could produce 1 trillion cubic meters per year.

Even today, Gazprom could easily meet any consumption upswing, given that it has around 150 bcm per annum of spare production capacity in its Western Siberia system, notably from Bovanenkovo, in the Yamal region. As Russia has no gas resource constraints, its future gas production will depend solely on the availability of markets and investments to build the new gas transportation infrastructure. Three different scenarios for the gas industry are regarded in this study, depending on the supply–demand balance on the global gas market:

- High Asian gas demand growth and on-time development of competing suppliers (US LNG, Australia, East Africa, Iran);
- Low Asian gas demand growth and slower development of competing suppliers;
- Low Asian demand growth and rapid development of competing suppliers.

Though the domestic market is consuming two-thirds of Russian gas output, it is very difficult to expect its radical expansion, as it is historically strongly correlated with GDP (which is projected to grow weakly). Thus, the major influence on Russia's upstream would come from abroad, depending primarily on external demand. So, Russian production outlook is particularly sensitive to international market developments. Any unexpected increase in global demand would likely trigger a large supply response from the region given the ample spare capacity available in the Russian upstream system.

In the high-demand scenario, all new gas would be absorbed by booming Asian markets. That means that more LNG would divert to Asia and slightly more Russian pipeline gas would be required by Europe. Such a call on Russian gas results in rather bullish production projections, rising from 650 bcm in 2010 up to 820 bcm in 2025 in the high-demand scenario (Fig. 9.6), though these figures are still much lower than the previous production targets of the General Scheme of gas industry development drafted in 2008–2010.

The speed and success of the development of alternative suppliers is critical. If their entrance to the market is postponed, or some of them fail to deliver gas, then Russia is always in a position to compensate for

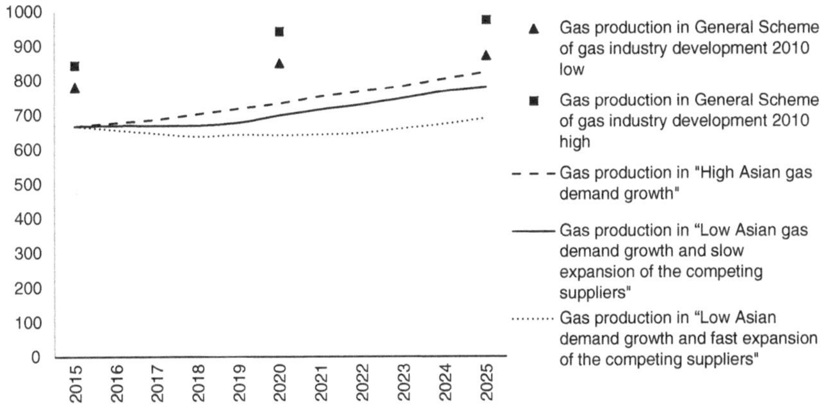

Fig. 9.6 Russian gas production outlook, bcm. *Source* Author's projections

their unavailability. But if they are successful in their project development, then Russia could face very difficult circumstances, especially in the period 2018–2023, when huge new volumes of gas are expected to enter the global market. In this case, Russia would have to struggle to protect its market share both in Europe and in Asia, and production volumes could stagnate for a decade. Much will depend on the pricing strategy chosen by the authorities and Gazprom: Russia has now huge spare gas production and transportation capacities, so it could theoretically follow Saudi Arabia's example and try to squeeze other out producers by flooding the market with cheap supplies. But, so far, there are no evidences of such a strategy.

9.2 Russian Gas Exports: Historical Background

9.2.1 Traditional Russian Gas Export Strategy in Europe

Gas export strategy of the Soviet Union, which was largely inherited by Russia, was based on the following few premises:

- Orientation on the single—European—market and expectations of stable gas demand growth in this export market.
- The USSR, and later Russia, had the cheapest gas in the market and was interested in maximizing export volume, not price (given the

serious deficit of hard currency, the country strove to increase gas exports to Europe to the maximum, and not infrequently resorted to loss-making in order to enter new markets).

- There were only a few competing suppliers in the European market, and market areas were clearly divided among them (Algeria, for all practical purposes, controlled southern Europe, Norway controlled the Northwest, and Russia controlled Central and eastern Europe).
- Gas was supplied solely on the basis of oil-linked long-term contracts with "take-or-pay" and destination clauses.
- The contracts only arranged for gas delivery as far as the national borders of the individual European countries. These bilateral contracts were usually supported by intergovernmental agreements, and they were the only real legal basis for the regulation of deliveries.

Beginning in 2002, after the appointment of new management at Gazprom, this "traditional" strategy was supplemented with a number of new features:

- Irreconcilable transit conflicts with Ukraine and Belarus, resulting in suspension of gas supplies to European countries, led to the appearance of strategies to bypass the transit countries (Nord and South Streams).
- Gazprom concentrated on maximizing export income, not volume. The government obliged the company to increase the value added of gas being sold, so the strategy of moving downstream and gaining access to end users in European countries was announced. At the same time, the government decided that Gazprom had good chances of becoming a champion of the Russian cause, i.e., an international player representing Russian national interests globally. Gazprom started to position itself as a transnational energy company (instead of the national gas company it had hitherto been) and began globalizing its activities, developing a large number of joint ventures, as well as involving itself in searching for European storage, transport and distribution assets.[8]

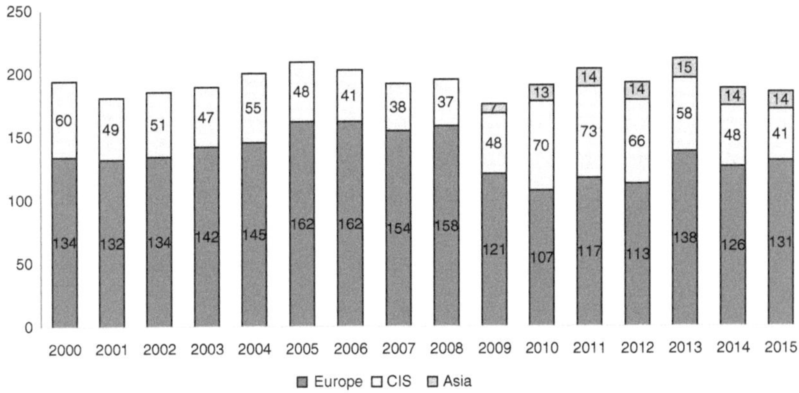

Fig. 9.7 Dynamics of the Russian gas exports in 2000–2015, bcm. *Source* Author's elaboration on Central Bank of the Russian Federation, Russian Custom Service data, Gazprom

From the start of the twenty-first century, Russia continued increasing its export volume and presence in the European market until the onset of the global economic crisis in 2008 (Fig. 9.7).

9.2.2 Russian Gas Exports in 2008–2016: Implicit Gas Export Strategy Adjustments

The economic crisis of 2008–2009 revealed and intensified a distinctive phased transition in the European gas market. All fundamental factors (demand, supply, pricing) have been undergoing a transformation, and there have been parallel drastic changes in market regulation. First of all, due to the economic downturn, market saturation and the diminished appeal of gas in the power sector (caused primarily by unfavorable conditions of price competition with coal and low demand for electricity), natural gas consumption in the European market dropped significantly in 2008 and in 2013 was estimated to still be about 10% below the level of 2008. In fact, it had fallen to the level of consumption in 2000, showing no signs of recovery. This was a product of a maturing market, low population growth, higher gas prices (in large part due to the oil price linkage for a large proportion of contracted imports) and the migration of manufacturing industry to other parts of the world.[9] Growing energy efficiency, low demand for electricity, and the development of renewable energy sources also served to limit the growth of gas demand, thus constraining the market niche for Russian gas. Noteworthy, that many of the mentioned factors are hardly irreversible. For instance, the inter-fuel competition is expected to remain one of the major drivers of gas demand. The investments in renewable energy technologies have already influenced development of the European energy mix.

Supply-side developments were also quite unfavorable for Russia: The volume of LNG supplies to Europe had sharply risen in 2009–2011 because of the bulk of new LNG capacity and drastic drop in the US LNG imports. The LNG glut contributed to the rapid development of gas hub trade in Europe and created huge pressure on spot prices (Cedigaz 2013). The market niche in Europe started to shrink compared to previous assessments, with Russian gas becoming less attractive for European consumers from both a commercial and a political point of view.

The regulatory framework has complicated the situation. The introduction of the European Gas Directive (98/30/EC), the Second Gas Directive (2003/55/EC), and the Third Gas Directive (2009/73/EC) have brought fundamental changes to the natural gas sector across many European countries (Growitsch, Stronzik, and Nepal, 2012). European Union Member States are actively liberalizing their gas markets to encourage tighter competition and the efforts of the European Commission to create a single European gas market are increasing the legal and political pressure on Russia. In September 2012, the European Commission initiated an investigation against Gazprom, accusing it of price discrimination on the basis of oil-indexed prices (CERA 2013). At the time of writing (mid-2014), the results of the investigation have not been announced, but it is already clear that European anti-trust regulation will require some amendments to Gazprom's existing contracts.

The liberalization policy also implies the separation of gas production assets from transportation networks within the EU. This could mean that, unless legally exempted, Gazprom would find itself unable to hold and control the transportation side of its operations, and this will also constrain investment initiatives in the EU. In order to respond to the new rules, Gazprom, in effect, was forced to abandon its vertical integration strategy and downstream movement, but the most important issue is that deregulation is putting more and more pressure on its traditional contractual model (eds Henderson and Pirani 2014).

The recent dramatic shift in the gas pricing model has probably been the most painful development of the European gas market for Russia. In 2008, spot-indexed gas supplies accounted for nearly 20% of total gas consumption, but in the course of the last 5 years, due to market oversupply, this share has reached 50% (IGU 2013). The hubs have become price benchmarks for a significant number of market participants. The gap between prices in oil-linked long-term contracts and spot prices has become a basis for revising contract terms. On 10 September 2013, the European Parliament adopted a report on the internal energy market, calling for the abolishment of oil indexation and conversion to "more flexible alternatives" (European Parliament 2013).

These fundamental changes have created a wide range of threats to the traditional Russian export strategy in the gas sector and are undermining its basic premises. In 2009, Russian gas exports dropped by a dramatic 23% (Fig. 9.7). Gazprom started to lose its market share (down from 30% before the crisis to just 23% in 2009), and it was becoming obvious that the traditional strategy, which had worked excellently for half a century, now had to adapt to the new reality.

Due to the weak demand, Russia has not restored its pre-crisis export volumes to Europe (162 bcm in 2006 vs. 131 bcm in 2015—Fig. 9.7), though 2016 might be more optimistic, as low oil prices made Russian gas one of the cheapest and most attractive options in the market. At the same time, Ukrainian conflict dramatically reduced CIS sales volumes (in particular—Ukrainian purchases of Russian gas have fallen almost 90% between 2011 and 2015),[10] while gas export to Asia is now limited to Sakhalin LNG exports, and the recently signed deal on pipeline gas supplies to China is only a longer-term prospect.

9.3 Russian Gas Export Outlook

9.3.1 Changing Gas Market Conjuncture for Russian Gas

Weak demand in Europe and slower consumption growth in China together with the approaching LNG glut are going to be the main factors, defining Russian gas outlook in addition to the price of oil. Abundant supplies of LNG lead to strong competition among producers: Ample spare regasification capacity allows both Europe and China to arbitrage between pipeline gas and LNG based on pricing. Russia's share of Europe imports is threatened.[11]

Europe has been Russia's core export market for five decades, and though it has mature and declining demand, its hydrocarbon import needs will nonetheless increase because of declining local European production. In the short to medium term, the major challenge for the Russian gas exports to Europe originates from the supply side. The large influx of new LNG volumes—particularly between 2016 and 2018—will result in more supply pushing toward Europe. Until today, LNG has never been a real threat to Gazprom's position in the European market. Over the past 5 years, European LNG imports ran at low levels, and the net effect of weak demand, weak production, and losses of North African volumes actually meant an increase in Gazprom's share of the European market. Oversupply in global markets will lead to fierce competition in Europe, with flexible US and Qatari volumes fighting hard to gain access to European customers. Outcome of this competition, however, is heavily dependent on Gazprom's commercial strategy and its response to the projected oversupply in the market.

Currently oil-linked Russian gas prices and hub prices are quite close, so in this situation, customers are more or less indifferent as to where they source their gas (through the spot market or through higher nominations of Russian gas). Nevertheless, the situation could change: The process of market rebalancing could take longer for gas than for oil. In this case, the gap between the value of oil-linked Russian gas and the price of spot gas would widen. European customers would then find cheaper-to-source gas from the spot market. This situation would likely trigger renewed tensions between Gazprom and its customers, including for those volumes delivered based on minimum take-or-pay obligations.

Nevertheless, as Russia has all of the necessary infrastructure in place (if Ukrainian transit is regarded as a viable option)—which fully depreciated long ago and has comparatively low upstream and midstream costs—it is in a very good position to compete with any newcomers to the European oil and gas market. Moreover, the existing portfolio of long-term contracts is also providing firm guarantees of the gas offtake by the customers.

At the same time, this stagnation on the European gas market accompanied by the declining sales in the CIS creates a clear commercial logic for Russia for seeking new gas sales to the world's fastest growing gas markets in Asia.[12] Indeed, Russia's export focus has shifted eastward because of unfavorable conditions in the core European market together with increasingly cool relations with the West, EU pressure on Gazprom, and Western sanctions, while strong dependence on energy export revenues drives new sources of economic growth in a difficult market. These factors are pushing Russia toward closer energy cooperation with Asia (primarily China), though this cooperation is not developing effortlessly.

It would seem that Russia and northeast Asian countries are ideal, complementary partners in energy trade: One is the holder of enormous hydrocarbon reserves, a leading exporter, and the others represent the largest consuming region and importer of hydrocarbons. But, for Russia, building energy relations with these countries is not a simple story at all. To lessen dependence on the European gas market, Russia managed finally to sign the gas deal with China, but "Power of Siberia" pipeline construction will take five years and an additional five years will be needed to bring the pipeline to its full capacity of 38 bcma, which means that at least until the mid-2020s; eastward gas exports will not be able to replace the reduction of supplies to Europe.

9.3.2 Russian Long-Term Gas Supply Contracts in Europe: Existing Portfolio and Its Renegotiation

Starting in 2009, Gazprom began receiving official notices from European buyers demanding that their contracts be reviewed. Under growing client pressure, and also because all other European suppliers were changing their contracts, Gazprom, wishing to maintain its market share, was also compelled to review its contracts in order to adjust to market conditions and improve the competitive advantage of Russian natural gas supplies.

The first such agreement was reached in 2009 after extended negotiations with the largest buyers: E.ON (Germany), ENI (Italy), Botas (Turkey), and GDF SUEZ (France), which together account for the purchase of about 40% of all Russian gas in Europe (Gazprom 2014). At that point, Gazprom had to make the following concessions:

- Consumers would receive a portion of the gas (specifically, the portion above the "take-or-pay" level—15% of the annual contract volume) at spot prices. Calling this a discount is a bit of a stretch, since under conditions where demand was dropping most of the buyers had difficulty in meeting their "take-or-pay" volume in any case. In the words of the Deputy Chairman of Gazprom, Alexander Medvedev, "the total cost discount amounted to only 3%, taking into account the spot component in the pricing formula for 10–15% of the combined contract volume."[13]
- Along with this, as indicated in the memorandum for the company's Eurobonds, Gazprom lowered the obligations of its European clients to accept delivery of gas from 2010 to 2012, three years altogether, to 15 bcm. But, the obligations of their clients under "take-or-pay" contracts were lowered on condition of larger deliveries later ("make-up gas"—volumes that were transferred to future years).

As Gazprom stated, "the introduction of the spot component and the transfer of obligations under the 'take-or-pay' conditions to a later period was done selectively, based on careful analysis of the client's reasoning in favour of such a transfer; in this way the basic conditions of the contract were not violated" (Gazprom 2011). But, Gazprom announced that even these measures were "temporary" and effective only between 2010 and 2012, once every three years Gazprom and its European clients can review the contract prices, and for this period, discounts are allowed.

After providing discounts in 2009, Gazprom took a more hard-line position in 2010. In response to Gazprom's refusal to allow further discounts, a number of European consumers began to resolve the issue by turning to arbitration, which is provided for in the contracts in the event that the parties cannot settle differences within 6 months from the start of negotiations. If the negotiations are not successful, the buyer can either go to arbitration or else agree to a bilateral termination of the contract (which obviously Gazprom would never agree to).

The Italian company, Edison, was the first to appeal to the Stockholm Arbitration Court for a ruling in the summer of 2010, but the case did not result in a ruling, with Gazprom preferring to settle the matter out of court. As a result, Edison secured a 200 million euro concession, a decrease in their obligations under "take-or-pay," and also the introduction of a number of additional factors in the accounting formula for the price of gas.[14]

In 2011, Gazprom started to give cash discounts on an individual basis to those buyers that were important to the company. VerbundnetzGas AG (Germany), Estonian and Latvian companies and also DEPA (Greece) and Botas (Turkey) received discounts, with the concessions for the latter two countries in fact being tied to negotiations on the South Stream pipeline. Other buyers that were not accorded the desired discounts in 2011—specifically E.ON (Germany), RWE Transgas (Czech Republic), and PGNiG (Poland)—followed Edison's example and brought suits against Gazprom before the international court of arbitration to demand a review of pricing on their long-term contracts. Their justifications for this step were the changes in the worldwide gas market and the substantial gap between spot prices on long-term contracts. The goal of these companies was to secure acceptable pricing levels and compensation for outflow from the previous period.

In 2012, the unfavorable situation in the European gas market forced Gazprom to make concessions once more. Prices were reviewed for only a few large clients—EconGas and Centrex (Austria), Sinergie Italiane (Italy)—whose volumes amounted to about 25% of Russian exports to Europe. These key clients received discounts through the review of the base price formula P0. According to Alexander Medvedev, the prices for these clients were reduced on an average by 7–10%.[15] There was no question of any spot component, and the price structure and volume requirements remained unchanged. Similar conditions were presented in March 2012 to one of the Gazprom's largest partners (and a participant in the South Stream pipeline project), the Italian company ENI. An agreement with German E.ON on price corrections was eventually reached only at the beginning of June 2012. E.ON managed to secure the same conditions that were presented to ENI, and furthermore the price for the German company was reviewed "retrospectively"—beginning from the fourth quarter of 2010.[16]

It is clear that Gazprom conducts negotiations on an individual basis, depending on the historical relationship and strategic significance of a given buyer; with a lot of companies, Gazprom is not just concluding contracts, but also major joint projects and joint ventures in the downstream sector. Hence, Gazprom's policy with regard to contract review has been based on the principal of delaying for as long as possible before providing the minimum discount acceptable to the buyer, under the terms of "special" bilateral agreements with various client countries. As a result, the difference between calculated oil-linked contract price for Russian gas and reported in Russian statistics export gas price appeared to vary significantly by importer country. The price adjustments also seem to have mitigated the differences in absolute levels of import price for Russian gas.

In 2013, Gazprom started to implement a new price discount model with so-called retroactive payments. According to this model, the company has to compensate its customers for the difference between contract price and spot price by the end of the year. This was an elegant way of executing a de-facto switch to spot indexation, while remaining formally within the framework of oil-indexed contracts (and to protect these contracts they were signed under the auspices of intergovernmental agreements). All these "compensations" are presented as temporary, and Gazprom has a right to remove them should the market become tighter.

The year 2013 saw the next round of negotiations with major customers (including ENI, GDF SUEZ, and Premium Gas). Negotiations with RWE Transgas were probably the most hotly contested and ended up in arbitration. The real outcome of the arbitration is still unclear as the details were not disclosed, while both Gazprom and RWE announced victory in their press releases.[17] But it was stated by the arbitrator that the prices should reflect the real market situation—so one can surmise that some spot component would have been a part of this decision (most likely based on the model of retroactive compensation for the difference in the spot price).

In 2014, contracts were renegotiated with SPP (Slovakia), Centrex (Austria), ERG and ENI (Italy), DEPA (Greece) and a number of Turkish companies—Akfel Gaz, Avrasya Gaz, Bosphorus Gas, Bati Hatti, Kibar Enerji, Enerco Enerji (Gazprom 2014). In 2015, contracts were reviewed with Engie S.A., Sinergie Italiane, PremiumGas, SPP with the introduction of larger spot component: According to Gazprom,[18] the share of direct spot indexation in the company's gas sales to the European customers in 2015 has reached 17.8% (compared to 16.5% in 2014). Moreover, in September 2015, Gazprom for the first time has launched gas auction for the supplies of gas during the winter 2015/2016 in several destinations in Europe. This gas was sold to 16 customers, total volumes reached 1.2 bcm.[19] Gazprom began to implement these auctions as a regular practice.

So, despite Gazprom's strident rhetoric in favor of traditional oil indexation, in actual fact, numerous adjustments and contract reviews have already been made in the course of the last 5 years. Analysis of Gazprom's official reports demonstrates a much more flexible negotiating position than has commonly been thought to be the case. During the period 2009–2015, as many as 60 gas supply contracts were reviewed with 40 clients, providing price discounts, easing of take-or-pay obligations, and a certain introduction of a spot component.

There are clear signals that Gazprom is opting for a more flexible marketing approach. The decision on September 2015 auction 3 on top of its long-term contracts marked a major departure from the company's previously stated strategy to stick to long-term sales and hints at a potentially more proactive (rather than reactive) pricing behavior. Gazprom conducted a second auction in March 2016 for the Baltic States, selling three-quarters of the offered volumes, or around 0.4 bcm. New auctions for both the Continental Europe and the Baltic States are expected to take place, particularly when and if spot prices are above oil-linked contract prices. If Gazprom volumes placed on the spot market grow, this could have a major impact on market dynamism in Europe. Taken together, these trends suggest that the period 2016–2018 will be very different for Gazprom, and the company might rethink its approach in light of weak demand in Europe, greater regional interconnection and large volumes of cheap LNG flooding the market. Additional supplies to Europe can conceivably displace Russian gas, but if there is a change in Gazprom's pricing policy aiming at defending market share, prices could be bid down to levels that trigger either coal-to-gas switching in the power sector or a significant supply-side response.[20]

Calculations using Russian Customs Service statistics, Gazprom reports and the Nexant World Gas Model (which allows the assessment of contractual prices based on the prices of oil products) shows that by 2015 Gazprom had already provided nearly a 25% average discount to its European customers compared to its pre-crisis traditional oil-linked price formulas. As a result of all, these price discounts and also the tightening European gas balance, already by the end of 2013 Gazprom managed to restore its market share to the pre-crisis level of 30% (Gazprom 2015).

At the same time, it is necessary to stress, that differently from all the other suppliers, Russia has a very stable and long-term portfolio of contracts, which will start to expire only in the late 2020s, still providing guarantees of quite stable export volumes to Europe. Just the existing portfolio of already-signed long-term contracts (even with the revised take-or-pay obligations) guarantees Russian sales of at least 128 bcma up to 2022 (Fig. 9.8). Additional volumes could be sold at European

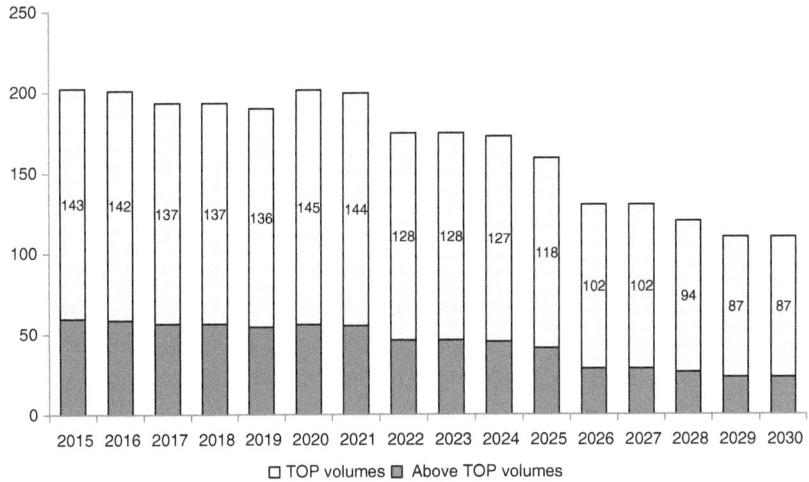

Fig. 9.8 Russian portfolio of long-term gas supply contracts. *Source* Author's elaboration on NEXANT, Cedigaz

hubs, thus providing an opportunity for Russia to protect current export volumes to the European market, though any significant growth of these exports does not seem likely. Nevertheless, any significant increase in Russian gas exports to Europe is unlikely in these circumstances.

9.3.3 Russian Gas Export Outlook

Russian medium-term gas export outlook is rather clear: It is defined by the long-term contracts in Europe and by the implementation of the Chinese deal in Asia.

In Europe during the next 5–7 years, Russian gas export strategy faces huge adaptation challenges. Gazprom has shown that it is willing to respond to competitive and regulatory pressure by adjusting its price level and its contractual terms, though some of these changes have been forced on it by arbitration cases. However, to date, Russian strategy has been very reactive, and once again it has looked for short-term solutions to immediate problems and to create multiple options for negotiating purposes rather than developing a coherent long-term strategy. Moreover, since 2014, exports have been strongly affected by Russia's geopolitical problems over Ukraine. This conflict has undermined European confidence in Russia as a secure source of gas supply. The interruptions of transit through Ukraine in 2006 and 2009 raised initial questions about the need for the European Union to reduce its perceived dependence on Russian gas, and these concerns have been amplified since the annexation of Crimea in March 2014 and the subsequent conflict in eastern Ukraine. Therefore, in order to protect its market niche, which could be targeted by the LNG suppliers, Gazprom will have to take more proactive position.

The positive news for Russia is that if it needs to compete to maintain its position in the European gas market then it has enough low-cost supply to meet its objectives. Although the full cost of developing greenfields is high (about 6–8 $/mmBtu), the short-run marginal cost of West Siberian supply is much lower, thanks to low upstream costs and benefits of Rouble devaluation (4–5 $/mmBtu). At this level, it can compete with US LNG imports, meaning that on a purely commercial basis, Russia could effectively choose its own market share in Europe.

Compared to most of its new competitors, Russia has a lower cost gas supply base and can thus engage in a price war if needed. Nevertheless, it would prefer to avoid "price war" with the US and Qatar LNG in order to maintain export revenues. Although Russia would prefer the status quo to persist, it is preparing to respond to change and competition by altering its pricing methodology and contractual terms. It seems inevitable that the continent will remain reliant on Russian gas for the foreseeable future, but it can avoid dependence on Russian gas by continuing its current strategy of increasing supply alternatives that are available via interconnections across Europe in order to ensure that countries that are most dependent on Russian gas can develop robust diversification strategies.

In Asia in the medium term, Russian gas export strategy is primarily related to the "Power of Siberia" project construction. It is under way now, but the key question is whether Gazprom will try to optimize its cost in the low price environment or postpone it, or indeed decide not to proceed with it at all. According to Sberbank CIB Investment Research, "at $30 per barrel oil price, the average price tag for gas delivered from 2021 at just north of $3/mmBtu. Even if oil prices would recover, gas prices in Asia are unlikely to exceed about $5.5/mmBtu in the long term. Even though Gazprom's China contract is tied to the oil price, we believe China will not take the gas unless it is competitively priced against alternative supplies. This sets a cap on the price. We have now cut our capex expectations for the project to about $40 bln from $55 bln previously, on the back of the depreciation. However, we have some doubts that the 'depreciation gains' could actually be skimmed. The capex cut pushes the breakeven price for the project down to $7/mmBtu, still way above what Gazprom could charge given the competitiveness of expected supplies to the Chinese market. The best course of action for Gazprom right now would be to find the least costly method of backing out of the contract. The second-best option would be to invite Chinese contractors to march into East Siberia and build the pipeline at what would likely be a much lower cost than Russian companies could offer."[21]

On the LNG side, Yamal LNG is the only project with a real chance to be launched in the medium term. All other LNG projects have been deferred, some indefinitely. Lower oil and gas prices and severe financial difficulties have dampened the "LNG fever" that had spread among Russian companies in recent years. Costs have also increased sharply due to the ruble depreciation, as Russia needs to buy much of its LNG equipment on the international market (where it is priced in US dollars).

Long-term projections of the Russian gas exports seem to be revised significantly downward, compared to the previous estimations, made in 2006–2008 (from 350 to 270 bcma), but they still remain the largest in the world. In the time horizon up to 2030, Russia will attempt to solve the twin problems of protecting its 30% market share in Europe while simultaneously substantially increasing gas supplies to Asia (Fig. 9.9). By 2025, total Eastern gas exports could reach 60 bcma (with the potential uplift to 85 bcma, depending on the success of the Altai pipeline negotiations and success in cooperation with the foreign partners in the joint LNG projects). This is one-third of the current exports

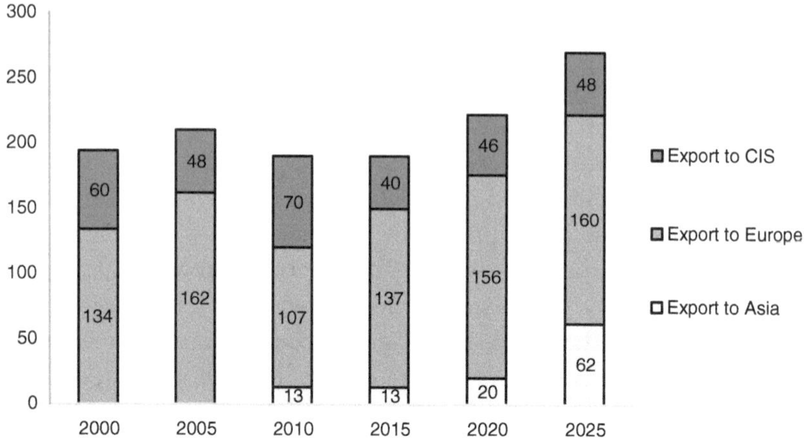

Fig. 9.9 Russian gas export forecast up to 2025 by destination, bcm. *Source* Author's projections

to Europe. Anyway, even in the long term, Europe will stay in Russia's focus.

9.3.4 Gas Pipeline Strategy

Reducing the role of Ukraine as a transit route has been a long-standing strategic priority for Russia. The country has made substantial progress toward this goal in recent years, with gas transiting through Ukraine halving since 2005, to around 67 bcm last year. Russia argues that Ukrainian transit risk can be solved once and for all only through the construction of bypass transit pipelines, and it has offered several alternatives including South Stream, Turkish Stream, and an expansion of the Nord Stream. The evolutionary and somewhat improvised nature of Gazprom's export strategy to Europe is perhaps best exemplified by its infrastructure plans, which over one year of 2015 have involved a commitment to end transit through Ukraine by building the South Stream pipeline; a switch from South Stream to Turkish Stream; a commitment to sell all of its Ukraine transit gas at a new Turkey/Greece hub; the announcement of an expansion of the Nord Stream; apparent uncertainty over the plans for the Turkish Stream, with contractor contracts being canceled, onshore lines postponed, and an intergovernmental agreement with Turkey delayed; and finally an apparent reversal, under the specific instruction of President Putin, of its original decision to eliminate Ukraine transit after 2019,[22] not to miss the recent reset of the relationship between Russia and Turkey and their joint statements on Turkish Stream.

Despite all the rhetoric, none of these pipeline options is guaranteed. The future of the Turkish Stream is strongly challenged by the fragile relationship between Russia and Turkey, while the Nord Stream's expansion faces a very negative attitude from the European Commission and a number of European Member States. It could eventuate that both options would be blocked and Russia would have to use Ukrainian transit without any new bypass construction. Other variants include construction of fewer Turkish Stream lines, additional Nord Stream lines, and different combinations of these two options. This gives the impression that neither Gazprom officials nor the Russian government knows how this game will end. It depends on too many factors, mainly political: relations with Turkey, the European Commission, and Russia's Western partners. Russia is, therefore, trying to create multiple choices for future developments: Having high hopes on a single project would be a mistake, and "improvisation" is the best term to describe the current short-term policy. An inability to make a long-term strategy in such an uncertain environment leads to multiple options to allow for flexible adaptation in the future, depending on market conditions and political barriers.

Aside from Ukraine, another difficult issue for Russia in Europe is the Third Energy Package, which makes it more difficult for major producers such as Gazprom to exercise dominance by controlling infrastructure or by monopolizing individual markets. The EU has also started proceedings against Gazprom with its competition authority, the EU Director-General for Competition (DG COMP), alleging unfair practices and pricing. The regulatory drive to ensure that Gazprom, as well as the other major gas market participants, adheres to new EU rules has added to the pressure on Gazprom stemming from the commercial issues mentioned previously (e.g., weak demand, increasing competition, and low prices).

Gazprom's success in launching Nord Stream 2 remains subject to resolution of a number of issues: Will it be able to agree to new supply terms with many of its European buyers before taking the FID? Will its partners in Nord Stream take a FID if Gazprom guarantees volumes and payment but has not successfully renegotiated its contracts and clarified the situation with Ukraine? All these answers depend largely on the EU position and relationship between the Member States.

Summing up, Russia has adopted a rather opportunistic export pipeline strategy in Europe, which is driven by an extremely high political uncertainty: Russia is trying to keep all the options open, hoping that at least one of them might work.

Notes

1. Tatiana Mitrova, *The Shifting Political Economy of Russian Oil and Gas* (Washington: CSIS, 2016), 1.
2. Fak, Alex and Valery Nesterov, *Russian Oil and Gas—Trimming the Belly Fat* (Moscow: Sberbank CIB Investment Research, February 2016), 41 (Fak and Nesterov 2016).
3. Andrey Movchan, *What's in Store for the Russian Economy in 2016?* (Moscow: Carnegie Russia, 2016), 1–2 (Movchan 2016).
4. IEA, *Key World Energy Statistics 2015* (Paris: IEA, 2016), 18 (IEA 2016a).
5. Tatiana Mitrova and Gergely Molnar, *Russian Gas Market: Entering New Era* (Paris: Cedigaz, 2015), 43 (Mitrova and Molnar 2015).
6. IEA, *Medium-Term Gas Outlook 2016* (Paris: IEA, 2016), 38–39 (IEA 2016b).
7. Medium-Term Gas Outlook 2016, p. 70.
8. Anna Galkina, Vyacheslav Kulagin, and Tatiana Mitrova, "*The transformation of Russia's gas export policy in Europe,*" Energy 168 (February 2015), 30 (Galkina et al. 2015)
9. Anouk Honore, *The Outlook for Natural Gas Demand in Europe* (Oxford: The Oxford Institute for Energy Studies, 2014), 21 (Honore 2014).

10. IEA, *Medium-Term Gas Outlook 2016*, 40.
11. IEA. *Medium-Term Gas Outlook 2016*, 53.
12. James Henderson, *The Pricing Debate over Russia Gas Exports to China* (Oxford: Oxford Institute for Energy Studies, 2011), 11 (Henderson 2011).
13. Alexander Medvedev, "*Revenue is the Important Thing*" (Gazprom Magazine 4 (2011), 22 (Alexander 2011).
14. Thierry Bros, *The future of oil-linked contracts* (Paris: Societe Generale, 2011), 2 (Bros 2011).
15. Alexander Medvedev, "*Revenue is the Important Thing*," 23.
16. Gazprom, *Annual Report 2014* (Moscow: Gazprom, 2015), 56.
17. RWE, "Analysis of the Arbitral Award Concerning the Price Revision Proceedings Between RWE Supply & Trading CZ and Gazprom Export Completed" 01 July 2013. Accessed on 1 December 2016, http://www.rwe.com/web/cms/mediablob/en/2402170/data/1360742/2/rwe-supply-trading/press/press-archive/Analysis-of-the-Arbitral-Award-Concerning-the-Price-Revision-Proceedings-Between-RWE-Supply-Trading-CZ-and-Gazprom-Export-Completed.pdf (RWE 2016).
18. Gazprom, *Annual Report 2015* (Moscow: Gazprom, 2016), 38 (Gazprom 2016).
19. Gazprom, *Annual Report 2015*, 104.
20. IEA, *Medium-Term Gas Outlook 2016*, 89–90.
21. Fak, Alex and Valery Nesterov, *Russian Oil and Gas—Trimming the Belly Fat*, 46.
22. James Henderson and Tatiana Mitrova, *The Political and Commercial Dynamics of Russia's Gas Export Strategy*" (Oxford: The Oxford Institute for Energy Studies, 2015), 45 (Henderson and Tatiana 2015).

Bibliography

Bros, Thierry. 2011. *The Future of Oil-linked Contracts*. Paris: Societe Generale.

Cedigaz. 2013. *Natural Gas in the World 2013 Edition*. Paris: Cedigaz.

CERA. 2013. *Challenge and Change in The EU-Russia Gas relations*. CERA.

European Parliament. 2013. *Report on Making the Internal Energy Market Work*. 2013/2005(INI). Strasbourg.

Fak, Alex, and Valery Nesterov. *Russian Oil and Gas—Trimming the Belly Fat*. Moscow: Sberbank CIB Investment Research, February 2016.

Galkina, Anna, Vyacheslav Kulagin and Tatiana Mitrova. *The Transformation of Russia's Gas Export Policy in Europe*. Energy 168 (February 2015): 30–40.

Gazprom. 2011. *Annual Report 2010*. Moscow: Gazprom.

Gazprom. 2014. *Annual Report 2013*. Moscow: Gazprom.

Gazprom. 2015. *Annual Report 2014*. Moscow: Gazprom.

Gazprom. 2016. *Annual Report 2015*. Moscow: Gazprom.

Henderson, James. 2011. *The Pricing Debate Over Russia Gas Exports to China*. Oxford: Oxford Institute for Energy Studies.

Henderson, James, and Simon Pirani (eds.). 2014. *The Russian Gas Matrix: How Markets are Driving Change*. Oxford: The Oxford University Press.

Henderson, James, and Tatiana Mitrova. 2015. *The Political and Commercial Dynamics of Russia's Gas Export Strategy*. Oxford: The Oxford Institute for Energy Studies.

Honore, Anouk. 2014. *The Outlook for Natural Gas Demand in Europe*. NG87. Oxford: The Oxford Institute for Energy Studies.

IEA. 2016a. *Key World Energy Statistics 2015*. Paris: IEA.

IEA. 2016b. *Medium-Term Gas Outlook 2016*. Paris: IEA.

IGU. 2013. *Wholesale Gas Price Survey—2013 Edition*. Oslo: IGU.

Medvedev, Alexander. 2011. Revenue is the Important Thing. *Gazprom Magazine* 4: 22–23.

Mitrova, Tatiana. 2016. *The Shifting Political Economy of Russian oil and gas*. Washington: CSIS.

Mitrova, Tatiana, and Gergely Molnar. 2015. *Russian Gas Market: Entering New Era*. Paris: Cedigaz.

Movchan, Andrey. 2016. *What's in Store for the Russian Economy in 2016?* Moscow: Carnegie Russia.

RWE. *Analysis of the Arbitral Award Concerning the Price Revision Proceedings Between Rwe Supply & Trading CZ and Gazprom Export Completed*. http://www.rwe.com/web/cms/mediablob/en/2402170/data/1360742/2/rwe-supply-trading/press/press-archive/Analysis-of-the-Arbitral-Award-Concerning-the-Price-Revision-Proceedings-Between-RWE-Supply-Trading-CZ-and-Gazprom-Export-Completed.pdf. Accessed 1 Dec 2016.

Author Biography

Tatiana Mitrova Ph.D., is Director of the Energy Center at SKOLKOVO, Moscow School of Management, scientific advisor at the Energy Research Institute of the Russian Academy of Sciences (ERI RAS), research scholar at the Center on Global Energy Policy at Columbia University and senior visiting research fellow of the Oxford Institute for Energy Studies (OIES). She has more than twenty years of experience in the analyses of the Russian and global energy markets, including production, transportation, demand, energy policy, pricing and market restructuring. She is also Head of the annual "Global and Russian Energy Outlook up to 2040," project member of the Governmental Commission of the Russian Federation on fuel and energy complex, and member of the board of directors at Unipro (former E.ON Russia). Dr. Mitrova is a graduate of Moscow State University's Economics Department, Assistant Professor at Gubkin Oil and Gas University, and Visiting Professor at the Institut d'Etudes Politiques de Paris (Sciences Po) Paris School of International Affairs. She has more than 120 publications in scientific and business journals and four books.

10

Earthquakes Shatter Dutch Gas Roundabout: Gas Dream to End Soon

Cyril Widdershoven

The Dutch natural gas landscape has been shattered by the negative effects of low-scale but continuing earthquakes in and around Europe's largest onshore gas field, the Groningen field.[1] After politically sensitive and sometimes subjective discussions, the Dutch government, majority shareholder in Groningen's operator NAM (Nederlandse Aardolie Maatschappij),[2] has decided to further cut overall production levels of main producing onshore gas field of Europe. At present, Groningen's gas production has been given a ceiling of 24 bcm, except in exceptional winter conditions, when production could be increased if needed.[3] The Dutch government, NAM (in which American oil major ExxonMobil and Dutch major Shell are 50% shareholder, respectively), and other stakeholders, however, will be in ongoing discussions the next years, as NGOs and several lobby groups in and outside of the Groningen Province are pushing for a further decline of overall production ceilings. Environmental NGOs are pushing for a level between 12 and 21 bcm, while lobby groups in Groningen want a total shutdown of the Groningen field forever.

C. Widdershoven (✉)
VEROCY, Elshardt 7, Heerenberg 7041SW, Gelderland, The Netherlands
e-mail: cyril.widdershoven@verocy.com

M. Hafner and S. Tagliapietra (eds.), *The European Gas Markets*,
DOI 10.1007/978-3-319-55801-1_10

The Groningen field issue, which has been a prime news item in Dutch news and papers, has been treated by all parties in the wrong way for years. Both sides, NAM and NGOs, have not been showing any real understanding for the other's position and predicament. At the same time, governmental decision-making process has also not been showing any overall statesmanship. All parties have been blaming the other for constraining the overall assessment process; political infighting and skull diggers have been leading the discussion without giving rational assessments the option to show other possible solutions. Without taking any side at present, the main NOT discussed item in the whole Groningen issue is the financial–economic impact of a lower Groningen production ceiling on government budgets and the welfare of the Dutch in general. The current discussion, which still is focused on the possible effects of earth quakes in the Groningen production area on houses and industry, has been kidnapped by anti-fossil lobbying organizations, Green parties, and alternative energy lobbyists. The effects at present are devastating to the oil and gas sector in general in the country.

Based on the subjective assessments being published by mainstream media and politicians, the total future of oil and gas (upstream–midstream–downstream) is under threat. After the successes by anti-Groningen groups, further actions are now been taken to block or end gas and oil production in the other smaller offshore and onshore fields. A full-scale anti-fossil offensive has been put in place, while the energy sector operators (oil–gas) are hanging in the rows as a KO-ed boxing champ. An objective assessment of the overall effects is needed; the below presented piece will give a first assessment of future and current gas operations and strategies in the Netherlands, while also looking at the possible effects for the Dutch Gas Roundabout and gas import–export plans.

10.1 Current Situation Dutch Gas Sector (Groningen Inc)

Dutch gas production is set to decline further in the coming years. Governmental decisions are expected to partly constraint a possible full-scale closure of Groningen, but public pressure is still high to assess the option to end natural gas production in the next decades. To support the latter, the Dutch gas sector is confronted by a multitude of problems as long-term gas export contracts have been signed with clients in Belgium, Germany, and even Italy. At the same time, investment is still needed to keep current gas-producing fields onstream and existing on and offshore gas pipeline infrastructure in place. Underestimated is also the fact that, based on the successes of Groningen and North Sea gas production, the Dutch economy and consumers are largely dependent on Groningen gas for production and domestic consumption. A further decline of domestic gas production volumes will not only necessitate a full-scale change in consumption patterns in the Netherlands, but also necessitate a direct increase of gas imports from third countries. Possible supply is available in Norway and Russia, but could have detrimental effects on government budgets and geopolitical risks. Some parties are currently calling for a diversification of gas supply, not only natural gas but also by substituting it by biogas or additional LNG imports. Energy experts agree almost unanimously that to fully remove natural gas from the energy equation in the Netherlands in the next 10 years is almost impossible.

Economically, based on the usage of gas as energy source for houses, etc., it will be also undesirable. Some scenarios predict that this is only an option around 2040–2050. First targets for the Dutch energy sector should be to remove coal from the energy mix, and substituting it by gas or alternative energy sources is feasible. Based on the current scenarios, imports of natural gas will become increasingly important. Gas analysts and traders expect that this will, in addition to geopolitical and economic issues, increase the importance of TTF (and NBP) pricing.

10.2 Groningen Debate

Groningen gas production future is still in a flux. At present, major decisions still have to be taken not only to address the concerns of the citizens of the production region in Groningen, due to earthquakes and induced seismicity issues, but also to support the future security of supply of the Netherlands and its contractual commitments to export partners. In 2015–2016, the total background and technical options to cope with the above have been under scrutiny, presenting the Dutch government (as major shareholder) and the operational partners, Shell and ExxonMobil, in the NAM (Nederlandse Aardolie Maatschappij), who is the acting operator on the field, with a long list of issues and challenges to be dealt with before the end of 2016. On June 24, the Dutch government has presented its strategy and ideas on the future of the Groningen field operations.[4] In its proposal, which has been approved in general on September 9, 2016, Dutch Minister of Economic Affairs Henk Kamp stated that based on all available information and advice of parties, he has decided to cap the Groningen production at 24 bcm per year.[5] The latter has one exception: production can be increased during a harsh winter if security of supply is under threat. The production ceiling has been set based on the available security and risk assessments. A possible 6 bcm increase of production to cope with harsh winter conditions is based on advice of Gasunie Transport Services (GTS), the Dutch gas infrastructure owner.[6]

At the same time, the Dutch government, based on advice of the Dutch Mining Authority, decided to decide on a new gas production strategy within the next 5 years. The NAM will have to submit a new gas production plan within the next years. At the same time, as was adviced by the Dutch Mining Authority, the government has put in place a new assessment round, to be held in 2 years, to assess if the current approach is still feasible.

In its report, the Ministry of Economic Affairs also reiterated that the current approach, combined with ongoing changes in L-gas (low caloric) demand in Germany, France, and Belgium, will lead to an overall lower demand in NW Europe after 2020.[7] The Dutch government already is holding talks with neighboring countries about the need for the latter to have their current L-gas systems converted into H-gas (high caloric).[8] For the Dutch domestic market, appropriate measures are needed to convert gas systems of industrial clients too. The above measures are necessary to quell ongoing discussions and fears within the population of the Groningen production area.

At the end of 2015, the Dutch government decided to follow the provisional decree of the Dutch Council of State (Raad van State) for the gas production year 2015–2016, which entailed the reduction in the gas production levels to 27 bcm per year.[9] The Council of State also decreed that before October 1, 2016, the government should have in place an assent for the gas production on the Groningen field. As a result of the latter, the NAM has submitted a new production plan for Groningen on April 1, 2016.[10] NAM's production plan proposal has been available from April 16 until the end of May for consideration and advise to the SodM, the Technische Commissie Bodembeweging (Tcbb),[11] provinces, city councils, and others. The Dutch Mining Commission has also been involved and has provided additional changes to the Dutch government's proposal.

Based on the NAM proposal, gas production is set at a level of 27 bcm for 2016/2017. In the planning, additional volumes are taking in, with a cap of 33 bcm per year. In reaction, the SodM has stated that a level of 24 bcm per year is more appropriate. According to the SodM, production fluctuations could increase induced seismicity or earthquakes. SodM reiterated that the speed of pressure loss in the field is playing a major role in the total earthquake issue. Even if production will be lowered or even totally ended, seismic activity will continue for the coming years. Based on her own assessments, SodM reiterates that a level of 24 bcm will be to keep seismic activity at the levels of 2015. The independent authority, however, has not looked at security of supply and demand.

The Dutch government also has taken a position on expected demand for L-gas (Groningen), taking into account domestic and international contracts. The latter has become a major issue, as the decrease of Groningen production not only has a direct effect on Dutch domestic gas supply but also negatively affects ongoing gas supply agreements with clients in Belgium, Germany, and France. Minister Kamp, however, has already on February 24, 2016 stated that the Dutch government will be pushing for faster end to existing gas contracts with international parties.[12] The latter will not be possible without investing in large-scale nitrogen capacity in NW Europe to convert H-gas into L-gas, which will be needed to counter existing demand for L-gas.

Still, a period of 4–5 years is available to set up the necessary frameworks to counter possible supply–demand issues and technological challenges. The Dutch government has given the NAM a five-year period to come up with a new strategy and production plan. For 2016/2017, a production cap has been set of 27 bcm. In a reaction, the SodM has stated that the NAM should submit a new production plan before March 1, 2021. The latter is based on the assumptions by SodM that a lower production level (24 bcm per year) will keep earthquakes in the region at the level of 2015. SodM also expects that NAM will need a 5-year period to comply with all new requirements. The Ministry of Economic Affairs has asked NAM to submit a production plan before October 1, 2020, taking into account time needed to react to the new plans.

NW Europe has used L-gas (Groningen) since the 1960s. Belgium, Germany, and France at that time signed long-term gas supply contracts, while setting up a gas pipeline infrastructure, build on Dutch supply specifics. At present, Dutch L-gas exports to Germany hover around 20 bcm per year, while Belgium and France take each 5 bcm. The impact of Groningen gas in these regions is high.

Still, since 2011, the Dutch government has started to discuss with these countries the options to decrease L-gas supply. The three European countries indicated in 2011 that they would be able to reach this in the next decade. According to 2011 analysis, Germany would be able to have all converted by 2020, while Belgium and France were expected to reach this level around 2024. However, the growing problems surrounding Groningen production, combined with dramatic changes in the NW European gas market, have pushed these dates forward. Kamp, forced by parliament and the Groningen parties, instigated further discussions with the other European countries. Information provided by GTS, who has been very active in talking with its fellow European infrastructure network compatriots, it now seems that Belgium, Germany, and France have already put in place a full-scale strategy by which it will start conversion not in 2024 but in 2021. Germany already is implementing a very aggressive large-scale conversion program. A fast-track, however, does not seem possible for Germany. Belgium at present is discussing its options (Table 10.1).

The Groningen decline will effect in Germany around 4 million households. With a total demand of 30 bcm of L-gas, the latter is around 30–33% of total German gas demand. In addition to L-gas

Table 10.1 Projections 2014–2024 NW Europe (Ministry of Economic Affairs 2013, "Groningengas op de Noordwest-Europese gasmarkt," November) in billion cubic meters (bcm)

	2014		2019		2024	
4	Demand	Production	Demand	Production	Demand	Production
Belgium	19	–	21	–	23	–
Denmark	6	6	6	6	7	7
Germany	92	9	90	6	92	4
France	52	–	52	–	54	
UK	91	47	91	38	86	28
Netherlands	45	71	46	57	47	39
Total NW Europe	305	133	306	107	309	78

from the Netherlands, additional gas is being produced in Germany. However, the latter also is in decline.[13] According to the Germans, domestic gas production will decline from 10 bcm per year in 2015 to 4 bcm in 2026.[14] The latter will need to mitigate partly by conversion of the total infrastructure. Since 2013, German gas network companies already are stepping up efforts to convert their total systems into H-gas supply. The total project is being coordinated by the German Bundesnetzagentur. *Total costs for Germany are expected to be around* €2 billion, which is excluding the costs made by companies to change processes to H-gas. Germany will need to convert from 2020 onwards 400–500,000 parts per year. In total, 4.3 million customers, with around 5.5 million installations, will need to be converted. Based on the current discussions between the Dutch and German government and gas network operators, it is expected that demand in Germany for L-gas will decrease around 2019–2020.

For the French, the issue is less important, but still entails conversion of a 5 bcm per year demand. L-gas currently makes up around 5% of total French gas demand and largely supplying the NW France, the region between Lille and Arras. Based on a French government proposal, the conversion from L-gas to H-gas will need to start between 2016 and 2020, entailing the conversion of around 85,000 installations in the Somme, Nord en Pas-de-Calais regions. A large-scale conversion is planned to start in 2021, reaching completion in 2030. The total costs of this operation are currently set at around €800 million.

L-gas supplies around 33% of total Belgian demand. Groningen supplies around 5 bcm per year, largely to the Antwerp-Brussel-Bergen (Mons) region. At present, plans are still being discussed within the Flemish-Belgian government. At present, a pilot already has been implemented, in which 14 large-scale industrial clients converted their systems between 2009 and 2015, with a total demand of 550 million cubic meters and one city (9 million cubic meters). Until 2030, 1.5 million customers will need to be linked to H-gas, but the start of the conversion project is still set for 2024.[15] Total costs for the Belgian conversion is set to be around €700 million.

The above conversion operation will directly lead to lower overall demand for L-gas. GTS already indicated in a report that current developments resulted in a 1 bcm lower gas demand from Belgium and France than was expected in 2015. Based on the current models, GTS expects that demand will decrease by at least another 2 bcm between 2019 and 2021. At the same time, these figures are still very optimistic, as the total effects of the so-called "Energierapport 2016" strategy are not yet clear.[16]

In the last weeks, the discussion on the Groningen field production strategy and future has been heating up. Not only Dutch parties have been involved in the discussion, but also international players such as Russian gas giant Gazprom or Norway have shown a direct interest, submitting their own ideas. Tor Martin Anfinnsen, Norwegian gas giant, Statoil's Gas Trading, stated to the press that the Dutch gas roundabout ambition in Europe will be under severe pressure if Groningen production will be lowered or even totally stopped.[17] He indicated that he questions the overall viability of this ambition in the future. The Norwegian also indicated that the Groningen issue is a major subject in the European security of gas supply discussion, as Brussels does not want to increase its dependency on Russia. Statoil will not be able to supply all gas supply needed if Groningen will decline, in combination with the already decline of European gas production in general. He also has questioned ongoing statements that global LNG supply will fill in the gaps. However, Statoil will be able to supply additional gas to the Netherlands if needed.[18] In 2015, the Netherlands has imported already 18.5 bcm of gas from Norway, in comparison with 10 bcm several years ago. Statoil has indicated that, based on available gas pipeline infrastructure, another 10 bcm could be delivered without real problems. The need for bilateral agreements is clear. Statoil has signed in 2015 two long-term gas contracts with the UK; this could be also possible for the Netherlands.

Internal discussions in the Netherlands, not only NGOs or Groningen lobby groups, also are heating up. Producers such as Engie are already in conflict with GasTerra, the Dutch gas trader. Both are in a conflict with relation to long-term L-gas supply contracts. Engie has indicated that their demand for L-gas is fledgling. In 2016, they expect to have several bcm of L-gas too much, without demand in France or Belgium available.[19] Engie now wants a change in contract, as GasTerra, as stated by Engie, still keeps to its contractual volumes. Engie also needs to convert the L-gas into H-gas, while the European market is flooded. In a reaction, Gert-Jan Lankhorst, CEO of GasTerra, stated that the suggestion that there is no demand for L-gas is wrong. The latter was said during a hearing in the Dutch parliament. GasTerra said that total demand for L-gas in NW Europe is not based on contracted volumes but on real demand. Lankhorst stated that the current discussion is not based on market demand, but purely a financial reason to get rid of existing contracts. At the same time, the Engie statement contradicts the fact, as the French company lately even contracted an additional 1 bcm from GasTerra.[20] Engie reacted in denial, openly asking for a change in contracts.

Dutch Shell has reiterated in the same discussion that they would opt for an optimal use of the Groningen gas reserves. In its own reaction to the Dutch parliament and the press, Shell Netherlands CEO Marjan van Loon reiterated that the latter is needed to use gas revenues for the development and expansion of alternative energy sources. A Delta Plan, as Shell said, will cost between €10–15 billion per year. In the Shell position paper, the company also said that this will be technically possible, taking into account all new safety requirements.[21] The latter is not new, as NAM director Gerald Schotman already stated the same in January 2016. To support the energy transition, in which the Groningen discussion is now playing a crucial role, the first step will be to close coal-powered electricity plants. Gas-powered plants, which are currently closed, can be reopened. gas as a transition fuel would be then put in place. It also would counter the impact of importing Russian gas. A report by Dutch consultancy CE Delft shows that the substitution of 10 bcm Groningen supply by Russian gas would result in an additional CO_2 emission of a 800 MW coal-fired power plant or 2 million cars.[22] Groningen gas production would mitigate the latter, while also supporting the gas sector's 16,000 jobs, of which 5400 are in Groningen.[23]

10.3 Government Revenue Discussion Groningen

The strategy for the gas production of the Groningen field also has a financial part. The Dutch government has been for years partly dependent on the revenues generated by Groningen. The term Dutch Disease has become a household name internationally due to the Dutch economy's dependency in the 1960–1970s on the gas revenues. This situation has dramatically changed, first due to economic growth in other sectors, but lately also due to the fact that the value of the Dutch gas assets has declined (Figs. 10.1 and 10.2). The Dutch statistical office Statistics Netherlands (CBS)[24] has stated in its latest report that the overall value

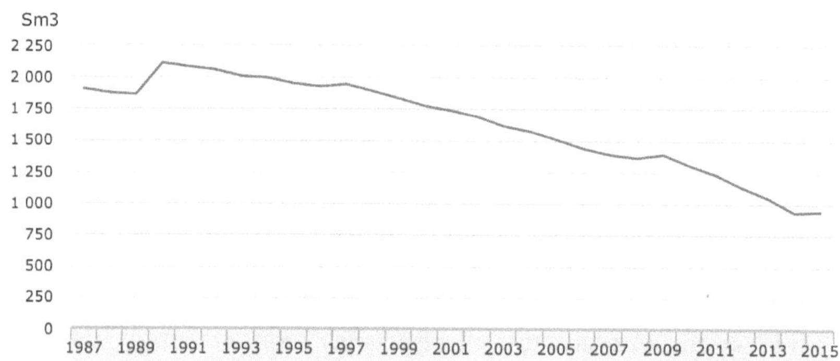

Fig. 10.1 Dutch natural gas reserves end 2015 (Centraal Bureau voor de Statistiek 2016, "Aardgas voor bijna 80 procent op," September 16) (Sm^3 = bcm)

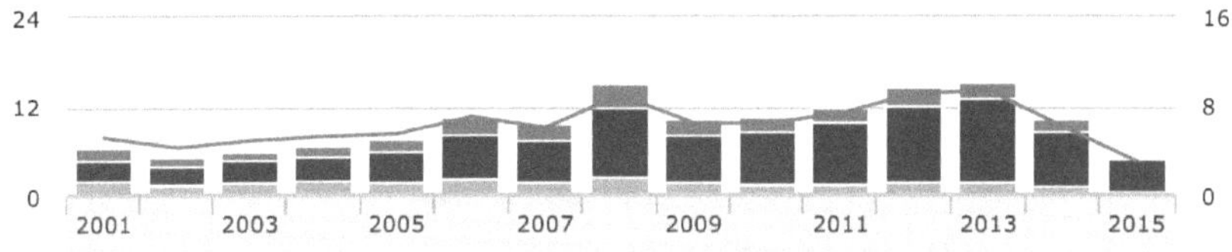

Fig. 10.2 Income Dutch government natural gas production *Left side Billion Euros, right side % government income* (Centraal Bureau voor de Statistiek 2016, "Aardgas voor bijna 80 procent op," September 16)

has again decreased substantially in 2015.[25] Due to lower production and demand for Groningen gas, combined with the earthquake and other unrest, the value[26] of this asset has almost diminished.[27]

10.4 Earthquake Costs

As the legal and commercial issues with regard to the Groningen gas production future are being settled, leading to a new normal, in which the Netherlands will be a net gas importer after leading the gas exporting world for years, a whole scale of societal issues still needs to be solved. The need to rebuilt or restore houses, churches, and industrial buildings in the region is clear. The costs of this major reconstruction program will be staggering, especially if gas revenues at the same time will be dwindling. The financial impact is still unclear. All parties are using different assessment models, financial structures and schemes. What is clear, it will be a multi-billion investment project, which will need to be paid by the NAM as main "culprit".

At the same time, evidence is not straightforward, so a long period of legal battles and political infighting is to be expected. The Groningen Province already has received around €800 million, but this is obviously not enough.[28] At present, more than 30,000 houses will need to be structurally adjusted to withstand the quakes. Some analysts even expect that this could increase to over 90,000 houses. First guess would be that this operation only will cost around €6.5 billion.[29]

Legal and political discussions are still going on to find the right framework to deal with these issues, bringing a right but workable scheme of liabilities and rights for all parties involved. If a solution is found in 2016, a real rational approach will be possible. Until now, emotions are heating up a gas production discussion which could negatively influence not only the Dutch economy but also the security of supply issues for NW Europe.

Notes

1. GEOExPro (2009).
2. Van Gastel et al. (2014).
3. Decision Dutch parliament September 14, 2016, publication at time of writing not yet available.
4. Ministry of Economic Affairs (2016b), June 24.
5. Dutch parliament majority VVD-PvdA for approval on September 14.
6. Gas Transport Services (2016).
7. WINGAS (2016).
8. See note no. 4
9. Raad van State (2015).
10. NAM Platform (2016).
11. For additional information on Technische Commissie Bodembeweging, look at tcbb.nl.
12. ANP Parlementaire Monitor (2016).
13. FNB Gas (2016).
14. Het Financieele Dagblad (2016b), "Afkicken van uniek Gronings gas is logistiek complex," September 8.
15. FluxEnergie (2016), "Gasbesluit Nederland kost België honderden miljoenen," September 1.
16. Ministry of Economic Affairs (2016a, b).
17. Het Financieele Dagblad (2016a), "Als Nederland Noors gas nodig heeft, kunnen we dat leveren," May 17.
18. Idem.
19. Het Financieele Dagblad (2016c), "GasTerra en Engie steggelen over gascontract," September 8.
20. Idem.

21. Het Financieele Dagblad (2016e), "Shell pleit voor 'maximale benutting' van Groningse gasvoorraad," August 31.
22. Vergeer et al. (2015).
23. See note no. 24.
24. Het Financieele Dagblad (2016d), "Overheid opnieuw armer door waardevermindering gasreserves," July 19.
25. For background information CBS, visit cbs.nl.
26. Value set by CBS at €105 billion end 2015.
27. Centraal Bureau voor de Statistiek (2016).
28. NU.nl (2015), "Verstevigen van Groningse huizen gaat vele miljarden kosten," February 6; FluxEnergie (2016), "Aardbevingsschade en de lusten en de lasten van het Groningse aardgas," April 20. For additional information on earthquake damage and payments, visit http://www.nam.nl/nl/nam-in-society/earthquakes.html.
29. Telegraaf (2015), "Stutten van Groningen kost 6, 5 miljard," February 6.

References

ANP Parlementaire Monitor. 2016. Lijst van vragen en antwoorden over Gaswinning Groningen en meerjarenprogramma NCG—Gaswinning.

Centraal Bureau voor de Statistiek. 2016. Aardgas voor bijna 80 procent op, September 16.

FluxEnergie. 2016. Gasbesluit Nederland kost België honderden miljoenen, September 1.

FluxEnergie. 2016. Aardbevingsschade en de lusten en de lasten van het Groningse aardgas, April 20.

FNB Gas. 2016. NEP 2016. Die deutschen Fernleitungsnetzbetreiber übermitteln den Entwurf des NEP Gas 2016 an die BNetzA, April 1.

Gas Transport Services. 2016. Groningen volume en leveringszekerheid periode 2017–2020, June 8.

GEOExPro. 2009. The Groningen Gas Field. Volume 6, No 4.

Het Financieele Dagblad. 2016a. Als Nederland Noors gas nodig heeft, kunnen we dat leveren, May 17.

Het Financieele Dagblad. 2016b. Afkicken van uniek Gronings gas is logistiek complex, September 8.

Het Financieele Dagblad. 2016c. GasTerra en Engie steggelen over gascontract, September 8.

Het Financieele Dagblad. 2016d. Overheid opnieuw armer door waardevermindering gasreserves, July 19.
Het Financieele Dagblad. 2016e. Shell pleit voor 'maximale benutting' van Groningse gasvoorraad, August 31.
Ministry of Economic Affairs. 2013. Groningengas op de Noordwest-Europese gasmarkt, November.
Ministry of Economic Affairs. 2016a. Energierapport. Transitie naar duurzaam, January 18.
Ministry of Economic Affairs. 2016b. Ontwerp-instemmingsbesluit gaswinning Groningen, DGETM-EO/ 16093085, June 24.
NAM Platform. 2016. NAM dient zienswijze ontwerpbesluit gaswinning Groningen in, August 12.
NU.nl. 2015. Verstevigen van Groningse huizen gaat vele miljarden kosten, February 6.
Van Raad State. 2015. Advies W15.15.0306/IV, October 22.
Telegraaf. 2015. Stutten van Groningen kost 6, 5 miljard, February 6.
Van Gastel, M., G. van Maanen, and W. Kuijken. 2014. Onderzoek Toekomst Governance Gasgebouw, ABDTOPConsult.
Vergeer, M., M.J. Blom, and H.J. Croezen. 2015. Maatschappelijke effecten van alternatieven voor gasproductie uit het Groningenveld, November, CE Delft.
WINGAS. 2016. How does natural gas form? Everything you need to know about natural gas.

11

Norway: A Reliable Long-Term Natural Gas Supplier for Europe?

Oeystein Noreng

11.1 Introduction

Norway is the second supplier of natural gas to the European market, second only to Russia. The Norwegian resource base has not been well explored, because of the huge maritime territory involved. For the past forty years, Norway has proved to be a reliable supplier. It has the potential for at least another forty years, but incentives are required. With lower oil and gas prices, Norway faces a dilemma about the future of its oil and gas industry, whether to continue investing in the petroleum activities in spite of the price risk, or whether to scale down oil and gas investment. The outcome is pertinent to gas supplies to Europe.

Security of supplyrequires security of demand because natural gas trade is a matter of reciprocity; for natural gas sellers to invest in extraction, there must be reliable buyers. The signals from the EU

O. Noreng (✉)
BI Norwegian School of Management, Oslo, Norway
e-mail: onoreng@online.no

M. Hafner and S. Tagliapietra (eds.), *The European Gas Markets*,
DOI 10.1007/978-3-319-55801-1_11

Commission do not point in that direction. The declared ambition to get out of all fossil fuels undermines investor confidence and discourages Norway.

Against this backdrop, the UK decision to leave the European Union, Brexit, appears as an advantage for Norway insofar as the UK would retain more independence in energy policy. Liberated from the Brussels energy agenda, the UK would be free to choose natural gas as the key fuel for power generation. That might permit the UK government to abandon plans for expensive nuclear power. Already in 2015, Norway supplied volumes to the UK market corresponding to 37% of consumption. With another major deal, the UK would be overwhelmingly dependent on Norway for natural gas supplies and indirectly for power generation stability. Price competition would be assured by domestic UK shale gas and especially LNG imports.

As has already been proved, Norway would be a reliable long-term natural gas supplier to the UK market. Insofar as the post-Brexit continental EU should have an interest in natural gas, Norway would also in this market be a reliable long-term supplier. The prerequisite is that the EU Commission gives incentives in the form of market opening, easy and inexpensive access to infrastructure and policy measures that welcome natural gas.[1]

11.2 The Issues

To question whether Norway is a reliable long-term natural gas supplier to Europe may seem preposterous, given the past record of dependable and rising gas supplies, but it is an issue of resources, technology, costs, industrial organisation, markets and prices, as well as on policy choices. The answers are not granted. They have relevance to Norway as well as to the rest of Europe. Security of supply requires security of demand because natural gas trade is a matter of reciprocity; for natural gas sellers to invest in extraction, there must be reliable buyers. The signals from the EU Commission do *not* point in that direction.[2] The declared ambition to get out of *all* fossil fuels undermines investor confidence and natural gas market stability. Increasing supplies of liquefied natural gas, LNG, from many sources integrate the various regional markets for natural gas, globalising gas trade, enhancing competition and flexibility. These factors improve the Security of supply in the sense of volume availability as well as price moderation. For these reasons, natural gas trade is expanding throughout the world, except in the European Union.

Norway is the second natural gas supplier to the European Union, after Russia. EU concerns about import dependency discredit the choice of natural gas a source of power generation, with a preference for costly renewable energy or cheap imported coal. In 2015, EU consumption-was 402 billion cubic metres (bcm) of natural gas, of which domestic output was 120 bcm and imports were 282 bcm; of which 133 bcm from Russia and 110 bcm from Norway.[3] Import dependency was 70%. Since peaking in 2005, EU natural gas consumption has declined by a quarter, in volume about offsetting the decline in output.

Declining output in the Netherlands and the UK points to rising import dependency. Currently, this prospect is seen as benefiting Russia, allegedly harming supply security. Norwegian gas could make a difference. Because of oil and natural gas exports, Russia is the backbone of Europe's energy supplies, a situation likely to prevail for many years.[4] Europe's dependence upon refined Russian motor fuel is higher than for natural gas, but it is not considered a problem due to presumed alternative supply sources.

Declining domestic EU production diminishes the attractiveness of natural gas as an energy option because just replacing lost output requires higher imports. The dilemma is that Russia, an immediate neighbour with spare capacity in production as well as in transportation, and thus capable of meeting Europe's needs, already has gas market power in Central and eastern Europe seen as excessive, practically being a monopolist, as discussed above. The reluctance to depend even more on Russia has been exacerbated by several supply transit crises involving Ukraine. From a natural gas perspective, Ukraine is a problem for both the EU and Russia. On two occasions, the Timoshenko government in Ukraine, in 2006 and 2009, provoked Russian deliveries to be cut due to non-payment, affecting EU customers. Since then, Russia has invested in gas pipelines to EU markets bypassing Ukraine. In 2014, this was topical once more after Russia halted deliveries to Ukraine. The prospect of larger gas volumes from Russia raises the anxiety of balancing supplies. That is pertinent to Norway.

On several occasions, concerns about finite, limited or unreliable supplies have weighed on European energy policy choices to avoid natural gas as a key fuel. Natural gas is particularly suited for heating purposes in stationary units. Therefore, it would fit European needs in room heating and especially in power generation. It is less polluting than coal, less costly than renewables and less controversial than nuclear power, and it is plentiful, with many sources around the world. As domestic supplies have waned, European energy policy makers have feared greater import dependence, especially from Russia, implicitly with a concern for blackmail and disruptions. Public attention has been on Russia, overlooking Norway as a second, complementary gas supplier.

In relation to Norway, the basic issue, the resource base, is an imprecise and flexible concept, as always in the petroleum industry. It is important to distinguish between ultimate resources in place, which are unknown, but subject to guesses that provide estimates for reserves, volumes that can be extracted under current prices and technology, meaning when costs do not exceed the market prices, whatever they might be. Consequently, price instability means that the notion of reserves is not a constant, but dynamic. It is also shifting by advances in technology and organisation able to reduce costs and offset price instability to secure an industrial continuity. This is particularly relevant to Norway.

11.3 The Resource Base

Natural gas is in geological terms the "sister" of oil; they are both petroleum and often, but not always located in the same formations. Therefore, insight on oil has relevance for perceptions on natural gas. Thus, the theory of *Peak Oil* leads to a theory of *Peak Gas*, although with a time lag.

Proponents of the *Peak Oil* theory often point to Norway as the evidence of an oil province, whose output at first soars, then peaks, and subsequently declines. Indeed, Norway's oil and other liquids production peaked in 2001 at 3.4 million barrels per day (b/d) and declined to 1.8 million b/d in 2013 before rising to 1.9 million b/d in 2015. Natural gas production, on the other hand, has increased almost continuously since 1993 and is expected to stabilise in coming years at about 120 bcm annually. The oil output decline since the 2001 peak is not only due to geological factors, but also due to restrictive licensing, costs and taxes.[5]

Estimates of future output are uncertain because of ambiguity concerning key parameters such as prices, taxation, technology and policies. From a historical perspective, estimates of future output have tended to prove conservative.

The world oil market tends to move in long cycles, with periods of scarcity and rising prices alternating with many years of surplus and falling prices. Capital investment and lead times make adaptation to changing prices, a long process marked by abrupt discontinuities as supply and demand trends precipitously intersect, causing brutal price shocks up or down. The 1973–1974 quadrupling of oil prices, followed by the 1979–1980 spike, caused investment in energy supplies as well as in energy conservation. The outcome was oversupply and by 1986 the bubble burst, followed by 15 years of declining real oil prices. Indeed, the real oil price of 1985, measured in constant US dollars, was only reached again in 2005. In the intermediary period, low oil prices had discouraged investment and encouraged consumption, preparing the market for years of rising oil prices. Low prices also provided incentives to efficiency and innovation in the petroleum industry, reducing costs so that it could reap a double benefit when prices rose from 2004. That bubble burst in 2014 as high prices once more had encouraged investment and discouraged consumption. By 2016, the outlook is for a protracted period of low oil prices, possibly disturbed by political crises in the Middle East. For natural gas, the outlook is for abundant supplies for many years, not only because of technological breakthroughs for exploiting shale gas in the USA, but also because of large deposits elsewhere.

Limited exploration means that the notion of resource base maturity should be applied with caution to the Norwegian continental shelf, especially as the little explored northern waters are opened. Some areas are well explored and appear as fully mature, with fewer, smaller and more adverse prospects. Other areas are hardly explored, and as virgin acreage, they have considerable potential.[6] Possibly, the Norwegian continental shelf has resources for petroleum activities for this century; the need is industrial continuity as well as more diversity and competition throughout the value chain.

The opening of new territories in the north and recent discoveries in the south indicate that Norway may be a less mature petroleum province than often assumed. Indeed, huge prospective areas are fallow, not subject to petroleum activity. After the agreements with Iceland and Russia, Norway's maritime territory, from the base line to the borders, 2 million kms^2, is larger than the entire Gulf of Mexico, which covers 1.6 million kms^2. According to estimates by the Norwegian Petroleum Directorate, about one-half of the area, one million kms^2, has rocks with a petroleum potential.[7] Most of that expanse has not been explored; it is fallow acreage. Fallow means uncultivated, idle, acreage not in use, but with an economic potential. More than one-half again of this territory, ca. 600 000 kms^2, has in principle been opened for petroleum activities, but it is far from fully explored. Areas not opened comprise parts of the Barents Sea, parts of the Norwegian Sea close to coasts, the territory around Jan Mayen, territories offshore the Lofoten and Vesterålen islands, as well as most of Skagerrak, the sea connecting Norway with Denmark and Sweden.

Since 1965, about one-half of the prospective acreage has been licensed, but most has been relinquished. Exploratory drilling has taken place on blocks representing a few per cent of the prospective territory, areas of less than 50 000 square kms.

The historical, cumulative finding rate has been 43%, against 23% on the UK continental shelf; in Norway, with less drilling, more resources have been found than in the UK.

Until 2010, altogether 884 exploration wells had been drilled in Norwegian waters, against 2366 in UK waters. Between 1965 and 2010, altogether 253 important finds were made, against about 550 in UK waters. The giant 2011 *Johan Sverdrup* discovery was made in an area considered mature, on a block that had yielded a dry hole in 1971. It may turn out to be one the largest finds ever made in Norway. The output may reach half a million barrels a day, for years. Total technical cost per barrel has been estimated at less than fifteen dollars.

Norwegian petroleum policy is marked by restrictive licensing, keeping much acreage off-limits to the oil industry, high taxes and a high degree of state participation, in addition to strict environmental safeguards. Labour costs and conditions add to the generally high-cost level. Nevertheless, the petroleum industry is thriving as measured by investment levels. Exploration is modest; at the moment, ten rigs are drilling exploration wells. Nevertheless, there are many successes, and the potential is substantial.

The Norwegian part of the *North Sea*, the southernmost part of the continental shelf, has been much less explored than the neighbouring UK side, but with more finds. The *Norwegian Sea*, the middle part, has indications of an oil and natural gas potential almost as large as those of the Gulf of Mexico, but with far less exploration. Technical challenges and costs are, however, substantially higher, partly due to a basalt layer. The *Barents Sea*, the northernmost part, has promising geology with both oil and natural gas finds in recent years. In geological terms, it is composite. The eastern slice has structures in common with adjacent Russian maritime areas, with a higher potential for natural gas; the western slice has structures in common with other Norwegian areas and higher potential for oil.

To sum up, in Norway, the limitation is not *"Peak Oil"* or *"Peak Gas"*, but labour costs, taxes, weather, environmental concerns and restrictive licensing. Historically, high oil prices have incited exploration, as in the late 1970s and early 1980s, while low prices have caused declining exploration activity, as in the 1990s. Figs. 11.1 and 11.2.

By the summer of 2011, the treaty establishing the boundary between Norway and Russia in the Barents Sea and the Arctic Ocean had been ratified by the parliaments of Norway and Russia. Contrary to the treaty between Norway and the UK, which establishes the median line as the border, the treaty between Norway and Russia builds on a compromise between the median line, advocated by Norway, and the sectoral line claimed by Russia. By marking a borderline, it brings legal clarity and predictability in the area, preconditions for a secure framework for economic activities.

The treaty also aims at ensuring the continuation of cooperation on fisheries regulation, which has been successful and beneficial to both

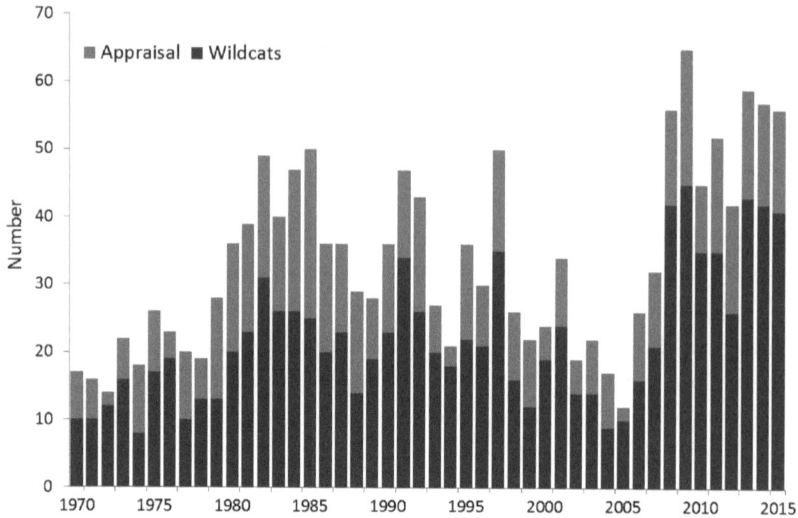

Fig. 11.1 Exploration wells spudded on the Norwegian Continental Shelf, 1970–2015 Updated: 10.03.2016. *Source* Norwegian Petroleum Directorate

countries. It introduces provisions for cooperation and settling disputes in relation to the exploitation of any petroleum deposits that may straddle the maritime border line. The essential point is unitisation of prospects that extend across the border line, to agree on the distribution. There is an agreement not only on the principle, but also on the procedure. This part of the treaty is modelled after the 1965 agreement between Norway and the UK on the maritime border in the North Sea, setting a precedent for cooperation and for settling differences within a legal framework agreed upon by both parties.

The Barents Sea petroleum resource potential is uncertain; it is mostly a separate geological formation, dissimilar from the other parts of the Norwegian continental shelf further south. Portions of the Barents Sea have been the subject of seismic studies as well as exploration drilling, but most of the area has not been subject to up to date seismic surveys and even less to exploration. The regional geology is characterised by large structures that in theory have an oil and gas potential. In practice, the oil and natural gas have been difficult to trace, as the resources seem to have migrated.

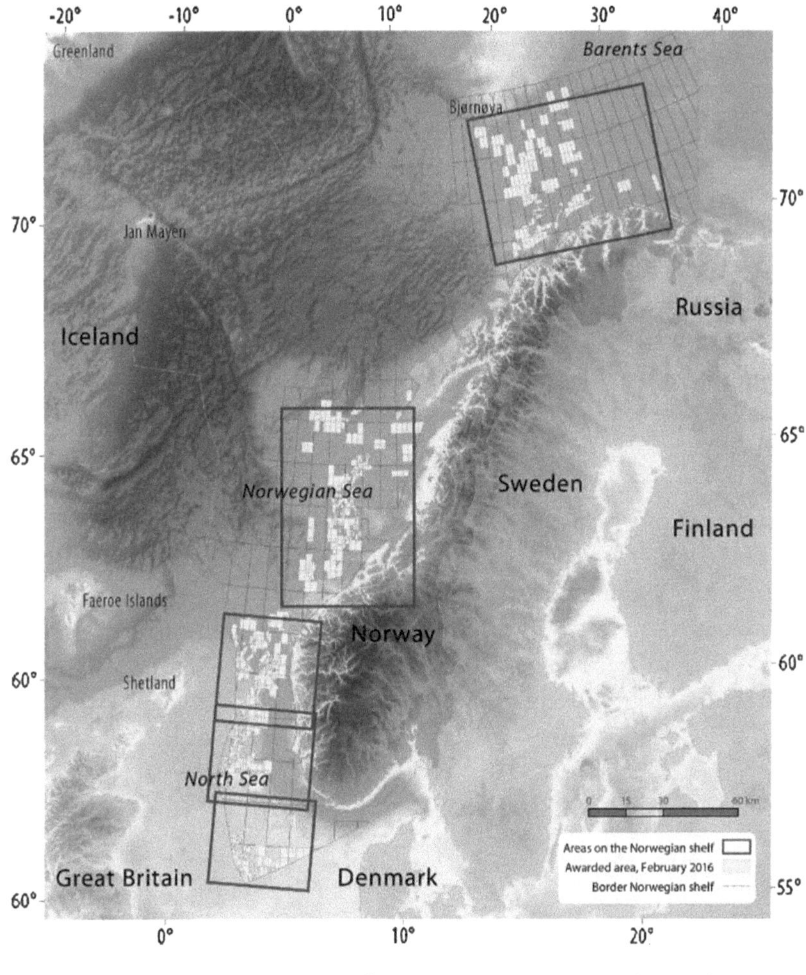

Fig. 11.2 Norway's continental shelf. *Source* Norwegian Petroleum Directorate

The conventional wisdom is that the most prospective parts appear to be in the eastern part, possibly extending into the disputed area. The USGS gives higher marks, however, to the Norwegian zone, the westernmost part of the area, than to the Barents Platform, which includes the formerly disputed region. This applies to natural gas and total reserves, not to oil. The USGS points to the Hammerfest Basin in northern Norway whose geology is oil-prone, but which has been poor in oil discoveries, so far. The basin has seen the *Snøhvit* natural gas field development and the *Goliath* oil field development. In 2011, oil and natural gas discoveries have been made in the basin. On the Russian side, several natural gas finds have been made east of the previously disputed area. More recent insight points to oil deposits in the eastern part and gas deposits further west, but there are also indications of huge structures of potential gas deposits in the northernmost part of the area.

Norway commenced seismic surveys in the newly acquired territory already in July 2011. The opening of the new territory and recent discoveries led to a new optimism. The conventional wisdom that oil and gas output would inevitably contract is yielding to a belief in a new oil and natural gas boom. That instead of declining slowly to about 2020 and the go down abruptly, Norwegian petroleum production might actually increase again and stay high for many years. That would require, however, a combination of a much higher level of activity and remarkable luck. A higher level of activity might require more extensive licensing and better fiscal terms for the industry. Luck might be enhanced by more diversity and competition. Much hope is being focused on the *Barents Sea*.[8]

The *previously disputed area* has been subject to seismic surveys since 2011. Finds have been made in neighbouring areas in both Norwegian and Russian waters. The seismic survey of the area has begun, and after reviewing the new data, the first licensing took place in 2013. New licensing in 2015 specifically targeted areas adjacent to the borderline with Russia.

For the *northern slice of the Barents Sea*, which is yet to be opened for the petroleum industry, knowledge is even more scant, with an incomplete database. The area to a large extent consists of sedimentary rocks that have a petroleum potential. During winter, large parts of the area are covered by pack ice.

For the adjacent *Lofoten-Vesterålen area*, knowledge is more complete due to extensive seismic surveys. Parts of the area appear highly prospective, but due to environmental concerns, petroleum activity in this area is a politically contentious issue in Norway. Therefore, the area has been put on hold.

The most optimistic assessment is that Norwegian oil and natural gas output will increase again until the 2030s and exceed the previous peak of 2000. The most pessimistic assessment is that the combined oil and natural gas output will stay at present levels until 2021 and then decline. These assessments are essentially based on what the *Barents Sea* including the formerly disputed area and the northern waters might yield, but not excluding surprises in the mature *North Sea* or the less mature *Norwegian Sea*. Figure 11.3

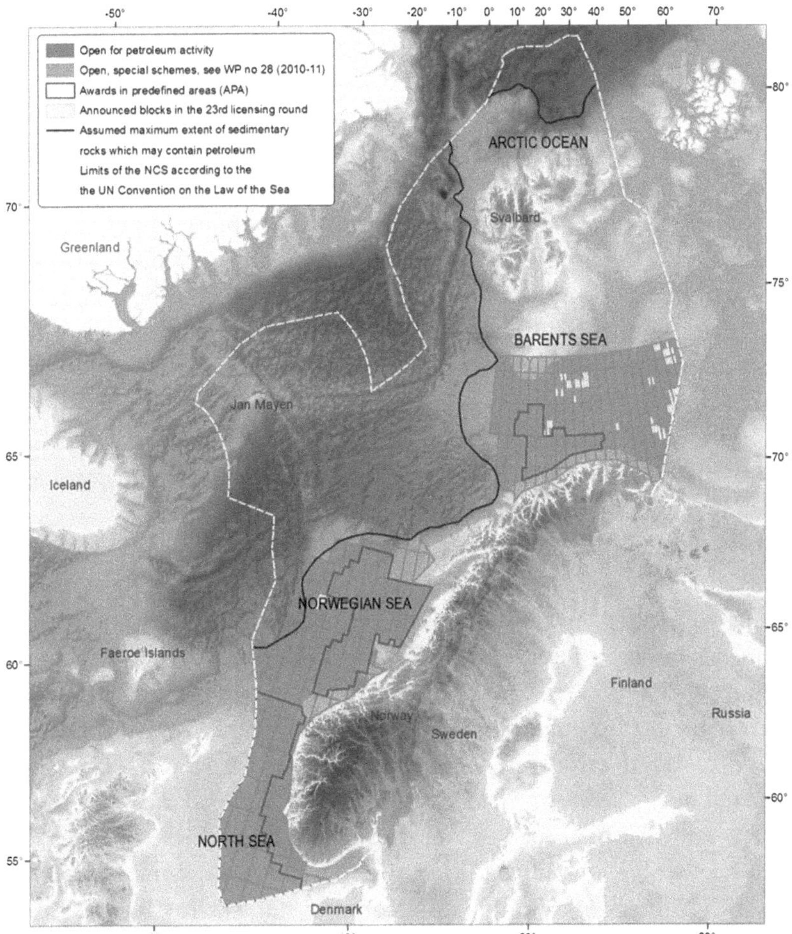

Fig. 11.3 Norway's petroleum provinces. *Source* Norwegian Petroleum Directorate

Assessing petroleum reserves is a complex matter; there is no single procedure. Reserve estimates vary by user and method. As mentioned, they are all price sensitive, and they embody subjective judgement on geology and technology. In Norway, as in the USA, they are primarily based on petroleum company reports to government agencies, but different government agencies have different needs and assess the data differently.

Oil and gas reserves make up the key part of oil company capital; their accounting is of critical importance to financial markets assessment. Therefore, investors and creditors need protection from potential oil and gas company optimistic exaggerations.[9] The US *Securities and Exchange Commission* (SEC) only recognises as proven reserves, volumes that can be commercially extracted at current prices, technology and regulation and that have a high probability, ninety per cent, of actually being lifted. Norway follows a largely similar procedure. Against this backdrop, public reserve estimates tend to be conservative, with lower figures than the reserves in place estimated by the companies.[10] Consequently, reserve data point to volumes that not only *can be* extracted at current prices and technology, but to volumes that reasonably *will be* extracted by commercial actors. The US *Department of Energy* and the US *Department of the Interior* have other needs and use different methods so that their estimates differ.

This commercial and legal dimension is often overlooked in the discussion of reserves available. In the case of Norway, with huge uncharted areas, it is pertinent. The Norwegian reserve accounting distinguishes between reserves already produced, reserves in place committed to production, and contingent resources in producing fields, in new discoveries and by potentially enhanced recovery, as well as an estimate of undiscovered resources. The figures other than for reserves already extracted and reserves in fields committed for extraction by commercial actors, petroleum companies, are uncertain, dependent on prices, technology and policies, in addition to commercial decisions. The bottom line is that reserves are a dynamic concept, subject to changing prices, technology, perceptions of geology and analytical perspective. Reserve expansion is a matter of both new discoveries and reassessment of what has already been found.

Reserves are remaining recoverable petroleum volumes subject to a commercial decision having been made, by the license holders, the companies, whether or not the government has approved. Conditional resources comprise proven volumes for which no production decision has been made, including potential future enhanced recovery, but excluding resources that currently are not considered commercially recoverable. Undiscovered resources comprise oil and gas probably in place and commercially recoverable, but which have not yet been proven through drilling. Against this backdrop, published reserve data are by their nature highly conservative and in no way exhaustive of the volumes that potentially could be extracted at current prices and technology. Therefore, reserve data tend to expand over time, and supplies prove to be more resilient than initially anticipated, at the field level as in petroleum provinces. Indeed, most oil and gas prospects tend to yield much higher volumes over their lifetime than foreseen at their discovery and investment decisions. Likewise, unexplored acreage in established petroleum provinces tends to have an upside potential.

Parts of the Norwegian continental shelf are mature in the sense that they have been subject to comprehensive seismic studies and exploration drilling, while other parts are virgin in the sense that they have not, or only scarcely, been subject to exploration. Even in areas considered mature, new and even large finds are made. Some estimates point to the possibility that the Norwegian oil reserve figure might double, but that much would be in smaller deposits.[11]

By the conventional conservative estimate, more than one-half of the liquids initially in place has already been produced, indicating Norway's oil output is set to decline, the question being how soon and how fast. By contrast, less than one-half of the natural gas initially in place has been extracted, indicating that Norway's natural gas production has the potential to stabilise at current levels for many years or even increase moderately. Estimates for petroleum initially in place and for undiscovered resources have been revised upwards from 2014 to 2015.

Recent discoveries justify a reassessment of the prospects for Norway's petroleum industry.[12] The conventional outlook has been that total extraction would stay stable until about 2001 and then decline. Alternatively, based on recent finds and new insight, there might be a potential for oil output to increase again until after 2030 and even exceed the previous 2001 peak. For natural gas, the potential may be even larger because companies essentially target oil in their exploration efforts, finding natural gas as a by-product. An assessment update by the Norwegian Petroleum Directorate in 2015 states "*There is more oil left to produce from fields and discoveries now than there was in 2005, and it is also assumed that there is more oil left to discover than was the case ten years ago*".[13] In the real world, prices, technology, policies and commercial decisions will decide. Table 11.1, Fig. 11.4

Among Norway's three petroleum provinces, the North Sea appears by far the most prolific in terms of resources originally in place. Here, about sixty per cent have been produced. The remaining proven reserves are still considerable, but the estimate for undiscovered resources is modest. In the Norwegian Sea, with a much smaller initial resource base, more than a third has been extracted so far. Reserves in place are limited, but the estimate for undiscovered resources is higher. Finally, the Barents Sea appears to have the smallest initial resource base, but the most undiscovered resources. Proven reserves are small.

Table 11.1 Resource categories

Status Sm^3	Oil	Condensate	NGL	Natural gas	Sum oil equivalents	Change 2015–2014
Produced	4075	114	179	2100	6630	229
Reserves	1023	28	116	1856	3128	167
Conditional resources in fields	328	2	22	222	594	11
Conditional resources in discoveries	375	13	15	323	739	−382
Potential enhanced recovery	155	0	0	60	215	−20
Undiscovered resources	1315	120	0	1485	2920	85
Total initially in place	7272	277	333	6047	14,227	90

Source Norwegian Petroleum Directorate 1 standard cubic metre, Sm^3, is equivalent to about 6.3 barrels of oil. See footnote about measuring gas

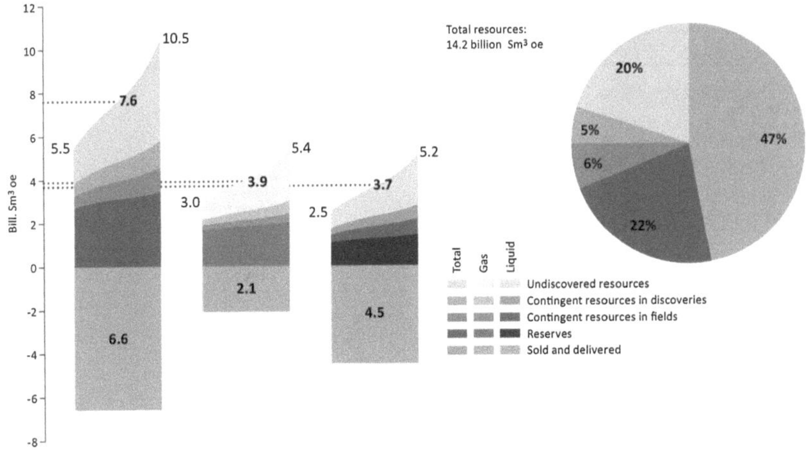

Fig. 11.4 Resource categories. *Source* Norwegian Petroleum Directorate

The North Sea is largely an oil province, but the Norwegian Sea and the Barents Sea are largely gas provinces. The largest proven gas reserves are still in the North Sea, but the estimated undiscovered gas resources are further north, the bulk being in the Barents Sea. The resource data indicate that the petroleum activity will move north, to the Norwegian Sea and especially the Barents Sea, presenting new challenges to technology and costs. Table 11.2, Fig. 11.5

11.4 Technology and Costs

Costs and taxes set the volume threshold for field development in Norwegian waters about three times the level in UK waters. The cost level is less related to geology than to labour costs, environmental, health and safety requirements, and distances, as well as bottlenecks and imperfect competition along the supply chain. It gives a strong incentive to technology development. The modern oil industry is based on technology as a tool to reduce risks and costs.[14] Assessments based on new finds and new technology change the outlook for Norway's petroleum industry. Declining output no longer seems inevitable;

Table 11.2 Resource categories by area

North Sea	Oil	Gas	NGL	Condensate	Total
Produced	3504	1650	129	75	5475
Reserves	874	1278	75	−1	2295
Conditional resources in fields	285	116	11	0	422
Conditional resources in discoveries	161	121	9	9	307
Undiscovered resources	455	250		40	745
Total initially in place	5280	3415	224	123	9244
Norwegian Sea	**Oil**	**Gas**	**NGL**	**Condensate**	**Total**
Produced	571	415	48	33	1111
Reserves	121	389	35	8	584
Conditional resources in fields	38	92	11	1	152
Conditional resources in discoveries	49	150	5	3	212
Undiscovered resources	320	410		45	775
Total initially in place	1099	1455	100	91	2834
Barents Sea	**Oil**	**Gas**	**NGL**	**Condensate**	**Total**
Produced	0	35	2	6	45
Reserves	28	189	6	20	249
Conditional resources in fields	5	14	1	1	21
Conditional resources in discoveries	165	52	1	1	219
Undiscovered resources	540	825		35	1400
Total initially in place	738	1 116	9	63	1934

some scenarios indicate sustained growth as the industry moves on. Most noteworthy, the recent giant find was struck by two smaller companies on acreage overlooked by big oil, led by senior geologists, whose advice had been disregarded by former employers, larger oil companies with a more conservative mindset.

The case of Norway throws critical light on the *"Peak Oil"* theory. It refers to US experience with private ownership of resources, seeking rapid maximisation of extraction and income, with no or weak government regulation. In sum, commercial considerations to cover immediate needs have driven US exploration for oil and the development of oil fields. These conditions do not apply outside the USA and parts of Canada. In North America, the advent of shale oil and oil sands indicates that the assumptions behind Hubbert's "Peak Oil" theory are no longer valid. In the USA, the oil reserve-to-production ratio has been

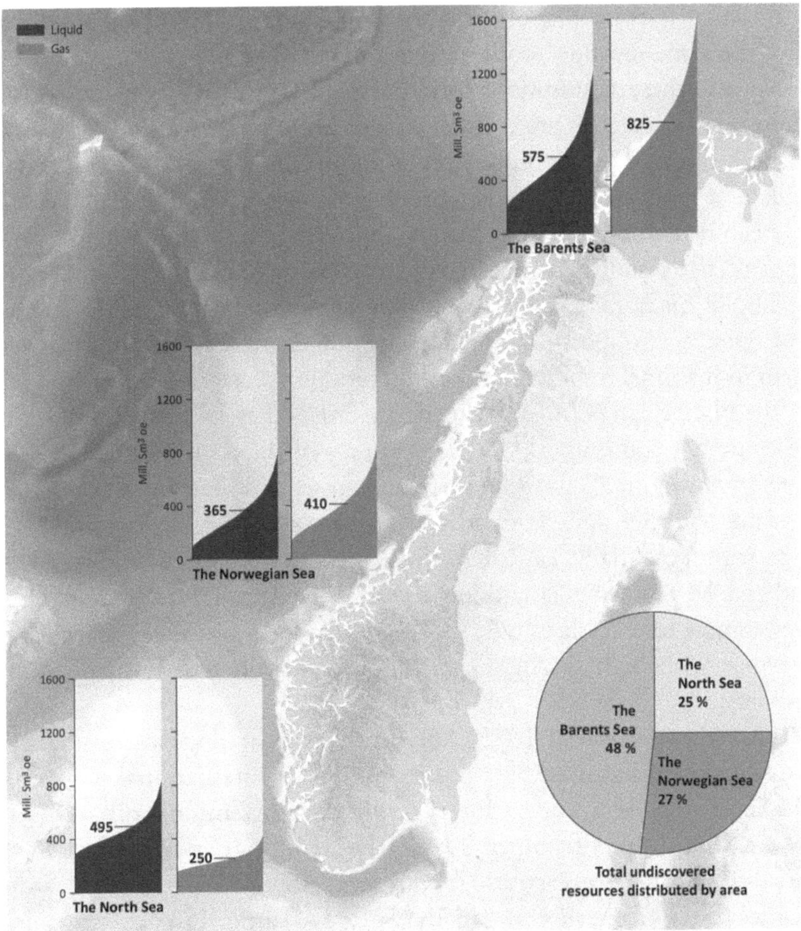

Fig. 11.5 Undiscovered resources by area. *Source* Norwegian Petroleum Directorate

about ten-to-one for the past 30 years, but the country is still one of the world's leading oil producers. In Norway, they never applied, given the vast area, restrictive licensing and high taxes.

The theory of "*Peak Oil*" rests on six key assumptions: knowledge of the world's oil reserves is complete, reserve estimates are constant, extraction inevitably takes the shape of a fairly symmetric curve, technology is constant, oil prices do not matter, and all oil producers have the same revenue/profit motive and goals.

None of these assumptions corresponds to reality. In the oil industry, all important parameters are dynamic. Knowledge of the world's oil reserves changes through exploration. Insight and technology cause a continuous update of reserves because of improving recovery rates. The actual extraction can take many different shapes, depending on the interests and strategies of the resource owner, in addition to technology and economics. Technology is not constant; new equipment and new methods lower costs and open new opportunities. Oil prices do matter, especially for investment in marginal resources. Not all oil producers have the same objectives.

The Norwegian petroleum experience proves these points. Technological development continuously lowers costs; a globalised oil industry disseminates innovation. New technology and lower costs make new resources available. Smaller and more adverse prospects become economical. New sites with potentially large resources open up. Experience lowers costs and enables new, smaller and less accessible fields to replace mature fields. Newcomers have fresh minds. In 1970, the North Sea was a marginal oil province with high geological, technical and economic risk; today it is ordinary.

Since the 1960s and 1970s, the need for technological and managerial innovation has been ever more pressing, as international oil companies have stepped up operations outside the conventional oil provinces, such as the plains of Iraq or Texas. The new frontier areas include the North Sea, offshore eastern Canada, the Arctic, and the deep waters of the US Gulf of Mexico and the South Atlantic. In conventional oil provinces, the need for technological adaptation and innovation was, and remains, most often moderate. Such was the context for the inception of the *"Peak Oil"* theory. On the other hand, frontier areas have presented and continue to present significant technology and management challenges. They include difficult weather conditions, the depth of waters and complex geology. With such challenges comes a corresponding necessity to innovate regarding costs, operational safety, health, and the environment, and to reduce risk. The payback is a supply elasticity; the higher the oil or gas price, the higher the volumes available, if not immediately, at least over time.

The rational practice for the petroleum industry has been to first develop and deplete the most accessible and largest prospects with the lowest unit costs, such as fields in the plains of Texas and Iraq. As such reserves are being depleted, international oil companies have moved into frontier areas in search of new reserves. The industry is caught in a race between resource depletion, which causes rising costs, and technology renewal, which aims at curbing such costs by enhancing efficiency, and will improve profits. This requires a sustained R&D effort. If there is insufficient R&D, there is a risk that the petroleum industry will be unable to meet this rising cost challenge as well as reasonable safety standards. Major international oil companies, which plan to explore for and eventually develop petroleum resources in Arctic waters,[15] must be ready for this challenge. This was the case in the North Sea in the 1970s and 1980s. Currently, it is the case in the Norwegian Sea and the Barents Sea.

To date, the move into new frontier areas has been facilitated by the spectacular reduction in technical costs that have occurred over the past decades, due to technological progress based on R&D.[16] Modern information technology has allowed for the use of 3D seismic exploration. This has reduced the number of wells needed for exploratory drilling, resulting in time gains and lower costs. Stratigraphic drilling allows for multiple targets to be reached from a single site, such as a platform. The result is a reduction of required drilling sites, positively impacting well productivity and the volume of oil handled per site, thus creating both time and cost efficiency gains.[17] A third example is the simplification of gathering and transportation systems as oil and natural gas can be shipped through the same pipe and no longer need to be separated at the wellhead. Finally, technological progress based on R&D has led to higher quality and lower cost of a range of production equipment, offshore as well as onshore.[18] The cumulative result is higher cost-effectiveness and improved operational safety. The industry has a constant need to renew technology and management strategies through R&D, especially as it moves to tap new reserves.

Lower oil prices since late 2014 raise the issue of the robustness of Norway's petroleum industry. From 2006 to 2015, exploration and development boomed, as did the costs. The concern is that oil prices, for example $40–60bl, will put an effective brake on exploration and development in Norwegian waters. The concern is justified, but there is hope.

A first remark is that Norway's maritime territory is huge, stretching from the Barents Sea border with Russia in the north-east to the North Atlantic border with Iceland in the north-west and to the North Sea border with Denmark in the south, not forgetting the North Sea border with the UK in the west. Conditions are different; the North Sea has the easiest conditions in terms of access, water depths and weather; as mentioned, it is considered a mature oil province, fairly well explored. The Norwegian Sea in the middle and the Barents Sea in the north are more challenging, and they have been little explored.

The North Sea is likely to stay attractive for the industry, but in the Barents Sea may suffer. Throughout the Norwegian waters, marginal projects are threatened by the oil price decline. Indeed, historically, petroleum activities in Norway have been highly sensitive to oil prices. In the low-price environment of the 1990s, exploration fell. It picked up only in 2006 as prices had risen markedly and new tax rules provided better incentives for newcomers.

A reduced pace in exploration and development was in the cards for the next few years even before the oil price declines. It causes several projects to be delayed or abandoned. Paradoxically, that may be an advantage. The cost increase in recent years has largely been driven by industrial factors rather than geology. In a Norwegian oil project, drilling typically represents 50–60% of total costs. Since 2002, rig rates have followed oil prices; the floater daily rate has risen from $100,000 in 2002 to $500,000 in 2013. The cost increase is due to bottlenecks in the supply chain, imperfect competition at many points and particular Norwegian rig specifications. Consequently, a lower level of activity is likely to moderate costs, as is already evident with rig rates. A more alarming sign is the loss of productivity in drilling. There is no good explanation why drilling operations have become slower. With lower oil prices, the problem is acute.

The Norwegian petroleum taxation permits a 90% deduction of capital costs against 70–80% tax bite of the net income. The disparity does not provide strong incentives for cost consciousness.

Nevertheless, even in the North Sea, a major find was made in 2011. It will possibly produce up to 650,000 bbls/day for decades. Technically, the find should provide incentives for more exploration. Economically, it could strengthen the argument for holding back. Nonetheless, that single find is likely to arrest or even reverse the decline in liquid production that has taken place since 2001. Technological progress is making Norwegian waters more accessible and more attractive for the petroleum industry. Since 1980, the secular trend has been a gradual increase in efficiency and productivity that can be summed up as a cost decrease on comparable projects of 3–4% a year. The trend contains some qualitative leaps such as three-dimensional seismic, stratigraphic drilling and joint transportation of liquids and gases. The outcome is a gradual expansion of the field of action of the petroleum industry. The major part of the Norwegian waters can still be defined as virgin territory from a petroleum point of view, with a potential that is difficult to quantify. Therefore, Norway's maritime territory represents a frontier for the petroleum industry. For Norway's actual output of oil and natural gas, the time horizon has probably been underestimated.

Nevertheless, major uncertainties remain. The level of exploration activity does not only depend on geology, but also depend on oil prices. With lower oil prices, the Norwegian petroleum frontier becomes more marginal and, of course, of less interest to the petroleum industry. Oil company interest in the Arctic is fading. Several marginal prospects are deferred or abandoned in a new low-price environment. Indeed, the Norwegian supply industry is likely to lose contracts both at home and at abroad; the oil price decline is universal and is hitting all producers. Lower oil prices affect the earnings of all oil companies, but more so those that are upstream only. In Norway, many of smaller newcomers and upstarts that have been successful in exploration in recent years will be badly affected and probably bought up by larger companies. A wave of consolidation will reduce the number of active explorers.

The question is what measures the Norwegian government can take to stimulate the activity. Reducing the special tax on petroleum extraction, currently at 51% in addition to the corporate income tax of 27%, may be cut, as well as accelerating the depreciation schedule from six years to one. Rig rates are falling quickly. A lower level of activity is creating spare capacity in the supply chains, reducing costs. Finally, lower oil prices again, as in the late 1980s and throughout the 1990s, force innovation and cost-cutting to the industry. The potential is considerable.

Historically, innovation in the Norwegian petroleum industry has been driven by joint efforts on the triple helix model between the government, research institutions and the companies. The government has initiated and sponsored comprehensive efforts in research and development. Low oil and gas prices since 2015 stimulate new efforts to cut costs, like before.

11.5 The Industry Record

Historically, the Norwegian petroleum industry has been reasonably robust to the oil price vagaries. This is pertinent to gas as well, since companies search for petroleum, in most cases hoping to find oil, but accepting gas as a by-product. The international oil industry at first caught in interest in Norway's offshore potential in the mid-1960s, when the nominal oil price was $1.80, about $13 in 2015 dollars. The oil price quadrupling in 1974–1975 caused a wave of interest in the petroleum industry, boosted by the subsequent doubling again of oil prices in 1979–1980. After the oil price crash of 1985, investor interest at first did not weaken much, but the persistent low oil prices during the 1990s led to less interest in Norwegian petroleum. Investor interest strengthened again as oil prices began to climb after 2003.

The suddenness and the amplitude of the 2014–2015 oil price decline surprised most investors, but so far, the depreciation of the Norwegian krone, NOK, versus the US dollar has softened the impact for Norwegian investors and the Norwegian government. Therefore, the current risk picture also includes a possible dollar depreciation. During the first six months of 2016, the average Brent crude spot price was $40.76/bl., with an average exchange rate of $8.45/NOK, giving an oil price of NOK334/bl. Indeed, for nominal oil prices, Norway is set back twelve years, to 2004, when oil was at NOK 351/bl. In real terms, in constant NOK, the oil price is back to the level of late 2004.

In hindsight, those were good times for the oil industry and for the Norwegian economy. In the meantime, the industry has internalised high prices and rising costs, even if technology has improved and the geology essentially remains the same. There is an evident parallel with the late 1980s.

Historical experience gives room for some optimism. From 1985 to 1986, oil prices fell by one-half after 12 years of then unusually high levels, from USD 27.56 to USD 14.43 (annual averages), accompanied by an abrupt fall in the dollar exchange rate from NOK 8.60 in 1985 to NOK 7.40 in 1986. The outcome was a decline in oil prices from NOK 237/bl., to NOK 107/bl., i.e. by 55%. Nevertheless, the Norwegian oil industry continued its activity, incited by a tax reduction and inspired by major discoveries made in the late 1970s and early 1970s. Costs had soared in earlier years, but came down. Oil companies reduced their exploration drilling, but capital investment continued to increase, also through the 1990s, as oil prices were low.

The robustness of the Norwegian oil industry to price adversity has its explanation in geology, technological progress and a stable regulatory environment. More recently, the regulatory and tax changes introduced in 2005 have proved been successful. The previous fiscal regime had discriminated between incumbents, i.e. companies with a cash flow from production, and newcomers without one. The tax reform benefited newcomers, able to write off exploration expenditures immediately and in case of failure, to withdraw with the government recompensing 78% of the costs, the standard petroleum tax level.

The change has attracted a multitude of newcomers, essentially smaller and medium-sized independent companies, foreign as well as Norwegian. They have brought fresh ideas. The result is an increase in exploration and a significant improvement in the success rate, which in recent years has been between 50 and 60%. The most spectacular case is the discovery of the Johan Sverdrup field in 2011, by Lundin Petroleum, a newcomer, at the location where Elf had drilled a dry well in 1974. Current insight and technology are far better, but the clue was a newcomer thinking in non-conventional ways and willing to take the risk. By a prudent estimate, the field should contain about 2500 million barrels of oil, possibly producing up to 650,000 bbls/day for at least

20 years, with a break-even cost of perhaps fifteen dollars a barrel. Even if this is a unique discovery, in recent years the industry has made many smaller finds in all three parts of the Norwegian continental shelf, i.e. the North Sea, the Norwegian Sea and the Barents Sea. Hence, the latest experience attracts the oil industry, even with lower prices.

The impact of lower oil prices on the Norwegian oil industry will not be uniform. Throughout the Norwegian waters, marginal projects are under threat by the oil price decline. Indeed, historically, petroleum activities in Norway have been highly sensitive to oil prices. In the low-price environment of the 1990s, exploration fell. It picked up only in 2006, as prices had risen markedly and new tax rules provided better incentives for newcomers. Most of the prospects commissioned for development are likely to be completed, but the industry could delay or abandon several smaller prospects unless they are close to infrastructure. The North Sea with infrastructure in place is likely to stay attractive for the industry, but the Barents Sea has impediments in remoteness and immature infrastructure.

In any case, a reduced pace in exploration and development was in the cards for the next few years even before this fall. Paradoxically, that may be an advantage. Over the past ten years, Norway's petroleum industry has experienced cost escalation, driven more by industrial factors than geology, meaning rising labour costs as well as rising supply chain costs. The investment boom has caused bottlenecks, reduced competition and rising costs; indeed, in this industry, the oil price is largely driving costs. In a Norwegian oil project, drilling typically represents 50–60% of total costs. Since 2002, rig rates have followed oil prices; the floater daily rate has risen from $100,000 in 2002 to $500,000 in 2013. The cost increase is due to bottlenecks in the supply chain, imperfect competition at many points and particular Norwegian specifications for rigs. A more alarming sign is the loss of productivity in drilling. It may be partly due to advancing automation, as workers operate faster than complex machinery on the platform decks, but with greater health and safety hazards. With lower oil prices, the problem is acute.

In Norway, the petroleum taxation amplifies this problem, as the government takes 90% of the capital costs but only 78% of the net profit. The disparity does not provide strong incentives for cost consciousness. The mechanism is an additional uplift on capital expenditure of 5.5% a year for 4 years against the special tax of 51%. Currently, a lower level of activity leads to lower rig rates and more idle capacity in the supply chains. Break-even costs are coming down. Therefore, the outlook is for a slower pace of exploration and development at lower costs.

The signal from the government is that Norway is not in a crisis. The prime minister has stated that the oil price decline provides a good occasion to diversify the economy. Her oil minister, from a different party, intends to boost petroleum investment, especially in the Arctic waters of the Barents Sea. Until further notice, fiscal measures to boost petroleum investment seem unlikely. The 23rd licensing round focusing on northern waters, especially the Barents Sea, has met considerable interest in the industry, in spite of low prices, long lead times and long distances. In May of 2016, the government awarded 10 licenses consisting of 40 blocks on the Norwegian continental shelf, of which three in the Barents Sea. In all, 13 firms are offered participating interest. In January of 2015, the government awarded 56 production licenses to 36 companies on the Norwegian continental shelf as part of its Awards in Predefined Areas (APA) announced in 2015, of which about half in the North Sea, five in the Barents Sea and the remainder in the Norwegian Sea.

In the wake of the oil price decline, natural gas prices have been halved in the UK as well as on the Continent. If prices stay low or fall further, over the next years, the industry response is likely to weaken. In that case, the government might consider measures to stimulate investor interest, including tax relief. The issue is also the cost of transporting natural gas from the Barents Sea to the UK and continental markets.

11.6 Marketing Norwegian Gas

Norwegian gas exports commenced in 1974 by the new pipeline from the Ekofisk field in the southernmost part of the Norwegian sector of the North Sea to the continent, servicing a market in Belgium, France, Germany and the Netherlands. At the time, gas from Norway supplemented gas from the Netherlands, whose volumes were being restricted in an effort to stretch out the lifetime of the reserves. There was no proper natural gas market in western Europe, and prices were negotiated, essentially indexed to oil. That also applied to Algerian gas sold to Belgium and France. The principle was retained for the next Norwegian gas deal, the Statfjord gas around 1980, but under the impact of the second oil price shock and the perception of resource scarcity, Algeria and Norway could negotiate exceptionally high gas prices on new contracts.

In the meantime, the late Soviet Union had concluded a major gas contract with France and West Germany. Contrary to Algeria and Norway, the Soviet Union did not exploit the oil market panic to renegotiate the contract, as that would have been contrary to its objective of getting a long-term foothold in the west European gas market and a large market share rather than high prices. By moderate pricing, the intention was to promote the use of natural gas in western Europe, but it did not wish to take the whole market.

In the erstwhile Soyuzgazeksport, there was a sophisticated understanding that the European buyers want and need supply diversity, so that other natural gas suppliers would not only be competitors, but also be partners in developing markets. From this perspective, there never was a conflict between the Soviet Union and Norway over gas supplies to Europe; the two never tried to outbid each other. As Norway's first gas deal was launched in the 1970s, the Soviets did not present themselves with competing bids. When subsequently, the first major Soviet pipeline deal with France and then West Germany was launched, against US pressure, Norway wisely remained on the sidelines. The US Reagan administration tried to wreck the Urengoi gas deal by banning the use of US compressors, but alternatives were available. The request to Norway was to advance the development of the giant Troll gas field found in 1979 in order to cut short the French and West German need for Soviet gas, but the Norwegian response was that this was technically impossible. Moreover, it was not in Norway's interest to appear as an obstacle to east-west trade in Europe and as a champion of US interests.

The next round was a Norwegian deal that raised Soviet eyebrows because of the high prices. Soviet fears that high prices for Algerian and Norwegian supplies would compromise the competitiveness of natural gas in the European energy market proved justified; the aforementioned

Norwegian deal was cancelled as the larger Troll gas deal was negotiated a few years later, with larger volumes at lower prices. Following tradition, the Soviets stayed on the sidelines, not making competing bids. In hindsight, this can be seen as a Soviet recognition of reciprocity with Norway.

The record is a tandem between Norway and the Soviet Union, later Russia, partly run by the two suppliers as they observe each other's moves, as a duopoly following unwritten rules of common interests and reciprocity. Historically, Norway and Russia have shared the north-west European natural gas market, not as a cartel, but as complementary partners. For decades, the natural gas market of north-west Europe has been the subject of a tacit bargaining game between the two major suppliers and the major buyers, aiming at stability and risk diversification. There is little evidence of price competition between Norwegian and Russian natural gas, as the buyers have been alternating contracts in order to balance supplies. Indeed, the Soviet Union, in spite of ideological opposition and military tension, was a reliable supplier of oil and natural gas to western Europe.

For Norwegian natural gas, which is more costly, this game has been helpful in securing a stable market at reasonable prices. Russia has lower costs as exports are the bonus of a huge home market; it could, in theory, undercut Norway, but realising that the buyers would prefer to rely on more than one supplier, it wisely lets Norwegian gas into the market in order to facilitate the sale of additional Russian gas. Not competing on prices for market shares is typical of a duopoly, a market dominated by two sellers. In this game, Norway has been the weaker part due to higher costs and limited spare capacity, but Russia as the stronger part, with lower costs and more spare capacity, needs to show restraint in order to keep Norwegian natural gas in the market, without which the buyers would be more reluctant to purchase Soviet or Russian gas. The game was facilitated by the monopsonies, single buyers, on the demand side, able to alternate supply contracts. Norway is the major supplier to France, with Russia leading in the German market. Nevertheless, Norway's major gas market is Germany, followed by the UK, the Netherlands and France. Russia's major gas market is Germany, followed by Italy, Belgium and Poland.

As the second supplier of natural gas to Europe, the question is to what extent Norway could substitute for Russian deliveries. For several reasons, calls for substituting Norwegian gas for Russian gas seem futile. There is no way Norway would be able to replace Russian supplies. Even if there might be some spare capacity in the short run, the long-term challenge for Norway is to sustain and moderately expand the present level of gas exports through an enhanced effort in exploration and field development as well as in pipeline construction. Norway might have second thoughts about challenging Russia because of the crisis over Crimea.

After protracted and successful negotiations over the maritime border, the Barents Sea is developing as a new petroleum province, with a prospective for cooperation between Norway and Russia.[19] For Norwegian industry, offshore petroleum activities on the Russian side offer a huge potential market, not only for oil companies, but also for the supply and services industry. Any Norwegian attempt to drive out Russian natural gas from the markets of continental Europe could easily trigger retaliation aiming at Norwegian firms, in addition to compromising long-term political relations.

Moreover, it is unlikely that the major buyers, especially Germany, would welcome a Norwegian move to push Russian natural gas out of the market. Even if Germany abandoning nuclear power would need more natural gas, the preference is to balance supplies between Norway and Russia, as has been the case in the past. For Germany, there is more at stake than energy supplies. Russia is an important export market for German industry. There is substantial direct investment by German firms in Russian industry. The complementarity in resource endowment, industrial prowess, human resources and markets, is remarkable.

In order to purchase German goods, Russia has to export. Against this backdrop, natural gas trade between Germany and Russia is a most natural occurrence, based on mutual needs for the common economic benefit. In German business as among German politicians, there is no interest in deteriorating relations with Russia. Norway should keep its historical role as the alternative natural gas supplier to Germany, indirectly helping Russia to sell more, rather than obstructing German–Russian trade. There is no way Norway with 5 million people could substitute for the Russian market with twenty-eight times as many consumers. The lower purchasing power of the latter indicates a higher potential.

In case of lasting tense relations between the EU and Russia at least in north-west Europe, the preference would be for more Norwegian natural gas at the expense of Russia's market share. As mentioned, Norway has the ability to moderately increase volumes, but at a high cost and with a long lead time. For Norway, such a prospect would also pose a dilemma, as any stabilisation of natural gas demand in Europe requires competitive prices, and only Russia has the volume capacity to moderate natural gas prices.

The advent of more plentiful supplies of LNG is also likely to moderate natural gas prices in Europe and enhance supply security, in addition to potential new pipelines from the Middle East and the Caspian region. In order to fully benefit from more diversified supplies, the EU natural gas market would need additional infrastructure investment in order to integrate flows, and preferably a single regulator to facilitate cross-border trade and impose free infrastructure access. That appears as a requirement for a multilateral European market for natural gas, in which larger numbers buyers and sellers would interact, reducing risks of supply as well as of demand. That would also reduce the threat of Russian dominance. After all, Russia invests in gas production facilities and pipelines in order to make money, not to make trouble, with markets in both Europe and Asia. There never was a direct supply crisis between Russia and west European buyers. The erstwhile Soviet exporter, Soyuzgazeksport, meticulously honoured supply contracts and had a reputation of price moderation. The interruptions of natural gas supplies to east European buyers were caused by price disputes. In any case, the European acceptance of rising imports from Russia would to some extent depend on Norway's ability and willingness to raise gas exports. In the Arctic, there is a large potential for cooperation between Norway and Russia to develop natural gas resources, but that requires détente, not confrontation. This underlines the reciprocity of Norway's and Russia's gas interests in resource management as in marketing.

The UK is the second market for Norwegian gas; in 2015, it took close to a quarter of gas exports. The backbone is the Langeled pipeline with a capacity of 25.5 bcm annually; with a length of 1166 kms, it was the world's longest underwater pipeline until the construction of the Baltic Sea Nordstream. The pipeline was commissioned without any volume contracts; the investors were confident that the natural gas would find customers in the open and competitive UK market. So it was. Figure 11.6.

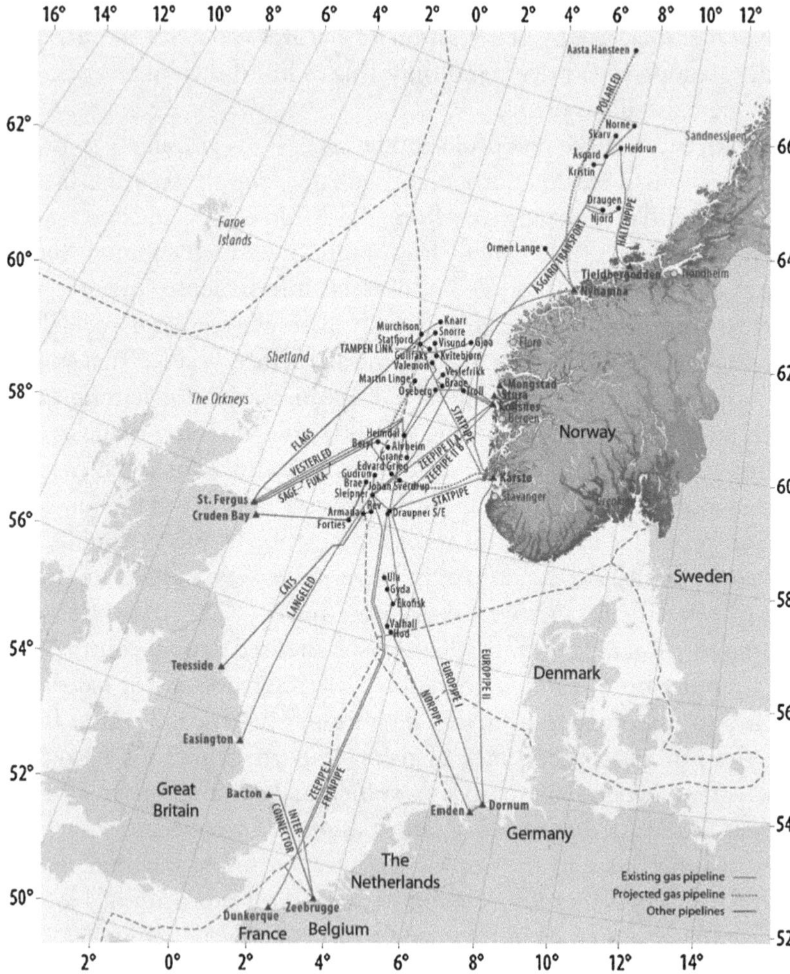

Fig. 11.6 Natural gas pipelines from Norway. *Source* Norwegian Petroleum Directorate

11.7 Norway's Dilemmas

Norway faces a recurrent dilemma about its petroleum industry, especially natural gas. Historically, in the 1970s as oil prices soared, fearing an overheated economy, Norway opted for a moderate pace of development, to keep more of the resources on the ground. Subsequently, in the 1980s, the issue was natural gas; the choice was not to flare associated gas and rather sell it at low prices. Paradoxically, at the same time as the EU Commission restricted the use of natural gas in power generation, Norway had problems finding outlets for the gas. Currently, the looming issue is what to do in case of major gas discoveries in the Barents Sea, with a long time horizon for production. As a starter, the location is remote and transportation costs are high, whether by pipeline or by ship.

At an earlier stage, before the advent of shale gas in North America, the USA appeared as a prospective market for Barents Sea gas. The *Shtokmanovskoye* project was advancing until it was literally put on ice. The US market is saturated. As mentioned, the EU Commission does not want any fossil fuels in the longer run. The Asian markets are served by natural gas from multiple sources, and distance means a competitive disadvantage for Norway.

Without sufficient markets for natural gas, investor interest would weaken and so would the level of activity in the Norwegian petroleum industry. Some Norwegian politicians would see that as an advantage. Apparently, Norway enjoys a robust macroeconomic position, but lower oil and gas prices make the situation more precarious. In 2015, with Brent at USD 53/bl., the petroleum industry accounted for about 18% of the gross national product (GDP) and 39% of total exports, with additional 2% from petroleum-related services. The budget had a surplus corresponding to 6% of GDP, including oil and gas revenues, without which it would be in deficit. That year, Norway's trade surplus was 6% of GDP; without petroleum exports, Norway's foreign trade would have had a deficit of 10% of GDP.

The pace of economic activity is slowing down. Low oil prices affect Norway's large exports of petroleum technology and services, in addition to the domestic supply industry. Low oil prices reduce the capital build-up of Norway's Sovereign Wealth Fund, but in principle not government revenues that are a percentage of the Fund's capital stock. This mechanism essentially shelters Norway's domestic economy from the vagaries of the oil market. Moreover, lower oil prices are likely to stimulate the world economy and hence the return on the Fund's investment abroad.

Indeed, the Fund is the buffer against adversity. With a value of almost USD 900 billion, it is the world's largest in the summer of 2016. By comparison, the 2016 estimate for Norway's gross national product (GDP) is USD 430 billion, and USD 335 billion for the mainland economy, excluding the offshore petroleum activities. The Fund is more than twice the GDP and almost three times the size of the non-oil economy, which the Fund is to shelter. Thus, the Fund permits expenditure without corresponding taxation, permitting a generous welfare state with comparatively low income tax.

Without the sizeable economic rent, the petroleum industry will no longer be the generator of huge surpluses that have marked the Norwegian economy over the past 10 years. Even if the economic rent essentially has accrued to the government, the oil industry and the supply chains, part of it has also benefited the public. In spite of channelling petroleum revenues directly to the Sovereign Wealth Fund, the indirect effects of using the return of the Fund have been important over the past 10 years. The Norwegian economy has internalised high oil prices and the ensuing economic rent in the sense that households have become used to ever-rising purchasing power, businesses have adjusted to an ever-expanding domestic market, and public authorities have become accustomed to ever-larger budgets. Indeed, for the past decade, high oil prices have caused a combination of pressures in the Norwegian economy. Rising oil investment has spurred escalating factor costs, with ripple effects throughout the rest of the economy. The mounting value of the Sovereign Wealth Fund has caused higher returns and triggered larger transfers to the budget. Norway's central bank has warned that this luxurious economy is ending. The challenge is to convince the public and especially the politicians.

Norway's labour market institutions contribute to macroeconomic robustness. Centralised wage bargaining mean that in good times the benefits are broadly shared, contributing to comparative income equality. In adverse times, it facilitates wages restraint. The trade union confederation, LO, in 2015 declared that there was little room for wages increases, a reasonable position in a situation with low unemployment after many years of rising real wages.

The downsizing of the oil and oil services industry will make human resources available for other industries. High-tech industries and infrastructure investment are potential new drivers of the Norwegian economy. Currently, there is a critical shortage of engineers and other skilled personnel. In addition, the depreciation of the Norwegian currency, NOK, is making other industries more competitive. Indeed, as raw materials exports are important, causing an economic cycle opposite to those of the trade partners, there is some wisdom in retaining an independent currency. Since early 2014, the Norwegian currency has depreciated by 20% against the US dollar, but by only 3% against the euro. In the current context, it might be tempting to let the Norwegian currency slide further against the euro, although there is a risk of higher inflation that might offset the gains.

To stimulate other export industries, the most urgent tasks are to step up research and development and to improve infrastructure. The preceding centre-left coalition neglected these tasks, even with huge revenues at its disposal, but gave priority to welfare and private consumption.

Reaping the benefits of rising raw materials prices, successive governments have underestimated the need for research and development, except for the offshore petroleum industry. Norway's transportation infrastructure is critically inadequate. Roads are insufficient to cope with traffic. The rail network is even worse with little investment and poor maintenance over many years. Freight traffic is increasingly going by road, not by rail, in spite of most politicians' verbal commitment to the environment and climate cause. Most of Norway's fish and seafood exports move by huge trucks on narrow roads, in a business whose export value is time critical.

Norway usually appears as a mature oil province, but most of the huge maritime territory is fallow, unexplored. The reasons for the slow pace of exploration are complex. Technology has been an important obstacle, but the desire to keep a moderate pace in licensing and exploration is also due to a combination of environmental concerns and economic needs. The Norwegian waters are some of the world's major fishing grounds, important to the world's food supplies. Moreover, Norway's small population of just 5 million has limited economic requirements. Living standards are among the world's highest; there are serious concerns about the economy overheating and the petroleum industry dominating. Low oil prices enhance concerns to diversify the economy.

The petroleum industry priority is to cut costs, as was the case after 1986. The petroleum taxation needs an overhaul to correct the disparity between incentives for capital expenditure and surplus taken by the government, in order to give stronger enticements for cutting capital costs. A possible measure would be to abandon the capital uplift mentioned above against an immediate depreciation, and perhaps a lower special tax reducing total government take. Petroleum activities in the Arctic waters, close to the ice edge, are politically controversial in Norway; lower prices make them economically more questionable. As an alternative, there is plenty of unexplored acreage further south. Indeed, the resource potential might provide a basis for sustained petroleum activities perhaps through this century, depending on markets, prices, technology and costs. Against this backdrop, Norway has the prospect of remaining a frontier area for the petroleum industry for decades to come. There is an evident need for pluralism and competition in order to ensure efficiency and innovation.

Norway is facing important dilemmas concerning the petroleum industry with serious repercussions for natural gas.[20] The key question concerns Norway's economic strategy and the trade-off between continued petroleum investment and diversification. Over the past 10 years, oil and gas exploration and development have boomed thanks to tax breaks and high oil prices, driving costs up. As a rule, the government take 78% of the net income, but companies are able to deduct 90% of capital investment. The supply and services industries have benefited greatly, partly due to bottlenecks and imperfect competition. In Norway, industry factors have driven the costs, more than the geology.

The oil industry cost escalation has spread to other sectors, as it has been able to pay high prices for goods and services, and high remunerations for labour. Thus, the oil industry expansion has weakened the competitiveness of other industries in a small, open economy, even if it has also spread technology. For that reason, the crowding-out effect, Norway's oil wealth is not an unmitigated blessing. Coexistence is difficult between a highly capital-intensive industry capturing economic rent and labour-intensive industries like manufacturing. Insofar as the oil companies should manage to squeeze costs, their earnings will be more robust in a low-price environment, and the supply and service industry will take the brunt. In brief, high oil and gas prices have for many years given Norway a generous portion of economic rent, excess earnings above the level needed to sustain the petroleum industry. The rent represents easy money. Although the state has captured the larger part through taxation and direct participation, much of the easy money has entered Norway's non-petroleum economy through the oil companies as well as the supply and service industry. Bottlenecks and limited competition in critical parts of the supply chain have driven up costs, with effects outside the oil and gas industry.

In Norway, a recurrent question is what the country should live off after oil, implicitly that it should prepare for a different export structure and deliberately scale down the petroleum industry, implying gradually falling natural gas exports. The issue is exaggerated, as Norway in all likelihood has a resource base able to sustain a petroleum industry for decades, depending on technology, costs, prices, tax conditions and markets. Indeed, the question is rather what Norway should live off together with the petroleum industry. Leaving aside the issue of business development based on natural and human resources, the question is how to treat the oil and gas industry. Should it be further promoted or should it be downsized? Opinions differ.

In current money, petroleum investment in Norway quadrupled from 2002 to 2014, while other business investment only rose by 76%. Thus, the boom years made Norway more exposed to oil and gas market risk. The outlook for petroleum investment is a decline by 34% after the 2014 peak, affecting new projects as well as enhanced recovery on producing fields. Already, the supply and service companies feel the squeeze, cutting jobs and shedding sub-contractors. Unemployment is beginning to affect local communities along the coast that until now experienced unusual wealth. The good news is that with lower level of activity, costs will decline. The bad news is the risk of losing industrial competence.

So far, the government seems unwilling to take any measures to stimulate exploration and development. The prime minister seems to think that through low oil prices, market forces will drive the structural changes needed to diversify the economy.

Nevertheless, oil and natural gas represent the country's major industry, essential to the level of economic activity, especially along the coast. The risk is that oil prices in the short term will fall further and cause additional decline in exploration and development, before eventually stabilizing, perhaps at current levels. The government's hands-off attitude seems inspired by general macroeconomic considerations, ignoring the special capital-intensive and knowledge-intensive character of the petroleum industry. So far, demand from the petroleum industry has given stronger growth impulses to the Norwegian economy than has the use of petroleum revenues. The risk is that reduced oil and natural gas activities will cause a stronger cooling-off of the economy than anticipated, with regional and sectorial depressions. Therefore, requests for measures to stimulate petroleum activities are likely to mount, in so far as low prices persist and investment drops. In Parliament, a majority would probably be in favour, but the government might have to seek new partners.

Even if Norway is not a member of the EU, it is partner closely associated and attentive to signals from Brussels. There is indeed a certain coincidence between the EU ambitions to phase out fossil fuels and the requests by Norwegian environmentalists to scale down and discontinue the petroleum activities. For Norway, it would make little sense to invest more in gas exploration and development if the major market, the EU, should not want any more of it. With the activity heading north, to the Barents Sea, with longer and costlier transportation requirements, this issue is pertinent. The counterpart is that if the EU should favour natural gas in power generation, e.g. through a carbon tax that would hit coal, Norway would respond favourably and be a reliable long-term future supplier.

More immediately, the markets for Norwegian gas are Germany, followed by the UK, the Netherlands and France. In 2015, Norwegian gas accounted for close to one-half of German demand, second to Russian supplies. In the UK market, Norwegian gas has a share of close to 40%. In the Netherlands market, Norwegian imports correspond to more than half the demand; in the French market, the figure is above 40%. Consequently, Norwegian natural gas is a key factor in the energy balance of north-west Europe. As discussed above, there is a likely potential to expand this role, but Norway would need incentives and assurance that markets are available. Paradoxically, Norwegian concerns about limited markets for its gas seem to mirror EU apprehensions about limited and insecure supplies.

Natural gas from Norway, a member of the European Economic Area, reasonably should be seen as secure as that from the Netherlands or the UK, with the latter about to leave the EU. In 2015, Norway's gas output of 117 bcm practically matched that of the EU at 120 bcm. Thus Norway accounted for one-half of Europe's natural gas extraction west of Russia. After Brexit, Norway's gas output will overshadow that of the rest-EU by about one-half.

Counting Norwegian gas as "European" would raise domestic output to 237 bcm and raise self-sufficiency from 30 to 57%. Consequently, just by a redefinition, an accounting alteration, import dependency and the assumed consequent supply risk would appear less ominous, politically facilitating the choice of natural gas in power generation, whether from Norway, Russia or other exporters. This would make sense, provided Norway shows a willingness to remain a key gas exporter, if possible with higher volumes. The matter is pertinent to the project of a European energy union. From a Norwegian perspective, the most imperative task is to simplify access rules and transportation tariffs in the pipeline system, so that natural gas could be easily and inexpensively transported to customers across the remaining EU, as has been the case in the UK for decades. Such a measure would also enhance supply security for the buyers. The requirement is overcoming incumbent vested interests.

Recent gas discoveries in the Mediterranean improve the supply prospects for southern Europe. The emergence of a world gas market based on LNG further dispel risks to supply security. These factors should enhance the competitiveness of natural gas also in the eyes of EU energy planners, but their preference seems to be renewables, regardless of the cost. The issue is the resilience of an energy policy provides Europe with the world's most expensive energy to the detriment of industrial competitiveness and employment. In the late summer of 2016, after the Brexit referendum, the new UK government signalled changes in energy policy, with more emphasis on cost efficiency and less on climate. From a Norwegian perspective, that is a positive sign for gas demand. The question is whether the UK will set an example for countries remaining in the EU.

Against this backdrop, the UK decision to leave the European Union, Brexit, appears as an advantage for Norway insofar as the UK would retain more independence in energy policy. Liberated from the Brussels energy agenda, the UK would be free to choose natural gas as the key fuel for power generation. That might permit the UK government to abandon plans for expensive nuclear power. Already in 2015, Norway supplied volumes to the UK market corresponding to 37% of consumption. With another major deal, the UK would be overwhelmingly dependent on Norway for natural gas supplies and indirectly for power generation stability. Price competition would be assured by domestic UK shale gas and especially LNG imports. As has already been proven, Norway would be a reliable long-term natural gas supplier to the UK market. Insofar as the post-Brexit continental EU should have an interest in natural gas, Norway would also in this market be a reliable long-term supplier. The prerequisite is that the EU Commission would give incentives in the form of market opening, easy and inexpensive access to infrastructure and policy measures that welcome natural gas.

Notes

1. The author is indebted to Morten Lindbaeck for critical advice and to the Norwegian Petroleum Directorate.
2. Energy 2020.

3. Figures are from *BP Statistical Review of World Energy*, 2015 Edition. Presumably, the data refer to *standard* cubic metres, measured at 15 °C, not *normal* cubic metres, measured at 0°C. The difference is appreciable as natural gas expands in volume at rising temperatures.
4. Westphal, Kirsten, 2014.
5. Mills, R. 2008, p. 142.
6. Ibid. p. 141.
7. Petroleum resources 2013.
8. "Northern exposure", 2014.
9. Osmundsen, P. 2010, p. 22.
10. Meling, L.M. 2006.
11. Carstens, H., 2012.
12. Nyland, B., 2011.
13. Evaluation of reserve growth for oil—2005–2014.
14. Bret-Rouzaut, N. and Thom, M., 2005.
15. ExxonMobil, 2008.
16. Oil and Gas Exploration and Production: Reserves, costs, contracts, 2004.
17. Gorelick, S. p.131.
18. Ibid, p. 156.
19. Norway, Russia exchange Barents Sea seismic data 2014.
20. Northern exposure 2014.

Bibliography

Arctic Leadership. 2008. Irving, Texas: ExxonMobil. Available at: http://www.exxonmobil.com/Corporate/Files/news_pub_poc_arctic.pdf.

Carstens, Halfdan. Vi kan doble oljereservene, Geo365, www.geo365.no

Energy 2020. 2011. *A strategy for competitive, sustainable and secure energy*. Brussels: EU Directorate General for Energy.

Evaluation of reserve growth for oil—2005–2014. 2015. Stavanger: Norwegian Petroleum Directorate.

Meling, Leif Magne. 2006. The origin of challenge-oil supply and demand. *Middle East Economic Survey*, June 12.

Mills, Robin. 2008. *The Myth of the Oil Crisis*. London: Praeger.

Nadine, Bret-Rouzaut and Michael Thom. 2005. Technology strategy in the upstream petroleum supply chain, Les cahiers de l'économie—no. 57. Rueil-Malmaison: Insti-ut Français de Pétrole.

Northern exposure, Petroleum Economist, March 2014, pp. 34–35. *Petroleum Economist.*

Northern exposure, Petroleum Economist, March 2014, pp. 34–35, London: Petroleum Economist.

Norway, Russia exchange Barents Sea seismic data, Oil and Gas Journal October 26, 2016, www.ogj.com.

Nyland, Bente, Fire framtidsbilder for Barentshavet og Norskehavet. 2011. Stavanger: Norwegian Petroleum Directorate.

Oil and Gas Exploration and Production: Reserves, costs, contracts. 2004. Paris: Institut Français de Pétrole Publications, *Éditions Technip.*

Osmundsen, Petter. 2010. Chasing reserves: Incentives and ownership. In *Energy, natural resources and environmental economics*, eds. Endre Bjørndal, Mette Bjørndal, Panos M. Pardalos, and Mikael Rönnqvist, 19–38. Berlin: Springer.

Petroleum resources on the Norwegian continental shelf 2016. 2016. Stavanger: Norwegian Petroleum Directorate.

Steven M. Gorelick. 2010. *Oil panic and the global crisis.* London: Wiley.

Westphal, Kirsten. 2014. *Russian energy supplies to Europe.* Berlin: SWP-Comments 16, *Stiftung Wissenshaft und Politik.*

Author Biography

Oeystein Noreng Ph.D. is Professor Emeritus at the BI Norwegian Business School, with teaching and research assignments at Norwegian and foreign universities and institutions, and Director at Norengconsult. He has been a research fellow at the Norwegian Humanities Research Council, research project manager at the Norwegian Book Club, counsellor at the Planning Department of the Norwegian Ministry of Finance, Research and Planning Manager at the Marketing Department of Statoil and professor at the Norwegian Business School from 1977 until 2012. Dr. Noreng has been an expert witness before the International Court of Justice and the US Tax Court. He was the recipient of the OPEC Award for Research 2012. He is the author of numerous books and articles in Norwegian, English and French.

12

The North African Gas Export Outlook Between Commercial and Political Challenges

Francis Perrin

North Africa can be defined in various ways, but we will consider in this chapter that this region includes Algeria, Egypt, Libya, Morocco, Mauritania, and Tunisia. Algeria, Egypt, and Libya are producers of natural gas and these three countries were until recently net exporters, which is no longer the case of Egypt and Libya. Tunisia is a small producer of gas, and there are interesting prospects in Mauritania following recent exploration work. Morocco imports gas by gasline from Algeria and intends to become a liquefied natural gas importer.

Algeria is able to increase its gas exports to Europe because it has adequate reserves and infrastructure. Libya is struggling with huge political and security problems, and it is very difficult to predict the future development of its hydrocarbon industry despite UN efforts to impose a government of national accord. Egypt has become a net importer of gas, but important discoveries and developments underway should allow it

F. Perrin (✉)
Stratégies et Politiques Energétiques and Energy Industries Strategies Information, Paris, France
e-mail: fperrin@stratener.com

M. Hafner and S. Tagliapietra (eds.), *The European Gas Markets*,
DOI 10.1007/978-3-319-55801-1_12

to turn the tide in a not too distant future. gas resources have been discovered off the coast of Mauritania and Senegal, and the two countries will have to agree to facilitate the development of these resources from both sides of their common maritime border.

Box 12.1 Morocco's LNG Import Project

Morocco evaluates its gas needs at about 5 billion cubic meters per year in 2025. In May 2015, the Ministry of Energy, Mines, Water, and Environment estimated at $4.6 billion the cost of the project, including the power plants which will be fuelled by natural gas. The terminal will be located at Jorf Lasfar, and the pipeline is expected to have a transport capacity of 7 billion cubic meters per year.

The first step is the selection of foreign partners, and it will be followed by the signing of contracts. From the date of signature, the construction of gas infrastructure should last 48 months. The development work on power plants could last 36 months. Commissioning of gas and power facilities could take place during the first half of 2021.

The Office National de l'Electricité et de l'Eau Potable (ONEE), which is a national company, will play an important role in this future LNG project.

This project was discussed with a lot of companies and organizations, including the Japanese group Mitsui, the Chinese company Norinco and with several US and other firms during a visit of the Minister to the USA (ExxonMobil, Shell, BP, Cheniere Energy, and General Electric). Previously, Russia had also been approached for this purpose. It is the same for the Siemens group. And the list is not exhaustive.

For Algeria, one of the main suppliers of the European market, the main unknowns concern the rapid increase in its domestic consumption and the appetite of the European Union for Algerian gas. The authorities and Sonatrach highlight the gas potential of the country, the revival of exploration, a great number of discoveries, the increase in the recovery rates in operating fields, gas developments, very substantial resources of unconventional gas, and ambitious programs for renewable energy and energy efficiency. All this should allow the country to meet its gas and power needs, which are rising rapidly, while remaining a significant exporter of gas to the European market in the long term. But the key issue of very low energy prices would also have to be dealt with and the sooner the better.

When we think of North Africa now, many political problems and risks immediately come to mind: war in Libya, the presence in this country of the Islamic State, a political power shared between Tripoli and the East, political tensions in Egypt, the impacts of the "Arab spring" and others. All these are well known and important and they have, or are likely to have, consequences in terms of energy but there are also a lot of progress on the ground through discoveries and developments and the conventional and unconventional potential in this region remains very significant. We will highlight in this section (completed in early October 2016) these positive elements, which are not always well known, before examining some problems that North African countries have to face to better exploit their gas potential in the medium and the long term.

12.1 Nine Good News About Gas in North Africa

12.1.1 Algeria on a Growth Dynamics for its Production and Exports of Hydrocarbons Again

The statistics published by Sonatrach in September 2016 confirm that Algeria is once again on a growth momentum for its production and exports of liquid and gaseous hydrocarbons. For the first eight months of 2016, exports totaled 71.5 million tons of oil equivalent (Mtoe), an increase of 9.3% compared to the same period of 2015 (65.4 Mtoe). Between January and August, the primary hydrocarbon production was 127.4 Mtoe, representing 99% of the objective of Sonatrach. The national company said the average monthly production was 15.7 Mtoe for the first six months and that it reached 16.9 Mtoe in late August. Furthermore, this average will continue to increase in September and over the following months. The year 2017 should mark the start of a further increase through the coming on stream of new fields, especially of natural gas.

Crude oil exports, however, fell by 8% in the first eight months of 2016 due to higher volumes processed in refineries in the north of the country (unless otherwise stated, all figures quoted in 13.1.1 are from Sonatrach and cover the first eight months of 2016). But exports of natural gas by pipeline increased by 43% compared to the same period in 2015, which is very significant. Sonatrach estimates that this increase confirms its “repositioning on the European market” and stresses that its gas exports to Italy grew by nearly 3 billion cubic meters in the first half of 2016. Italy is the main consumer of Algerian gas. Exports of refined products were 2% higher than the January–August 2015 volumes.

At the Algeria-European Union Business Forum on Energy, held in Algiers in May 2016, the then Energy Minister, Salah Khebri, said that primary production of hydrocarbons in Algeria, which was almost 200 million tons of oil equivalent in 2015, could reach 241 Mtoe in 2020, an increase of 20%. It was 195.2 Mtoe in 2014. The return to growth is expected from 2016. It will come from the commissioning of new fields and higher reserves recovery rates in the producing fields.

Mr. Khebri said that Algeria spared no efforts to increase its contribution to meeting the world’s energy needs. Its export capacity of natural gas has increased from 50 billion cubic meters per year to 90 billion cubic meters per year, of which about 50 bcm/year by pipeline and the rest in the form of liquefied natural gas (LNG). The country is connected to Europe through three pipelines, one to Italy and two to Spain.

Sonatrach’s long-term plans are fairly ambitious. It is planning a production of 1.28 million b/d of crude oil by 2040 as against about 1.1 million b/d currently. The increase will be greater for conventional natural gas, whose output could rise from 128 billion cubic meters in 2015 to 165 bcm in 2040. Gas exports could return to their historic level of 60 bcm in 2020.

Box 12.2 Africa’s Oil and Gas Reserves

According to the 14th edition of the World Oil and Gas Review, an annual statistical report on hydrocarbons in the world published by the Italian group Eni, Africa’s proven oil reserves increased by 37% between 2000 and 2014, from 91.75 billion barrels to 125.76 billion barrels over this period. The increase was 2.3% per year over these 14 years, a growth rate slightly higher than that registered for the whole world (+2.1%). According to the World Oil and Gas Review, stronger growth was recorded for Russia and Central Asia (+5.4%) and the Americas (+3.7%). Data collected by Eni mainly come from Enerdata and the US Energy Information Administration (EIA).

At the end of 2014, Africa represented approximately 8% of world oil reserves, behind the Middle East (48%), Latin America (20%), and North America (13%). Among the top ten holders of oil reserves, there is a single African country, Libya, which is in ninth position ahead of the USA.

After Libya, whose proven reserves are estimated at 48.36 billion barrels, the other major African countries are Nigeria (37.07 billion barrels), Algeria (12.20 billion barrels), and Angola (8 42 billion barrels). Then come the Sudan (five billion barrels—in its publication, does not distinguish between Sudan and South Sudan, Editor's note), Egypt (4.4 billion barrels), Uganda (2.5 billion barrels), Gabon (two billion barrels), Congo (1.6 billion barrels), Chad (1.5 billion barrels), and Equatorial Guinea (1.1 billion barrels).

According to the same source, Africa's share in global proven gas reserves is around 7%. Between 2000 and 2014, Africa's gas reserves increased by 16.2%, from 12,460 billion cubic meters to 14,478 billion cubic meters, or +1.1% per year. For gas reserves, Africa is behind the Middle East (40% of world total), Russia and Central Asia (32%) and Asia/Pacific (8%). Two African countries, Nigeria and Algeria, are among the top ten holders of gas reserves. They are ranked ninth and tenth, respectively.

In Africa, Nigeria was at the end of 2014, according to the World Oil and Gas Review, the first holder of gas reserves with 5,111 billion cubic meters. It was ahead of Algeria (4,504 billion cubic meters), Egypt (2,168 billion cubic meters), and Libya (1,505 billion cubic meters).

12.1.2 Gas Developments Underway in Algeria

Several developments are underway in Algeria and many of them are gas projects as this country has more a gas profile than an oil profile. Sonatrach and Total are thus developing the Timimoun field which will produce 1.6 billion cubic meters of gas per year. Its forecast start-up is 2017. Eni continues to develop MLE-CAFC, which is already in operation. Eni, which is active in the upstream in Algeria since 1981, says that it is the first foreign producer of hydrocarbons in Algeria and the second in the country after Sonatrach.

The development of the Ain Tsila gas and liquids field on the Isarene permit (Illizi basin) is also underway and production is expected to begin in the fourth quarter of 2018 or early 2019. According to Petroceltic International plc (Dublin, Ireland), the operating cost is estimated at only $2 per barrel of oil equivalent (boe). Capital expenditures up to start-up would be $1.6 billion. The plateau production of Ain Tsila would be 81,500 boe/day approximately. The duration of this future plateau is estimated at 14 years based on the recoverable reserves of this field (proven and probable reserves—2P).

During the peak phase, gas production will be 355 million cubic feet per day. The completion of the development project will require the drilling of 124 wells, including 30 before the start of production. According to Petroceltic International (38.25%), which is associated with Sonatrach (43.375%) and Enel (Italy, 18.375%), 2P reserves of Ain Tsila are estimated at 2,148 billion cubic feet of gas, 110 million barrels (Mb) of LPG and 69 Mb of condensate, or a total of about 530 Mboe. Ain Tsila was declared commercial in 2012. The gas, which will be exported, will be marketed entirely by Sonatrach, and liquids will be sold separately.

Another gas field, Touat, will come on stream soon. It is developed by Sonatrach and Engie, formerly GDF Suez (France). The goal is to reach an output of 155 billion cubic feet/year (4.4 billion cubic meters). Reggane Nord should start-up in 2017. The expected peak is 155 billion cubic feet/year. For this project, Sonatrach is associated with Repsol (Spain), DEA (Germany), and Edison (Italy).

A joint venture between Sonatrach (35%), BP (33.15%) and Statoil (Norway, 31.85%), In Salah Gas (ISG), began to set in operation the In Salah Southern Fields (ISSF) project covering the start-up of four gas fields in the region of In Salah. These start-ups will allow ISG to keep at 9 billion cubic meters per year the production of the In Salah gas project. In Salah is one of the largest gas projects in Algeria. The final investment decision for ISSF was taken in 2011.

BP is one of the largest foreign investors in Algeria. Besides In Salah, the British group is involved in another gas project, In Amenas, and its partners are also Sonatrach and Statoil. Both projects have the same production capacity, 9 billion cubic meters per year, but In Salah is producing dry gas and In Amenas wet gas (gas plus liquids). In Salah has been on stream since July 2004 and In Amenas since 2006. This second project was targeted by a terrorist attack in January 2013 that led to the death of 40 people.

The In Salah project includes seven gas fields. Three of them, in the northern part of this area, Krechba, Teguentour, and Reg, are in production since 2004. The ISSF project focuses on four fields south of these ones, Gour Mahmoud, In Salah, Garet el Befinat, and Hassi Moumene. For the development of the southern fields, 26 wells must be drilled and the work started in 2014 will continue until 2018. ISG expects production will reach 14.1 million cubic meters per day (5.15 billion cubic meters a year) in a few months as and when the wells will be drilled on Hassi Moumene and Garet el Befinat. Recoverable resources of In Salah are estimated at 159 billion cubic meters. The exported gas is sold to the Italian company Enel.

Box 12.3 Algeria's Transportation Capacity of Hydrocarbons

The overall transportation capacity of hydrocarbons (crude oil, condensate, LPG, and natural gas) in Algeria has significantly increased between 2005 and 2014, rising from 365.8 million tons of oil equivalent to 418.8 million toe (Mtoe), up to 14.5%. At the same time, transport volumes fell from 244.18 Mtoe to 219.11 Mtoe (-10.3%). In 2014, the utilization rate was therefore 52.3%.

Transport capacity increased for all categories of hydrocarbons. For crude, they were 165.9 Mtoe in 2005 and 172.9 Mtoe in 2014. For natural gas, the corresponding figures are 145 Mtoe and 170.5 Mtoe. For condensate and LPG, these capacities have increased from 32.2 Mtoe and 22.7 Mtoe in 2005 to 44.9 Mtoe and to 30.6 Mtoe, respectively, in 2014.

The decrease in volumes transported derived from crude (98.72 Mtoe in 2005, 78.64 Mtoe in 2014) and LPG (21.5 Mtoe and 16.74 Mtoe, respectively). The quantities of natural gas and LPG transported were rather stable.

The hydrocarbons transportation activities are performed by Sonatrach, which holds 22 concessions, including 10 for the transportation of natural gas, seven for the transport of crude, three for condensate and two for liquefied petroleum gas. The national company operates 21 oil pipelines (9,946 km long), 16 gaslines (9,677 km), 82 pumping and compressor stations, 127 storage tanks (4.2 Mtoe), and two dispatching centers.

Sonatrach said that the hydrocarbons transport network in Algeria consists of two parts, the southern and northern networks. The first transports crude oil and condensate to Haoud el-Hamra (Hassi Messaoud), where the dispatch center for liquid hydrocarbons (CDHL) is located, and gas and LPG to Hassi R'Mel (national dispatching center for gas—CNDG). The northern network allows the transport of crude and condensate from the CDHL to refineries and export ports; of gas from the CNDG to the domestic market, export pipelines and liquefaction complexes; and of LPG from Hassi R'Mel to the separation complexes.

12.1.3 Egypt: Zohr Potential of 30,000 Billion Cubic Feet of Gas is Confirmed

The appraisal of the Zohr supergiant gas field continues. Eni completed the drilling of a fifth well on this field in the Mediterranean, Zohr 5x, and a sixth was planned by the end of 2016. Zohr 5x, which was drilled successfully, allowed Eni to confirm in September 2016 its estimate of a potential of 30,000 billion cubic feet of gas (850 billion cubic meters) for the whole field. The company does not refer to reserves but to gas originally in place.

Drilled to a total depth of 4,350 m by 1,538 m of water, Zohr 5x also confirmed the very good deliverability capacities of the reservoir. The well could produce up to 250 million cubic feet per day under normal conditions. Eni intends to put Zohr on stream in late 2017 on a fast-track basis at a rate of 1 billion cubic feet per day of gas. Ultimately, Zohr's production ceiling could be 2.7 billion cubic feet per day.

Zohr is located on the Shorouk block, which is 100% owned by the Italian group. The company is active in Egypt since 1954, and its share of production in the country was about 200,000 barrels of oil equivalent per day in 2015.

12.1.4 The Potential of Gas in Place in the Greater Nooros Area in Egypt is Estimated at 3,000 Billion Cubic Feet

The successful drilling of the Baltim Southwest-2X appraisal well on the offshore Baltim South permit allowed Eni and BP in September

2016 to estimate at 3,000 billion cubic feet (3 trillion cu ft) the gas potential in place in the Greater Nooros area. The Nooros gas field currently produces 700 million cubic feet per day (mn cu ft/d) and, according to BP, the flow could reach 880 mn cu ft/d by early 2017. Nooros accounts for about 2 trillion cu ft of this potential and Baltim South-West for the rest. The two companies are studying development options for this latest discovery, which was announced in June 2016.

The Baltim Southwest-2X well showed a column of gas of 102 m. The net thickness was 86 m. The rocks date back to the Messinian and reservoirs are of excellent quality. This drilling has confirmed the great potential of the Messinian play in this area.

Egypt needs a lot of natural gas in order to satisfy its domestic market. BP said that drilling would be accelerated on the Nile Delta Offshore, Temsah and Ras el-Barr permits for which amendments to existing contracts have been signed with the Egyptian Ministry of Petroleum.

Eni and BP each hold a 50% stake on Baltim South. The operator is Petrobel, a joint venture between IEOC, a subsidiary of Eni, and the Egyptian General Petroleum Corporation (EGPC). Nooros is located on the Nile Delta Offshore permit (Eni 75% and BP 25%). The two companies are also associated (50–50%) with Temsah and Ras el-Barr.

12.1.5 Development of the West Nile Delta Gas Project in Egypt

For BP, the West Nile Delta gas project (WND) is strategic. The British group is the operator of the project, and it increased to 82.75% of its stake in this scheme, which is extremely high taking into account the huge investments which are planned. The development cost of WND is indeed estimated at $12 billion. BP's partner is the German firm Deutsche Erdoel DEA AG.

Egypt is a key area for BP, which has invested more than $25 billion over 50 years in this country. The company is a major player in Egypt since the joint ventures in which it is involved control nearly 30% of domestic gas production (BP, Eni and two Egyptian national companies, EGPC, and EGAS) and 10% of oil and condensate production (BP and EGPC). WND will mobilize reserves of 5,000 billion cubic feet of gas (5 trillion cubic feet) and 55 million barrels of condensate. WND gas production will be approximately 1.2 billion cubic feet per day in the plateau phase, which represents 25–30% of Egypt's current gas production. BP believes that an additional potential of 5–7 trillion cubic feet of gas could be identified there through future exploration.

WND is a strategic project for Egypt too due to the importance of this future production—from 2017 normally—and gas shortages in the country. Egypt, which was not long ago a gas exporter, is now a net importer. The Egyptian government looks forward to the start-up of WND as well as the coming on stream of other major gas projects in which BP and the Italian group Eni are involved. WND gas will supply the domestic market. It will not be exported.

The area covered by the WND project comprises two permits, North Alexandria and West Mediterranean Deep Water.

12.1.6 Fast-Track Development by BP of Atoll Discovery in Egypt

BP announced in June 2016 that together with the Egyptian Natural Gas Holding Company (EGAS), it has sanctioned development of the Atoll Phase One project, which is an early production scheme that will bring up to 300 million cubic feet a day gross of gas to the Egyptian domestic gas market starting in the first half of 2018. BP has a 100% interest in the concession.

According to BP, the acceleration of the Atoll project "will bring critical gas to the Egyptian market and establish a new material hub offshore East Nile Delta". BP recently completed multiple transportation and processing agreements accelerating the development of the Atoll field which contains an estimated 1.5 trillion cubic feet of gas and 31 million barrels of condensates. Onshore processing will be handled by the existing West Harbor gas processing facilities. BP announced the Atoll discovery in March 2015 on the North Damietta Offshore concession.

Atoll Phase One is an early production scheme (EPS) involving the recompletion of the existing exploration well as a producing well, the drilling of two additional wells and the installation of the necessary tie-ins and facilities required to produce from the field. Success of the Atoll Phase One EPS could lead to further investment in the Atoll Phase Two full field development.

Thanks to these various projects Egypt is looking to boost its gas production to 5.5–6 billion cubic feet/day (57–62 bcm/annum) by the end of 2019, said its Oil Minister, Tarek El Molla. Egypt currently produces some 3.9 bcf/d of gas (40 bcm/annum), as against 4.4 bcf/d in 2015. Egyptian gas production is no longer high enough to meet the country's demand, which now stands at 5.3 bcf/d (55 bcm/annum).

12.1.7 Egypt Could Become an LNG Exporter Again

Thanks to these discoveries and developments and some others, Egypt thinks it will be able in some years to meet its internal consumption and resume its LNG exports through its Damietta and Idku plants (it is now importing LNG). The priority was given to the domestic market, and feed gas for the LNG plants was redirected to help meet national consumption. The future import of gas from Cyprus and/or Israel could also be used to feed the LNG units. It is likely that the LNG plants could be back at full capacity around 2020.

Egypt is no longer exporting gas by pipeline to Jordan and Israel.

12.1.8 Significant Gas Discoveries by Kosmos Energy off the Coast of Mauritania/Senegal

Kosmos Energy, which focuses its exploration on the Atlantic margin, believes in the potential of the Mauritanian and Senegalese offshore and facts seem to prove it right. After a gas discovery in May 2015 with the Tortue-1 well on the block C8 in the Mauritanian offshore, the Marsouin-1 well also resulted in a gas discovery on the same permit. The sites of these two wells are separated by 60 km.

Several important points should be noted about this:

- In both cases, the discoveries are in very deep sea: 2,400 m of water depth for Marsouin-1 and 2,700 m for Tortue-1 (this discovery was renamed Ahmeyim).
- In both cases, the geological formations date back to the Cenomanian. For Ahmeyim, there is also a smaller section in the Albian.
- Kosmos Energy described in both cases these gas discoveries as "significant."
- Net gas pays are of a good thickness: 70 m for the new well and 117 m for the first one (of which 107 m in the Cenomanian).
- Kosmos Energy's work seems to confirm the potential of the Cretaceous off Mauritania and Senegal (the Cenomanian is a stage of the Cretaceous) and to validate the geological model of the US company, which is very positive for the future.
- In addition to the C8 block, Kosmos Energy holds a 60% stake on Blocks C12 and C13. These three permits, which cover a total area of 27,000 km^2, are adjacent and water depths are between 1,000 and 3,000 m.

These three permits are held by a consortium led by Kosmos Energy (60%) and also including Chevron (30%) and the Société Mauritanienne des Hydrocarbures et de Patrimoine Minier (10%), which is the national company of Mauritania. These blocks were obtained under the framework of production-sharing contracts.

If this exploration and appraisal work results in the discovery of large reserves, Kosmos Energy would consider a liquefied natural gas export project that could come on stream beyond 2020. This would require reserves of at least 15,000 billion cubic feet.

After a program of five exploration and appraisal wells off Senegal and Mauritania, Kosmos Energy estimates the potential gas resources in the fairway between the Marsouin-1 well in Mauritania and the Teranga-1 well in Senegal (offshore Senegal Kosmos also made a gas discovery with the Guembeul-1 exploration well) at more than 50,000 billion cubic feet (50 trillion cubic feet). As of today, the gas resources discovered through this drilling program are estimated at about 25 trillion cubic feet (median estimate). According to the company, the gas fairway which was discovered has a length of about 200 km between the sites of the Marsouin-1 and Teranga-1 wells through the Greater Tortue area.

12.1.9 Very Significant Unconventional Gas Potential

In a paper on "Shale Oil and Gas: Developments in Key Countries," the Center hydrocarbures nonconventionnels (Center for unconventional hydrocarbons, CHNC, Paris) reviewed the status of 13 key countries for unconventional oil and gas, including four in Europe, three in the Americas, two in Africa, one in Asia, one in the Middle East as well as Australia and Russia.

Algeria is considered as a country with huge shale gas resources. They would be among the largest in the world. The CHNC stresses that the country has several advantages, including a favorable geology with qualities similar to those of the best basins in the USA, good thickness, a good organic matter content of good maturity and a likely good productivity. Frasnian and Silurian formations in major Saharan basins are particularly interesting in this regard. Sonatrach has started some pilot production and cooperates closely with leading international operators. For the CHNC, the main challenges are access to water, social acceptability, and the security aspects.

According to Alnaft (part of the Algerian Energy Ministry), the potential of tight gas in Algeria is very important in the basins of Berkine, Illizi, Hassi Messaoud, Ahnet, Bechar, and Timimoun. The formations of Lower Devonian, Ordovician, and Cambrian in the Saharan platform are most suitable for tight gas discoveries. The Saharan platform also presents a big potential for shale oil and gas, especially for Frasnian and Silurian source rocks. According to studies from the US Energy Information Administration/ARI, Algeria is a country with considerable potential for shale gas. The study of several basins has led to estimates of technically recoverable resources of 707,000 billion cubic feet of gas (707 trillion cubic feet) and 5.7 billion barrels of oil. It must be recalled that these estimates are the product of geological studies and are not derived from drilling, which means they must be treated with extreme caution. Moreover, technically recoverable does not mean necessarily economically recoverable.

- Algerian estimates are as follows:
- Volumes of shale gas: 6,500 trillion cubic feet.
- Volumes of shale oil: 170 billion barrels.
- Technically recoverable shale gas resources of 975 trillion cubic feet with the assumption of a 15% recovery rate.
- Technically recoverable shale oil resources: 20 billion barrels with a 10% recovery rate.

For shale gas, the potential of Libya and Egypt would be significant, though far less than that of Algeria. But in any case, only exploration work will allow us to have a clearer view of this potential. The key issues are thus to know which are the governments that will authorize or not this exploration and which are the companies that will or not take the risk of drilling. Algeria and Egypt seem to be in the starting-blocks.

12.2 Egypt Could Become Part of A Regional Hub for Eastern Mediterranean Gas

At the beginning of September 2016, the Republic of Cyprus and Egypt signed an intergovernmental agreement in Nicosia with the aim of building a gasline, which would allow Cyprus gas extracted from the Aphrodite field (Block 12) in the Eastern Mediterranean to be exported to Egypt. This is one of the first results of numerous discussions and negotiations between high officials in several Mediterranean countries, especially Cyprus, Egypt, Greece, Turkey, and Israel. Discussions are continuing and some other agreements could be announced in the future but it is a first step. Among the large oil companies, Eni was particularly active in trying to convince the governments in the region to cooperate in order to facilitate the development of gas projects in this part of the world. After this political agreement, companies will have to negotiate a commercial agreement and it will take some time before construction work will start-up.

The plan is to deliver Cypriot gas to Egypt that would either be consumed in Egypt or re-exported in the form of LNG. As explained above Egypt has two LNG plants which are no longer working because there is not enough gas to feed them (13.1.7). Cyprus wanted to export LNG directly to Europe and Asia but Aphrodite's resources are smaller than hoped, and it would not be commercial at this stage to choose the LNG option, which is costly. If exploration work offshore Cyprus resulted in new discoveries the LNG option could be revived.

For Cyprus, the choice of a pipeline to Egypt is thus in the present context one of the most efficient ways to monetize its gas resources. As far as Egypt is concerned it will strengthen its efforts to meet its national demand (see 13.1.3 to 13.1.6) and to resume its gas exports in the future. It also strengthens Egypt's hand in its discussions with Israel about the sale of gas from Leviathan, discovered by the US company Noble Energy and its partners in the Eastern Mediterranean offshore Israel. Noble Energy also discovered Aphrodite. At the end of the day, however, it will very likely not be Cyprus gas vs. Israel gas for Egypt but Cyprus and Israeli gas. But Turkey could try to spoil the game due to its disagreement with Cyprus about the northern part of the island.

Box 12.4 Toward More Cooperation Between Eastern Med Countries?

After approving the sale of gas from the Tamar field in the Eastern Mediterranean, the Israeli Minister of National Infrastructures, Energy, and Water Resources, Yuval Steinitz, said that it was a first sign of the development of a regional cooperation in the energy sector which will include in the future, besides Israel and Egypt, Jordan, Turkey, and, perhaps, Europe. After years of debate and delay, we begin to move forward and to position Israel as a gas superpower in the region, the minister added.

Yuval Steinitz is right on one point at least. After much hesitation, Israel has begun to move toward becoming an exporter of natural gas, which is important for this country, for the region and, potentially, for Europe. But it is excessive to speak of a future regional gas "superpower."

As for the desired cooperation, several obstacles must be overcome to achieve it.

The favorable decision of the Minister relates to the sale by the consortium operating the Tamar field of 5 billion cubic meters of gas to the Egyptian company Dolphinus Holdings. This is an important but not sufficient condition. It will be necessary for the Egyptian government to approve the agreement and for East Mediterranean Gas, which operated the pipeline carrying Egyptian gas to Israel (Egyptian deliveries to Israel ceased in 2011), to also give its green light. Some work will have to be carried out on the pipeline which was designed to bring Egyptian gas to Israel and not the reverse.

The first condition is the most crucial. The Egyptian authorities are not opposed to this project, but its successful implementation was complicated by arbitration in favor of the Israel Electric Corporation (IEC) in its dispute against the Egyptian General Petroleum Corporation (EGPC) and the Egyptian Natural Gas Holding Company (EGAS), two Egyptian national companies. They must pay $1.8 billion to the IEC but Egypt disputes this arbitral decision by the International Chamber of Commerce. The Egyptian government ordered the freezing of negotiations on the purchase of Israeli gas by Egyptian entities as long as this issue is not resolved.

Although relations between Egypt and Israel are not characterized by great heat, the two countries continue to exchange and negotiate on topics of common interest, including natural gas, in a regional and international political context (Syria, Iran, Islamic State, Hamas, etc.) that facilitates a rapprochement between them. Despite the serious difficulty of the arbitration process that followed the cessation of Egyptian gas exports to Israel, the diplomats of the two states are maneuvering to find a compromise.

Dolphinus Holdings signed two letters of intent, one to buy gas from Tamar and another to import gas from Leviathan. Furthermore, two foreign companies active in the liquefied natural gas sector in Egypt, Union Fenosa, and BG, had also each signed a letter of intent to buy Israeli gas to feed liquefaction plants in which they have stakes and which are no longer supplied with Egyptian gas.

On December 17, 2015, Noble Energy (Houston) had welcomed the establishment and the beginning of execution by the Israeli government of the "Natural Gas Framework," which facilitates the development of Tamar and Leviathan, both operated by the US firm with holdings of 36% and 39.66%, respectively. Other interest owners in Leviathan are Delek Drilling with 22.67 percent, Avner Oil Exploration with 22.67%, and Ratio Oil Exploration (1992) Limited Partnership with the remaining 15%. Leviathan has an estimated 22 trillion cubic feet of recoverable natural gas resources. The final investment decision for this field could be taken at the end of 2016 and gas sales could begin three years later. The consortium signed contracts with Israeli buyers and with a Jordan company, and negotiations are ongoing with potential buyers in Israel, Jordan, and Egypt.

12.3 Some Political and Commercial Obstacles Ahead

12.3.1 Politics and Security

On security, or rather insecurity, in North Africa, Libya is in the front line. Foreign oil companies have largely withdrawn from the country since the war of 2011 because of the conflict, the presence of numerous militias and of the Islamic State and the existence of two governments, two parliaments (in Tripoli and in the east—Tobruk and Baida) and two national oil companies (NOC). Libya is rather an oil country than a gas country, unlike Algeria and Egypt, but its gas potential is very significant and much underutilized. The country therefore has a role to play in international gas markets, particularly in Europe.

We must be very cautious but, in 2016, some political progress has been made even if, at the end of the writing of this chapter (early October 2016), it had not generated significant improvements in the oil and gas landscape. Under the aegis of the UN, a government of national accord has been formed and a Presidential Council moved to Tripoli, at first timidly and then with a little more confidence. Headed by Fayez al-Sarraj, the Council got the allegiance of two key organizations, the National Oil Corporation and the Central Bank, and began to take control of various ministries and institutions in Tripoli. The parliament setup in the east, however, continues to make resistance and refuses to lower its flag.

In this dynamics, the historical NOC and the NOC created by the authorities in the east of the country have decided to merge. Its main headquarters would be moved from Tripoli to Benghazi in a clear aim of national reconciliation. The process was announced but it remains to be confirmed and implemented, which will not be easy.

In September 2016, the forces of General Khalifa Haftar, which are close to the authorities in the east, took control of what is often called the Libyan oil crescent, which contains several key terminals on the Mediterranean, an essential element of the Libyan oil value chain. As General Haftar still has not rallied to the banner of the Presidential

Council, this event was not well received by the international community that supports the full implementation of the agreements concluded under the auspices of the UN. But this occupation broke the domination of the Petroleum Facilities Guard, a force created to protect oil facilities and infrastructure but which often used its position to block them for its own benefit rather than allow them to run normally. The situation is thus complex and full of ambiguities but some hope was palpable after years of political crisis without any positive outlook.

If these developments were confirmed, Libya's oil production and exports could grow significantly in the short to medium term. In the gas sector, improvements will require more time and the effective return of foreign companies.

12.3.2 Relationship with the International Oil Companies

The investments in major gas projects are very heavy and North African countries need to work with major international energy players, especially oil companies. To develop more fully the gas potential of the region, it is necessary to attract these investments.

For Libya (see 13.3.1), the extent of security problems does not currently allow the country to fulfill that condition even if there is a hope of improvement due to certain political developments. For Egypt, the key problems are the delay in the payments due to foreign companies, which has led some of them to slow down their investments, and the price of gas (level and indexation) but compromises were found and the main operators, Eni and BP, are going ahead at full speed.

For Algeria, contractual terms are perceived as not attractive enough and low oil prices do not help the situation. The latest international tenders launched by Alnaft, an agency which is part of the Algerian Ministry of Energy, have not been very successful despite the signing of various contracts, and it will be necessary for Algeria to show some more flexibility. The adoption by the country of a new law on hydrocarbons in February 2013 was an undeniable progress but several companies continue to believe that results fell short of expectations. Only four exploration permits were awarded (out of 31 available) at the end of the fourth international bid round, ended on September 30, 2014, the first tender after the adoption of the 2013 law. Algeria is committed for at least 30 years to partnerships with foreign oil companies (the 1986 law on hydrocarbons was a very important step in this regard) and this policy of association is always a priority in its strategies for the exploration, development, and exploitation of hydrocarbons.

The 2013 law had complemented that of April 2005. The latter one was amended a year later by executive orders that established a tax on exceptional profits. This tax applies when the price of the Dated Brent is above $30 per barrel. Anadarko Petroleum Corporation (USA) and Maersk (Denmark) launched an arbitration procedure to seek financial compensations. Their argument was as follows: if a state has the right to modify certain tax provisions, our contract with Sonatrach includes a stability provision that protects us against the negative impact of such changes. These two companies were successful in this procedure, and Sonatrach had to give them the required compensation.

In May 2016, Total and Repsol, which are associated with Sonatrach on the gas and liquids field of Tin Fouyé Tabankort, indicated that they had also requested arbitration because of this tax on exceptional profits that downgraded the economic balance of the project. The two companies said they had negotiated with the Algerian national oil company but that these discussions did not yield any satisfactory results.

12.4 Algeria Wants to Retake the Initiative

Alnaft, the National Agency for the Monetization of Hydrocarbon Resources (Agence Nationale pour la Valorisation des Ressources en Hydrocarbures) stresses the importance of partnerships. The reserves discovered by associations between Sonatrach and foreign firms since the mid-1980s represent about 4 billion barrels of oil equivalent, or 17% of total discoveries and about 30% if one excludes the two biggest Algerian fields, Hassi Messaoud (oil) and Hassi R'Mel (natural gas), the agency said (Alnaft refers to P1 reserves, meaning proven reserves, but adds "in place" while the industry normally distinguishes reserves and volumes in place). Total cumulative discoveries are thus estimated at 23.5 billion barrels of oil equivalent and "reserves" of Hassi Messaoud and Hassi R'Mel at about 10 billion boe. In the first nine months of 2015, oil and gas production in partnership covered a quarter of domestic oil and gas production.

Also according to Alnaft, in July 2015, the acreage in partnership consisted of three prospecting licenses, eight research contracts (exploration), 15 development contracts (59 fields, of which 14 oil fields and 45 gas fields), and 14 contracts in the operating phase (53 fields, including 47 for oil and six for gas).

As emphasized repeatedly by the Ministry of Energy and Sonatrach Alnaft recalled at the end of 2015 that Algeria remained an underexplored and underexploited country. Over 10,000 wells have been drilled since 1948, date of the first commercial discovery in the country, 500 fields have been discovered and 250 of them are in operation, according to the agency. In the Berkine basin, the density of drilling is 60 wells per 10,000 km^2 but this proportion falls to 4/10,000 km^2 in the frontier areas.

Alnaft explained that vast underexplored areas or themes are of great potential. This is partly the case of mature basins as far as stratigraphic traps and deep horizons are concerned, of some frontier basins, of certain areas with a complex geology and of unconventional resources (tight formations and shale oil and gas).

In the first category, the exploration of deep horizons located under certain operating reservoirs in the Berkine basin is considered promising. These horizons include the Silurian AG, Hamra quartzites, and the Cambrian (producing horizons in this basin are particularly TAGI, the Strunian, and the Lower Devonian). In addition, stratigraphic traps have been little explored in Algeria even though their potential worldwide is well known. The basins of Berkine, Illizi, and Oued Mya seem interesting in this regard.

Alnaft stressed on this occasion that two-thirds of the national acreage were available for new research (exploration) and exploitation projects.

The assessments above on Algeria's hydrocarbons potential are not disputed in substance, but the attractiveness of the country is not what it was. Current legislation is not about to be amended soon and Alnaft anyway has not the power to do so. The agency is working on ways and means to improve research and exploitation contracts in the context of existing legislation. The main orientations have been identified:

- More operational flexibility, which means the ability to adapt the provisions of the operations agreement between Sonatrach and its foreign partners within the various consortia to allow this increased flexibility. Alnaft also proposes to provide budget overrun thresholds without prior recourse to its approval, which will increase the flexibility of the contractor in its budget management.
- Alnaft could strengthen its assistance to the contractors. In case of disagreements between Alnaft and the contractors, "consensual solutions" would be preferred.
- The number of annexes to research and exploitation contracts would be reduced by consolidating various provisions in a single document.

These ideas are interesting and they deserve to be deepened and implemented. However, it is not sure these steps ahead would be sufficient in the current context marked in particular by low oil prices, which greatly complicates the situation for Algeria as for all oil and gas producers.

Box 12.5 Future Oil and Gas Investments in Africa

WEO 2015: investments in the oil and gas supply in Africa would total $2,400 billion between 2015 and 2040

In order to meet oil and gas needs in Africa a lot of investment will be required. According to the World Energy Outlook 2015 (projections related to the New Policies scenario), a publication of the International Energy Agency, cumulative investment in the 2015–2040 period for the supply of hydrocarbons in Africa would total $2,400 billion in 2014 dollars, or about $92 billion per year.

By investments related to the supply of oil and natural gas, the IEA means investments in the upstream (exploration, development, and production) and in the transport of hydrocarbons. For oil, refining, which allows consumers to get petroleum products, is also taken into account.

These $2,400 billion in investments break down into $1,533 billion for oil and $868 billion for natural gas. The shares of these two fuels would therefore be 63.8% and 36.1%, respectively. The upstream would get the lion's share with $1,990 billion (82.9% of the total), or $76.5 billion annually, of which $1,356 billion for oil and $634 billion for natural gas. For transport, the projections are $90 billion for oil and $233 billion for gas. Investments in refining are projected at $87 billion over the period 2015–2040 (all figures are in 2014 dollars).

According to the WEO 2015, global investments in the supply of hydrocarbons as defined above would exceed $25,000 billion over 2015–2040.

Bibliography

Algerian Ministry of Energy, Law of 20 February 2013 relating to hydrocarbons.

BP, *BP Energy Outlook: Transition to a Lower Carbon Future*, 10 February 2016.

BP, *BP Statistical Review of World Energy*, June 2016.

Cedigaz, *First Estimates 2015*, April 2016.

Eni, 2016–2019 Strategy Presentation, 18 March 2016.

International Energy Agency, *World Energy Outlook 2015*, 10.November 2015.

International Energy Agency, *World Energy Investment 2016*, September 2016.

OPEC, *Annual Statistical Bulletin 2016*, June 2016.

OPEC, *World Oil Outlook 2016*, October 2016.

Sonatrach, Annual Report 2014.

Sonatrach, Annual Report 2015—Abstract.

Sonatrach, *Une dimension gazière internationale.*

Author Biography

Francis Perrin is Senior Research Fellow at the OCP Policy Center based in Rabat (Morocco). Senior Research Fellow at the French Institute for International and Strategic Affairs (IRIS, Paris). Chairman of Stratégies et Politiques Energétiques and of Energy Industries Strategies Information. Contributing Editor for Petrostrategies, a weekly energy newsletter based in Paris. Contributing Editor for the Cyclope World Commodities Markets published every year (Economica, Paris) and for Cyclope Africa, published by the Cercle Cyclope and the OCP Policy Center. Courses at the OCP Policy Center, at IRIS and at the University of Lyon (France). Presentations in many professional conferences and interviews to many media on various energy issues.

13

Toward A New Mediterranean Gas Hub?

Dario Speranza and Daniela De Lorenzo

13.1 The Southern Mediterranean's Role in Future Energy Scenarios

As a geostrategic category, the "Southern Mediterranean" has traditionally denoted a complex Gordian Knot of seemingly irreconcilable interests and intractable conflicts. Nevertheless, the recent discoveries of huge gas reserves in the Eastern Mediterranean—in the offshore of Israel, Egypt, and Cyprus—are redefining its meaning, even heralding scenarios of energy cooperation in relation to the concept of creating a Mediterranean gas hub.

With due caution, we can say that momentous changes in its landscape—fueled by new discoveries and region-wide energy "plays"—have the potential to bring about a paradigm shift with long-standing political and economic implications for the region and beyond. There are objective interests which might act as catalysts for cooperation between

D. Speranza (✉) · D. De Lorenzo
Eni, Rome, Italy
e-mail: dario.speranza@eni.com

M. Hafner and S. Tagliapietra (eds.), *The European Gas Markets*,
DOI 10.1007/978-3-319-55801-1_13

the countries on the Levantine Basin and, as we will see below, even Turkey and Greece. And the change in the “Southern Mediterranean category” based on an abundance of natural resources (particularly gas and renewables) is happening as new demand and consumption patterns emerge, particularly but not only, in its most populous countries such as Egypt.

The south Mediterranean countries share a growing need of energy resources for the internal market and a high reliance on fossil fuels. In this regard, the challenges for the energy sector in the region will be to satisfy increasing internal demand with affordable and reliable supplies, to create conditions to efficiently export excess resources, and to craft a sustainable energy mix (i.e., more gas and renewables).

The southern shore is, in fact, a fast-growing and dynamic area, with a 1.6% demographic and 3% energy consumption annual growth rates. It needs a rapid push toward a sustainable energy transition, primarily to strengthen local economies and social cohesion. The north Mediterranean, however, is also set to benefit from the new energy transition.

The north Mediterranean, on the contrary, is now a mature region, not only in demographic terms (population growth rate of approximately 0.8% a year) and economic growth, but also in energy consumption (which is flat). In 2014, the European gas demand stood at 410 billion cubic meters (bcm) and projections for 2030 are above 480 bcm. Internal production—which covers 34% of demand—will decrease, while imports—which cover the remaining two-thirds—are viewed as uncomfortably vulnerable to disruptions to interconnections and supplies.

So, there is a strong will on the part of the EU to further geographically diversify sources of gas and concurrently increase energy autonomy and reduce environmental impact by strongly encouraging the use of renewable energy sources. To achieve these goals effectively, the consolidation of a fully integrated EU domestic energy market is the key: “Effective interconnection of electric grids and pipelines would allow member states to trade energy more flexibly than today, mitigating the impact of supply disruption and dependence on non-EU suppliers.”[1]

Moreover, new European commitment against climate change and the target of reducing CO_2 emissions by 40% in 2030 could further boost natural gas consumption and consequently increase energy demand.

In this respect, the recent discoveries in Eastern Mediterranean, together with the prospect of further hydrocarbon finds off the Levant Basin, have the potential to set the region on a positive trajectory in terms of economic gains, energy security, and diversification, in terms of both sources and suppliers.

Europe indeed has an urgent need to diversify energy suppliers in order to increase the resilience of its internal gas system and create a more competitive regional market. In this context, strengthening the south–north corridor could contribute to the recovery of the Mediterranean and boost the security and the development of the whole region.

13.2 Levantine Landscape

According to a March 2010 USGS report,[2] the Levant Basin area holds an estimated recoverable gas of 3453 bcm and an estimated 6310 bcm (mean estimated) of technically recoverable natural gas in Egypt's Nile Delta Basin (three times the presently proved reserves of the country).

The discovery of the Zohr supergiant gas field (850 bcm), in August 2015, represents a lifeline for Egypt and has structurally altered the energy landscape of the region.

Although Egypt was in the past a major gas exporter, gas production dropped from 62 bcm in 2008 to 46 bcm in 2014, while energy demand rose sharply in the same period reaching 50 bcm in 2015. As a consequence, in April 2015, the country began to import LNG.

As far as the future scenario, Cairo's energy demand is expected to grow by a further 37% by 2030. More specifically, by then, the country will need an additional 34 bcm of natural gas to meet demand.

With the discovery of Zohr, Egypt is set to achieve self-sufficiency, which will eliminate the need for LNG imports for the years to come. Rising Nile Delta gas production also offers the promise of relief from current ongoing shortages and of export the gas production surplus.

At the same time, the country's close proximity to other discoveries in Israel and Cyprus and the possible synergies with them could result in the creation of a strategically important gas hub.

The prospect of synergic cooperation with its neighbors is strengthened by the fact that Egypt has existing infrastructure for production (such as flow lines and treatments plants) and export (LNG terminals and pipelines), for an overall capacity of 35 bcm.

As for pipelines, the Arab Gas Pipeline (AGP) from Egypt goes to Jordan, Syria, and Lebanon. In 2008, an extension of the AGP was built under water from al-Arish in Egypt to Ashkelon in Israel. Now, Egypt could import gas from Israel via reverse flow from the same pipeline.

Egypt has also two liquefaction plants with a combined capacity of 17 bcm per year.

Consequently, the Egyptian gas industry could both meet local gas demand and also allow Cairo to become an export hub, resuming LNG exports from Damietta and Idku (channeling potentially even Israeli and Cypriot gas through its unexploited infrastructures), possibly as early as 2022.

Discovered in 2010, Israel's Leviathan field (620 bcm)—which is double the size of Tamar (280 bcm), discovered just 1 year before—has been a game changer for the country which is currently an energy importer but is firmly intending to become a net exporter. In perspective, this change, according to Israeli officials, is the most important energy news since the founding of the state.

Regarding the future energy landscape in Israel, gas demand in 2016 stood at 9.4 bcm, while energy demand is expected to grow 30% by 2030. At present, Israeli resources (around 1100 bcm) could allow the country not only to meet local demand, but also to become a net exporter.

The management of the country's natural resources generates in Israel strong sentiments and is traditionally a sensitive political topic, especially about the way revenues are allocated and redistributed. After a long period of political debates and divisions, the new gas regulatory framework, finally approved in a revised form in May 2016, should create a more benign environment for the Leviathan partners, Texas-based Noble Energy and Israel's Delek Group, to resume investments.

Yuval Steinitz, Israel's Minister of National Infrastructure, Energy and Water Resources, referring to the concept of creating a Mediterranean gas hub, said: "Now we have a golden opportunity, for the first time, to create something significant in the axis of peace of Israel, Egypt and Jordan".

As for Cyprus, so far the country has discovered a limited amount of gas in the Aphrodite field (128 bcm), but following the Cyprus third offshore licensing round—which attracted also new international operators and in which Nicosia assigned three more Blocks in the south of the island (6, 8, and 10)—new discoveries could be fast-tracked. In this case, given the small size of the island and its limited domestic market, most of the potential gas could be exported.

In addition, Israel—which seems to have cleared most of the regulatory hurdles—had launched a new international bid round for 24 offshore blocks, in November 2016, while recently the Lebanese council of ministers decided to put on offer, after four years of delays, five blocks in the country's first bid round, which it hopes to award by the end of 2017. The government claims potential for up to 700 bcm.

If these countries are able to define common strategies and share the existing infrastructures, the overall gain in terms of cost reduction and efficiency will be very significant.

This could boost the development of the entire region from North Africa to the Middle East and—to borrow a keyword from the political debate and even the new *EU Global Strategy for MENA countries*—contribute to making states and societies more "resilient."

This would render them more able to withstand and recover from "internal and external crisis," providing the tools "to strengthen endogenous force" and "move the region into a more cooperative order."[3]

13.3 Maritime Borders and Export Routes

The political challenges surrounding the ongoing disputes over maritime borders need to be tackled before any long-standing investment decision can be made.

On the positive side, in recent months, there has been important movement on the "Cyprus problem"—Greek and Turkish island communities have been willing to acknowledge each other's existence.

The leaders of the two communities, Nikos Anastasiades, President of the Republic of Cyprus (RoC), and Mustafa Akinci—elected in 2015 as President of the Turkish Republic of Northern Cyprus (TRNC)—agreed to intensify negotiations and to meet more frequently in order to solve outstanding issues.

However, the negotiation process is moving slowly, primarily because relationships between Nicosia and Ankara are fraught with new tensions. The Cypriot (RoC) Minister of Energy, Georgios Lakkotrypis, said that Nicosia would pursue its offshore development projects, keeping them completely separate from the negotiation process.

In March 2016, immediately after the launch of the licensing round by the Cypriot authorities, the Turkish government issued a statement warning Cyprus against proceeding with the tender. Ankara accused Nicosia of challenging the rights of the Turkish community in the island and of violating Turkey's sovereign rights.[4]

Similarly, Turkey recently contested the assignment of the new exploration licences by Cyprus, saying it would violate turkish sovereign waters, and on May 2017 issued three NAVTEX, reserving areas within Cyprus EEZ for seismic research. Moreover, Ankara, after the diplomatic reconciliation with Israel, could potentially gain the unique opportunity of becoming a Mediterranean gas hub to Europe.

Indeed, exporting Israeli (and Cypriot) gas to Turkey via pipeline could be one of the most economically viable options. According to the state company BOTAS, Turkish gas demand could grow from a current 50 bcm to over 76 bcm by 2030, and the two countries could produce as much as 25 bcm per year, that is to say, half of the current Turkish demand, although, recent studies however reported a possible stagnation of turkish future demand for gas.

However, without a settlement of the Cyprus problem, this solution appears difficult to implement.

Another possible option, the Eastern Mediterranean Gas Pipeline—one of the EU's Projects of Common Interest—which aims to connect by 2025 Israeli and Cypriot gas to the shores of Greece and Italy. Energy ministers from Greece, Cyprus, Italy and Israel signed in Tel Aviv a joint declaration on April 2017 to promote construction of the 2,200-km deep-sea pipeline, a privately funded $6–7 billion plan.

However, doubts were stirred by some experts[6] who have challenged the economic viability of the pipeline at least without other major gas discoveries, because currently "Russian gas is selling in Europe from \$4.7 to \$5 per mmbtu (million British thermal units). By comparison, the proposed East Med gas, factoring in the cost of the pipeline, would in the best-case scenario likely go for around \$9 to \$10, making it uncompetitive."

The gas subsea pipeline could also pass through Lebanese waters, but there is a historical dispute regarding 1000 km of the Israeli–Lebanon maritime border, which directly impacts the rights on Lebanese Blocks 8 and 9.

In 2010, Beirut presented the United Nations with its own map on EEZ coordinates, which unilaterally defined Israeli–Lebanon maritime borders and modified those signed with Cyprus in 2007 (never ratified by the Lebanese parliament).

Subsequent steps to solve the dispute with the mediation of the USA have failed. Moreover, a recent study by the Lebanese Petroleum Administration (LPA) showed that Lebanon and Israel might share large natural gas reservoirs (particularly in Blocks 8 and 9 on the southern maritime border with Israel).

Consequently, some Lebanese politicians expressed their concerns about the possibility that Israel could exploit Lebanon's southern energy resources, and called for immediate action. They urged the Lebanese council of ministers to speed up decisions about the development of energy sector. In addition, Syria and Lebanon have not reached an agreement on 1100 km of maritime borders, which affect ownership rights on Lebanese Blocks 1 and 2.

Finally, it is important to note that even Israel and Cyprus have unsolved issues regarding gas. They have been negotiating on a unification agreement for the Aphrodite gas field for a long time: "it is estimated that up to 10% of its reserves extend into Israel's exclusive economic zone (EEZ). The situation with that field is entangled because two of the gas field's owners, Noble Energy and Delek Group are partners in Aphrodite gas field and are opposing any claim by Israel to part in the field, which, on the Israeli side, is licensed to another business group."[7]

In any case, at present, energy experts seem to share the view that Egypt represents the best regional destination for Israeli and Cypriot gas, which through the LNG facilities of Damietta and Idku, currently not utilized, could reach Southern Europe (mainly Italy and Spain).

13.4 Competitive Advantage and Accountability

The Middle East and North African (MENA) countries are facing a challenging regional landscape, fraught with political risks. Their vast oil and gas reserves, characterized by low breakeven and operating costs, offer the potential to improve their trajectory in terms of growth and stability.

Their low-cost production base gives them an important competitive advantage, which will allow them to gain market share at the expense of higher cost producers, thus becoming even more crucial and critical in future energy scenarios.

However, energy as utility must remain affordable to avoid inequality and to preserve competitiveness, topics which raise important issues.

Firstly, revenue streams in resource-rich countries—as a share of production not received in cash—remain frequently unquantified. Thus, the promotion of accountability and transparency is an important issue, especially in those countries where energy is the major or almost exclusive source of national wealth.

Given the emergence of improved *transparency and resource management* practices, international and financial institutions are especially important in promoting the adoption of international standards in environmental and social practices as well as in corporate governance.

Secondly, these institutions can foster the transition a market-based management of the energy sector. Countries such as Algeria, Libya, Egypt, and Lebanon allocate about 10% of GDP to energy subsidies, which are considered by many to be economically inefficient and unsustainable in the longer term.

In the southern Mediterranean, the adoption of an effective energy mix based on low-carbon energy sources (i.e., natural gas and renewables) is a key driver for sustainable economic growth. The region should carefully avoid the development pattern followed in Southeast Asia, whereby the strong increase in the use of coal produced an uptick in pollution and CO_2 emissions. Natural gas emits around half the carbon dioxide of coal when burnt for power generation; it is also very flexible as it can be brought online quickly to meet fluctuating power demands and is thus the best partner for intermittent renewable energy sources.

A reform of the energy sectors in these countries is urgent. The EU is the most logical partner to work with in order to launch sectoral reforms, which should address policy planning, legal and regulatory design, infrastructure development, and technology transfer.

Thirdly, other important tools to foster the energy transition on the southern shore might be the public–private partnership, which could involve local governments, international energy companies, and financial institutions, such as the European Bank for Reconstruction and Development (EBDR).

Lastly, rethinking the role of multilateral cooperation is also a core enabler of East Med energy development, and in this respect, it would be meaningful to consider the potential role of the Organization for Security and Cooperation (OSCE) in offering a framework, which includes not only the economic and environmental dimensions but also the political, security and humanitarian spheres.[8]

The Arab Spring and its aftermath have added an element of instability to a region traditionally fraught with political tensions. The breakup of the social contracts between citizens and states has fostered a sense of fragility, exacerbated by extremism and terrorism.

To this end, public and private stakeholders have an important role to play in promoting stability and preventing conflicts.

The broader Mediterranean region will benefit enormously from cooperation at a multilateral level, while from a pure energy perspective, a gas hub framework could make them more interdependent, while boosting their endogenous political, economic, social strength and resilience and therefore increasing cohesion and limiting conflicts.

Notes

1. http://www.ecfr.eu/rome/post/salzano.
2. http://pubs.usgs.gov/fs/2010/3014/pdf/FS10-3014.pdf.
3. http://www.iai.it/sites/default/files/menara_fn_1.pdf.
4. Indeed, on September 2011, Turkish Prime Minister Erdogan and Turkish Cypriot President Eroglu signed an agreement in New York on the delineation of the continental Shelf between Turkey and Turkish Republic of Northern Cyprus and exploration licenses were given to TPAO to explore for hydrocarbon around the island. According to this delineation, Blocks 1, 4, 5, 6, and 7 claimed by Nicosia overlap with Turkish continental shelf and shall be treated as Turkish territories. In this regard, according to the Turkish government, they can be defended on the principle of national sovereignty.
5. http://cyprus-mail.com/2016/08/07/turkey-blame-negotiations-fail-says-kasoulides/.
6. http://cyprus-mail.com/2016/02/07/tripartite-gas-would-be-double-the-price-experts-say/.
7. http://www.naturalgaseurope.com/israel-and-cyprus-to-resume-energy-talks-30718.
8. http://www.iai.it/sites/default/files/newmed_energy.pdf.

Bibliography

Descalzi, C. 2016. *I rischi energetici se non si torna ai fondamentali,* Conference: "Middle East and North Africa Energy Power, Security and Energy Markets", Chatham House, London, (published also in ASPENIA, n. 72).

Dreiski, P. 2016. *The OSCE as a Player in the Regional Energy Context.* In "The OSCE's Contribution to Energy Governance in the Mediterranean Region"—New Med Research Network. ed. Silvia Colombo and Nicolò Sartori. http://www.iai.it/sites/default/files/newmed_energy.pdf.

Ellinas, C, J. Roberts and H. Tzimitras. Foreword by Koranyi D. 2016. *Hydrocarbon Developments in the Eastern Mediterranean.* Washington, DC: Atlantic Council.

Gürel, A. 2016. Eastern Mediterranean Gas: Source of Prosperity for the Region? In *The future of natural gas,* IAI-OCP Policy Center, ed. Silvia Colombo, Mohamed el Harrak and Nicolò Sartori.

Hazou, E. 2016. *'Tripartite' Gas Would be Double the Price, Experts Say,* Cyprusmail http://cyprus-mail.com/2016/02/07/tripartite-gas-would-be-double-the-price-experts-say/.

Salzano, P. 2014. *The Mediterranean is Back,* Longitude.

Soler i Lecha, E, and N. Tocci. 2016. *Implications of the EU Global Strategy for the Middle East and North Africa,* Middle East and North Africa Regional Architecture- Menara, Future Notes http://www.iai.it/sites/default/files/menara_fn_1.pdf.

The EITI, Nocs & the First trade. 2015. Extractive Industry Transparency Initiative. https://eiti.org/sites/default/files/documents/EITI_Brief_NOC_FirstTrade_March2015.pdf.

Turkey to blame if negotiation. 2016. Cyprusmail. http://cyprus-mail.com/2016/08/07/turkey-blame-negotiations-fail-says-kasoulides/s *fail says Kasoulides.*

US Geological Survey. 2010. *Assessment of Undiscovered Oil and Gas Resources of the Levant Basin Province, Eastern Mediterranean.*

Zalel, Y. 2016. *Israel, Cyprus Resume Energy Talks,* Natural Gas Europe http://www.naturalgasworld.com/israel-and-cyprus-to-resume-energy-talks-30718.

Authors' Biography

Dario Speranza is a Chartered Mechanical Engineer and Master in General Management, member of the Society of Petroleum Engineers since 2008. Almost 20 years in the oil and gas industry and in Eni since 2002, Dr. Speranza has competencies in market analysis, benchmarking, gas supply, investment control, medium- and long-term strategic plan, oil market scenarios, and upstream strategic development. He is head of political and institutional analysis and scenarios within the International Affairs Department of Eni.

Daniela De Lorenzo is graduated in political science from Istituto Universitario Orientale of Naples with specialization in political and economic geography and master in international development. In Eni since 1998, Ms. De Lorenzo is Middle East/Levantine Basin Activities Manager within the International Affairs Department of Eni with competencies in geopolitical and institutional analysis, international relations, and eastern Mediterranean and Middle East scenarios.

14

The Southern Gas Corridor

John Roberts

14.1 Introduction

14.1.1 The Attainment of the Southern Gas Corridor

The Southern Gas Corridor is a series of projects carrying an initial cost of close to $40 billion and intended to serve as a mechanism for bringing gas from a variety of sources in the Caspian and the Middle East to Europe.

The core elements of the SGC are already under development, with pipeline construction under way in Azerbaijan, Turkey, Greece and Albania that will initially carry 6 bcm/y of Azerbaijani gas to Europe and around 10 bcm/y to European countries beyond Turkey. Crucially, more than 90% of the work on developing the giant second stage of the Shah Deniz gasfield—commonly known as the SD2 project—in

J. Roberts (✉)
Atlantic Council, Washington, DC, USA
e-mail: John.roberts@methinks.org.uk

M. Hafner and S. Tagliapietra (eds.), *The European Gas Markets*,
DOI 10.1007/978-3-319-55801-1_14

the Azerbaijani section of the Caspian Sea, which will provide this initial 16 bcm/y, has already been completed. Furthermore, the initial infrastructure under development between the gas processing centre at Sangachal on Azerbaijan's Caspian coast and the junction in southern Italy where the system will connect into Italy's Snam Rete pipeline network is designed so that it can eventually carry 32–33 bcm/y as far as western Turkey and 20–22 bcm/y onwards to Italy.

Although Azerbaijan hopes that in time it will be able to provide the bulk of the gas required to ensure the system is fully filled, in an era of relatively low gas prices and consequent reluctance to invest too much, too quickly in new upstream projects, the issue of how and when the initial SGC system will be filled to capacity remains open. So, too, does the issue of whether any other groups of producers and consumers will come together either to expand the initial SGC system by building a parallel system utilising the SGC's rights of way or by emulating the SGC partners and developing complementary projects, such as a 900-km east–west pipeline across the Black Sea or a 1,600-km pipeline from the Eastern Mediterranean to the Greek mainland. Such brand new projects—which would, like the SGC itself, cost several tens of billions of dollars to build—would be required if the European Commission were ever to attain its apparent goal of developing a Southern Gas Corridor capable of bringing not 10 or 20, but 80 or even 100 bcm/y of gas to Europe.

The SGC's history so far relates to two imperatives: Azerbaijan's need to find a major commercial export market for Shah Deniz gas and the European Union's wish to diversify gas imports so that it becomes less reliant on Russian gas supplies.

To succeed, the project had to meet both commercial and political requirements. In general, the commercial issues have proved to be the most crucial, since the majority of the funding for the system has come from the private sector, although funding arranged by Azerbaijan's state-owned oil company, SOCAR, and by European institutions has also played an important role.

On the political front, securing the approval of both national and local governments in the various transit countries has not always been easy. But the projects that constitute the initial Southern Gas Corridor will make it possible both to bring important substantial volumes of gas from a wholly new supply source to major European gas markets and will serve to provide a reasonable basis for European hopes that, in time, additional infrastructure can be developed so that gas from a variety of Caspian and Middle Eastern producers can reach European markets.

14.1.2 The Current State of the SGC

At the time when final investment decisions on the key elements of the project were concluded towards the end of 2013 (see Table 14.1: Costs & Shares in the SGC), international energy prices were high. Even so, there were concerns among some of the original participants that one core component, the Trans-Anatolian Pipeline (TANAP) across Turkey, might not be commercial. However, for two of the three major financial backers of the project, the State Oil Company of Azerbaijan (SOCAR) and BP, the sheer necessity of both profiting from and safeguarding their upstream interests in Azerbaijan, coupled with deep-rooted concerns about relying on the existing pipeline network operated by Turkey's BOTAŞ, overrode any possible doubts. Moreover, if they did not proceed with the integrated development of the giant second phase of the Shah Deniz field (SD2) and with the infrastructure it required, they would stand to lose not only the value of the gas from the field, but the value of condensate which could be pumped into the existing Baku–Tbilisi–Ceyhan oil pipeline.

The subsequent fall in international energy prices has impacted on the development of the SGC. But this has not so much affected the development of the initial projects to ensure delivery of 6 bcm/y of gas to the Turkish market and a further 10 bcm/y to European customers beyond Turkey as on the longer-term development of the system's expansion. There are two main reasons for this. One is that the system is already underpinned commercially by firm contracts concluded in 2013 for the sale of gas both along the route and further afield. The other is that falling energy prices have reduced the cost of construction, not least by contributing to reduced prices for steel pipe. The overall cost for the SGC project chain had thus fallen from around $45 bn at the time that

Table 14.1 Costs and shares in the Southern Gas corridor

(Principal shareholders only)

	SGC Project Costs ($ bn) Original Dec-16	($ bn) Revised June 2016 **	Share SOCAR* Stake 2016	in SGC BP Stake 2016	Project Turkish Stake 2016	Share ($ bn) SOCAR Share	of ($ bn) BP Share	Costs ($ bn) Turkish Share	
Upstream SD2	23	18.9	16.67%	28.80%	TPAO 19%	3.15	5.44	3.59	
(26 wells offshore)	(6)								
(Offshore facilities)	(15)								
(Sangachal onshore terminal)	(2)								
SCP-X	5	4.9	16.67%	28.80%	TPAO 19%	0.82	1.41	0.93	
Tanap	11–12	9.3	58%	12%	BOTAS 30%	5.28	1.12	2.79	
Tap	c.5	6	20%	20%		1.2	1.2		
Total:	c.44–45	**39.1**				10.45	9.17	7.31	26.93

* *Note* Socar's shares in all the major projects have changed significantly since 2013

** These are figures given by Azerbaijani Energy Minister Natiq Aliev in Baku on 2 June 2016. The TAP consortium has not provided any official figure for the cost of TAP

Source Methinks Ltd

the key FIDs were taken in late 2013 to $39.1 bn as of June 2016.[1] The longer-term consequence, however, is that lower gas prices will mean much tighter margins for new suppliers seeking to enter European markets, particularly if their gas has a relatively high production cost or has a long way to travel.

Since the final investment decisions were taken, there has been very rapid progress indeed in terms of progressing upstream SD2 field development, constructing a new pipeline across Azerbaijan and laying a brand new pipeline across Turkey.

Progress on the last stage of the system—the Trans-Adriatic Pipeline (TAP) from the Turkish–Greek border to southern Italy—has been slower, but this largely reflects the fact that it is not required for use as early as the lines to and through Turkey since the build-up of actual gas production at Shah Deniz will only allow for deliveries to Italy to start in 2020, whereas the current intention is that deliveries to Turkey will start in the second half of 2018.

As of late September 2016, the BP-led Shah Deniz consortium had awarded no less than $18 bn worth of contracts for the $23.8 bn worth of work required to develop the $18.9 bn SD2 upstream element and the accompanying $4.9 bn expansion of the South Caucasus Pipeline, commonly dubbed SCP-X. (Both elements are the direct responsibility of the Shah Deniz consortium and are often treated as a single element). In December 2016, BP Azerbaijan Vice-President Bakhtiyar Aslanbeyli told the author that work on the Azerbaijani and Georgian sections of the project (the upstream SD2 development and the associated expansion of the South Caucasus Pipeline from Sangachal to the Georgian–Turkish border) was 82% complete.[2] With regard to the key central infrastructure section, the $8.5 bn, 1,802-km Trans Anatolian Pipeline (TANAP), Saltuk Duzyol, TANAP's CEO, said on 8 December 2016 that construction work was 55% complete and that Turkey's state gas company, Botas, would receive its first deliveries from TANAP in June 2018 and that deliveries from TANAP to TAP would start in June 2019 deliveries from TANAP to TAP would start in June 2019. By June 2017 almost all of TANAP's 1,334-km, 56-inch section from the Georgian border to Eskişehir had been laid. This left some 450 km of 48-inch pipeline between Eskişehir and the Turkish border with Greece still to be completed, along with the 18-km subsea crossing of the Dardanelles. In July 2016, TANAP awarded Malaysia's SapuraKencana TL a $125 m contract to lay twin- 30 inch pipelines across the Dardanelles, along with the associated fibre-optic cables.

On the final section, the Trans-Adriatic Pipeline (TAP) contracts for laying the 215 kms of onshore line in Albania and the 545 kms of onshore line in Greece (together with the river crossing connection to the TANAP system at Kipoi/Ipsala) were awarded in March 2016, and the following month Italy's Saipem was awarded the contract to lay the crucial 105-km subsea section between Albania and Italy and the accompanying eight-km connection to the Italy's Snam Rete pipeline system.

However, although onshore construction work in Greece and Albania proceeded apace throughout 2016, TAP still appeared to face some problems with regard to actual landfall in Italy. The TAP consortium has already secured full governmental authority from Rome for a landfall at San Foca on the heel of Italy and onward connection to the Snam Rete system, but this continues to be strongly opposed by local authorities in the Lecce region. As of mid-2017, changes in the Italian Government resulting from the resignation of Prime Minister Matteo Renzi in December 2016 had not led to any fall in central government support for the project. However, there were still ongoing disputes with local and regional authorities in southern Italy who were worried about the need to temporarily displace olive trees along the line and who argued that the pipeline's landfall in southern Italy should be shifted from San Foca, near Lecce to a point near Brindisi, some 80 kms further north.

Despite the problems concerning the development of landfall facilities in southern Italy, there is no commercial or direct governmental reason to suppose that the system will not be up and running in or around 2020. This is due to the fact that the system is underpinned commercially by firm contracts for the sale of gas both along the route and further afield and politically by guarantees from all governments along the route concerning project implementation. On the commercial side, gas sale-and-purchase agreements covering the delivery of around

10 bcm/y over a 25-year-delivery period were signed in 2013 with nine European companies: Bulgargaz (Bulgaria); Depa (Greece), Enel (Italy), Hera (Italy), GDF Suez (France), Gas Natural Fenosa (Spain), Axpo (Switzerland), E.ON (Germany), and Shell (the Netherlands/UK). A similar agreement covering the delivery of six bcm/y to Turkey's BOTAŞ was also agreed.

Normally, such commercial and governmental underpinning should be more than enough to yield the conclusion that even such a complex set of projects as the SGC would be delivered more or less on schedule. But there is one non-commercial issue that might yet pose a problem: instability within Turkey through which the crucial TANAP pipeline will run. This issue is addressed subsequently.

14.1.3 Expanding the SGC

From a European perspective, the Southern Gas Corridor is actually more than the sum of the projects that are generally called the SGC. This is because the infrastructure currently under development is regarded as the start of a process, and not simply the completion of a finite set of specific projects. This view considers that one of the major developments of current SGC project implementation is the establishment of a new route to reach Europe and that eventually—and this may be decades away—other pipelines will be able to follow the same or similar routes to bring gas from the Caspian and/or the Middle East to the European Union. Thus, the European Commission, in a one-page summary of oil and gas supply routes that was updated as recently as 20 September 2016, states that: "Initially, approximately 10 billion cubic meters (bcm) of gas will flow along this route when it opens in 2019/2020. Given the potential supplies from the Caspian Region, the Middle East, and the East Mediterranean however, the EU aims to increase this to 80 to 100 bcm of gas per year in the future".[3]

Most of the prospective suppliers were listed in an interview given by Azerbaijan's Minister of Energy, Natiq Aliev, shortly after the final investment decisions (FIDs) for SD2 and various infrastructure components of the SGC were taken in late 2013. "In the future, Turkmenistan and Kazakhstan may use the infrastructure we are currently building to export their resources to the European market. As such, the Shah Deniz 2 FID will have a major positive impact on the economies of many countries, including Azerbaijan, Turkey, Georgia, Italy, Greece, and Albania". Aliev added: "I think this is just the beginning because many resources will be added to existing production. There are talks about adding Iranian and Iraqi gas reserves to this corridor".[4]

The list of prospective suppliers continues. In May 2016, after listing most of the sources mentioned above, Socar's Vice-President for Southern Gas Corridor Development, Vitaly Baylarbayov, said: "....and there is a lot of gas in Iraq and Syria. And there is now gas in Israel, and there is gas in Cyprus".[5]

The prospect of adding gas from all these prospective suppliers is considered below, along with the issue of Egyptian resources and the question of whether Russia, too, might ever contemplate seeking access to at least part of the system.

14.1.4 The Centrality of the SGC

In today's low-price atmosphere, with the specific exception of Russia, there is no immediate prospect for new pipelines reaching Europe without using the TANAP or TAP pipelines currently under development. In considering SCG expansion, this chapter will therefore focus on what further use may be made of the SGC infrastructure that is currently under development, rather than the more theoretical issue of which suppliers might seek to access European markets by means of wholly new infrastructure along the general route pioneered by the developers of the current SGC.

The pipelines that are at the heart of the SGC's infrastructure are designed so that they can eventually carry twice the initial volumes intended for Turkey and European countries beyond Turkey. Overall, this means the system will be able to convey some 32–33 bcm as far as western Turkey and some 20–22 bcm onwards to Greece and Italy.[6]

In technical terms there will, of course, have to be further work to ensure such an expansion. This will principally involve additional compressors to push the gas through the line and, in Georgia, where initial deliveries are to be achieved through increased compression alone through a section of the existing South Caucasus Pipeline, a second pipeline will be required for some 240 km.

The core issue in terms of expansion, however, is the availability of gas in a cluster of potential suppliers and the interest of those suppliers in utilising the system to access markets in Turkey and European countries beyond Turkey.

When the European Union began considering diversification of gas imports in the early 2000s, it looked at an arc of countries from the Caspian and Iran to the Gulf and Egypt. Given their proven resources and relative lack of consumption at the time, this led to projections that one day Europe might indeed receive as much as 100 bcm/y from these countries.

In 2016, despite the protestations of the official website cited earlier, EU ambitions are more modest. Perhaps there is a greater degree of understanding about the fact that public funding is limited and that it is either private companies—usually international oil companies—or state energy enterprises in producer countries that principally have to foot the bill for both upstream development and the infrastructure required to deliver upstream output to international markets. More realistically, there is an increasing awareness of just how fast gas consumption has grown in many of the countries which the EU once appeared to regard as potential suppliers. Indeed, one of the most striking facts in recent gas history has been that Iran, which possesses the word's largest gas reserves (34 trillion cubic metres—tcm), has commonly been a net importer of gas and is currently only a marginal exporter. Saudi Arabia, with the world's sixth largest reserves (8.3 tcm), has a no-export policy for gas while the United Arab Emirates, which holds the world's seventh largest reserves (6.1 tcm) and which pioneered gas liquefaction, is now a net importer. And while Egypt, with 1.8 tcm in proven reserves, was once a significant provider of liquefied natural gas (LNG) to western Europe, notably France, its move to prioritise domestic supplies over the maintenance of export commitments resulted in BG, the prime developer and operator of the Idku LNG export facility, declaring force majeure in January 2014 and suspending operations.

The result is that although there are a wide variety of potential additional sources that could provide input to the SGC, it is remarkably difficult to say which of them will overcome their own development problems, some of which also involve transit issues, so that they can utilise the new SGC infrastructure.

14.2 Potential Suppliers: Azerbaijan

14.2.1 Azerbaijani Options

The corporate developers of the SGC, while open to additional input from other suppliers, are clearly looking primarily at Azerbaijan as the source for most, if not all, of the additional input required if the system is to operate at its full 32/20 bcm/y capacity.

BP Azerbaijan President Gordon Birrell made this clear in an interview conducted around the time of the December 2013 SD2 FID. Birrell said: "We expect this ability of the Southern Corridor to bring new sources of supply to European markets will extend as additional supplies become available. When I say new sources I definitely mean additional supplies that can be anticipated; with several gas opportunities in Azerbaijan including Shah Deniz Deep, Shafag-Asiman, and Azeri-Chirag-Gunashli (ACG) Deep that are being evaluated by BP and its co-venturers".[7]

Azerbaijan has two options for additional supply. The first, on which most attention is currently focussed, concerns the development of what Azerbaijan terms its "Next Wave" of offshore gasfield development, with the fields named by Birrell playing an important role, along with a cluster of fields being developed by other companies. The second, which is dependent on factors outside Azerbaijan's control, concerns the possible redirection of Shah Deniz Phase 1 output away from Turkey and towards other European customers beyond Turkey.

14.3 The "Next Wave" of Azerbaijani Offshore Gas

14.3.1 General Potential

Azerbaijan possesses great potential in terms of additional offshore gas development, but harnessing this potential will not be easy. Furthermore, Azerbaijan has a somewhat delicate gas balance, so SOCAR will have to pay considerable attention to the need to provide gas for the domestic market as well as for export. This poses a political as well as a practical problem since almost all of the 10 bcm/y currently being produced under the at Shah Deniz Phase 1 programme and the entirety of the planned 16 bcm/y gas to be produced by SD2 is bound for export under contracts whose abrogation seems unimaginable. Yet Azerbaijan is not only running short of gas for domestic consumption—so much so that in mid-2016 it was engaged in negotiations with Gazprom to import as much as 3 bcm/y of Russian gas—but it has ambitious plans for gas to provide much of the feedstock for a planned \$7 bn Oil and Gas Petrochemical Complex.

SOCAR's hopes for Next Wave development, as shown in Fig. 14.1, largely rest on two main sets of project: development of the Absheron

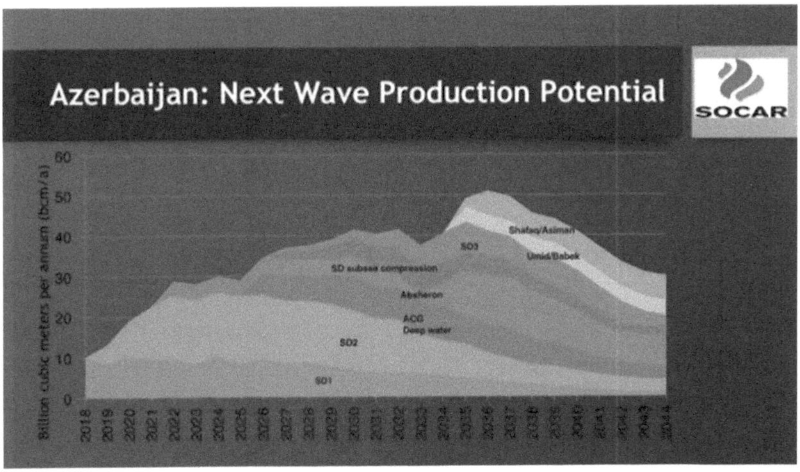

Fig. 14.1 SOCAR's hopes for the next wave of Azerbaijani production. *Source* SOCAR

field and the exploitation of deep level gas resources under the Azeri–Chirag–Guneshli (ACG) oilfield complex and a third phase at the existing Shah Deniz gasfield (SD3). Initially, the hope is that Absheron can be brought on line early, along with extra subsea compression to boost output at Shah Deniz. Such developments would result in Azerbaijan raising offshore production from around 26 bcm in 2021, when the SD2 project should be fully operational, to around 29–30 bcm a year or two later and ensuring a production plateau of around 35 bcm/y from 2026 onwards.

However, actual development of the Next Wave of Azerbaijani fields will be dependent on a variety of factors, notably the commercial environment within which corporate developers will be expected to make their initial investments, the technical complexity of dealing with each field (some projects involve gas that is both extremely deep and under extremely high pressure) and the availability of rigs to carry out all forms of drilling, from exploration and appraisal to actual production. Uncertainty concerning all these factors raises considerable doubts concerning the timing of individual Next Wave contributions to Azerbaijani input into the SGC, throwing both the extent and the timing of Azerbaijani input into an expanded SGC in doubt.

The core elements of the Next Wave are as follows. The reserve estimates in each field come from SOCAR presentations and, although considered to be reasonable, are based on seismic studies and have not necessarily been backed up by actual drilling and therefore cannot be considered as estimates of proven reserves. The estimates for potential output and for production start-up are the author's.

14.3.2 Absheron

This is the most important prospect, since its development is currently under active development. SOCAR originally estimated its prospective reserves at 350 bcm but after the operator, France's Total, drilled the first successful exploration well in 2012, estimates were raised to around 500 bcm. A second well, currently being drilled in a different part of the Absheron structure, may well result in a further substantial reserve increase. However, the initial focus is firmly on Azerbaijan's domestic market and there is no indication as to when, or even whether, a second phase might be initiated. As of mid-2017, the plan is for gas output from the field to be purchased by Azerbaijan while condensate production will be exported to international markets via the Baku-Tbilisi-Ceyhan (BTC) oil pipeline. In August 2016, SOCAR First President Khoshbakht Yusifzade said: "An initial production well is planned for the third quarter of next year. First gas is planned for the third or fourth quarters 2019." By June 2017, Total was preparing to issue a tender for the laying of subsea pipelines and communication cables.

If there is a subsequent phase, then it will be export-oriented. The Azerbaijani press reported on 6 September 2016 that the Azerbaijani government had taken up with the TAP consortium an approach from Total for access to the line – with TAP stating in its reply that the line was, indeed, designed to carry 20 bcm/y, of which only 10 bcm/y was currently booked. Commercial sources consider that Total requires European gas prices to be significantly higher than they were in 2016 in order to justify any export-oriented development. Moreover, as SOCAR has acknowledged, there are issues concerning the availability of rigs to carry out field development. So long as international gas prices remain relatively low, it seems reasonable to expect that if Absheron is eventually developed as an export-oriented project,then first gas would probably not be available for input into the SGC before 2025 or thereabouts.

14.3.3 BP's Deep Level Prospects

BP has proven the existence of substantial deep level gas reserves at both ACG and Shah Deniz, with SOCAR saying that ACG possesses some 300 bcm and deep level Shah Deniz some 500 bcm. BP has long been negotiating a production-sharing agreement (PSA) for what is commonly called Deep Level ACG (a special purpose PSA is required as the existing ACG oilfield PSA signed in 1994 does not cover resources found at the depth of the deep level gas reserves). In August 2016, SOCAR officials said that these negotiations—which had originally been expected to yield an agreement in mid-2014—were almost completed but that there were still some commercial aspects that had to be agreed.

If all goes well, BP should secure an ACG deep level PSA in 2017 or 2018. However, it will take time to develop this resource. On 1 June 2017, Energy Minister Natiq Aliev (in perhaps his last speech before his death from a heart attack a week later), said: "In case an arrangement with BP is reached, work could be started quickly," whilst adding that "in 2026, deep gas output may be started at the Azeri-Chirag-Guneshli block." The noted Azerbaijani energy analyst Gulmira Rzayeva envisages production of around 4–5 bcm/y starting in 2027–2028, a target that appears to be genuinely attainable.[8]

As for SD3, BP has publicly spoken of its ability to undertake such a project but there is no indication that it is currently a priority. Shah Deniz, according to operator BP, contains 1.2 tcm in proven reserves while Azerbaijani officials have spoken of it containing 2 tcm. In April 2016, Energy Minister Natiq Aliev declared: "After Shah Deniz-2 there will be the third stage of development of this mega-field".[9] Although SOCAR would clearly be happy to see SD3 come on stream in or around 2026, current commercial conditions make it unlikely it will start up before 2030.

14.3.4 The SOCAR Fields and Other BP Prospects

Azerbaijan has several other fields which, in time, will likely be developed with a view to exporting their output. Thus, SOCAR's Yusifzade has declared that fields such as Babek, Mashal, Asiman, Sharg, Nakhichevan, and Zafar, as well as Absheron and Umid, "enable us to increase our production".

One of these fields, Umid, is already in production, with SOCAR itself as an operator. But while the field has so far produced around 1 bcm, its development has been problematic and although SOCAR had spoken of its hopes that it would be able to sign contracts with foreign partners by late 2016 for the development of both Umid and the nearby Babek gasfield, as of late December 2016 there had been no such development. These are expected to be risk service contracts rather than full PSAs. SOCAR considers Umid to possess some 200 bcm and Babek 400 bcm. Rzayeva considers that Umid/Babek might come on line around 2026–27 and could produce around 5–7 bcm. This seems reasonable, so long as firm contracts for their development are signed in the near future.

At Shafag-Asiman, for which BP has a PSA, BP Azerbaijan plans to launch exploration drilling operations in 2019. SOCAR considers Shafag-Asiman to contain some 500 bcm, but until the results of exploration drilling are known, this cannot be confirmed. The most important element here is that BP is proceeding with an initial programme. When, or even whether, the field will be developed will depend on the outcome of the exploration drilling campaign.

Overall, although SOCAR considers that Azerbaijan possesses as much as 2.7 tcm in offshore gas resources still to be developed—in other words, excluding SD1 and SD2—the timing for this development remains highly uncertain. This is frankly acknowledged by Yusifzade who said in April 2016 that while the fields noted above (together with Nakhichevan, with 300 bcm in assessed resources) "contain great gas volumes" also acknowledged that "we lack floating drilling devices"—his term for drilling rigs—to ensure their development.[10]

If Total can deliver the rapid development of Absheron, then Azerbaijan would get a kick start in terms of managing to utilise the SGC's expansion capacity. But to this author, at least, it looks as if Azerbaijan's Next Wave of offshore gas development may not truly get into gear until the latter half of the 2020s.

14.3.5 Azerbaijan—Redirection

Azerbaijan currently exports around 6.5 bcm/y of gas of to Turkey from first-phase production at Shah Deniz (SD1). Deliveries were begun in 2007, but the contract is due to end in 2021. The general expectation is that the contract will be prolonged and that no further infrastructure adjustments will be required to ensure continued delivery. However, if Turkey should secure significant volumes of gas from other sources, so that it did not automatically need to maintain all or part of the SD1 deliveries, then SD1 gas could be added into SGC flows to the European Union. All that would be required would be, in effect, for gas currently delivered into the BOTAŞ system at the Georgian–Turkish border to be inserted instead into TANAP.

Such a development really depends on two factors. The first is whether other producers whose output might usefully be consumed within Turkey will indeed manage to get their production and export plans up and running by 2021. This issue essentially relates to output from northern Iraq and the Eastern Mediterranean and is addressed in the specific sections of this chapter devoted to these supply sources.

The second is whether Azerbaijan will feel sufficiently confident in 2021 that it has the supply issue for its own domestic market well in hand. If this is the case, then SD1 output can primarily be channelled to European markets; if it is not, then it is quite possible that a substantial proportion of current SD1 exports will be redirected for internal use.

14.4 Potential Suppliers: Turkmenistan and Kazakhstan

14.4.1 Turkmenistan's Options

Turkmenistan is an obvious potential source for long-term SGC input. It clearly possesses enough gas and has the advantage that its own offshore field infrastructure extends to within 100 km of Azerbaijan's offshore field infrastructure. Onshore, the discovery of Galkynysh, the world's largest onshore gas field, has raised standard estimates of its proven reserves to 17.5 tcm. Yet its production is only around 70 bcm/y, and its exports are barely 40 bcm/y. For its part, Azerbaijan has stressed it is fully prepared to be a reliable partner in the transport of other countries' natural gas to markets in or beyond Turkey. Speaking with particular regard to Turkmenistan, but in a remark that is equally applicable to Kazakhstan, SOCAR first Vice-President Khoshbakht Yusifzade told the author in late August 2016: "As soon as they have the inspiration and desire to export, we are ready to transit".[11]

Turkmenistan's problem is that both Russia and Iran, which used to import around 40 bcm/y and 7–8 bcm/y of Turkmen gas respectively, no longer import any gas from Turkmenistan. In the first case, Gazprom has simply decided it no longer needs to import Turkmen gas; in the second, Turkmenistan itself decided at the end of 2016 that it was no longer prepared to export gas to Iran, which was largely arranged on a barter basis, so long as Iran remained unwilling to settle a $1.8 bn bill which Ashgabat said was owed for previous gas deliveries. This means Turkmenistan is now solely dependent on China for its gas export revenues (see Table Two) and, since these are insufficient to cover government expenditures, it is also dependent on Chinese loans. A third factor is that the difficulty of doing business in Turkmenistan has resulted in China being the only significant source of external investment at onshore gasfields.

Offshore the situation is significantly better because foreign companies have been allowed production-sharing agreements, whereas onshore only the Chinese have been permitted to secure a PSA in the last twenty years or so. This means there are two very different approaches that should be considered concerning potential Turkmen exports to Europe. The first, on which the EU was engaged for the best part of a decade, concerns the transport of 30 bcm/y or more of gas from Turkmenistan's giant onshore fields by means of what would have to be a wholly new pipeline system. In the last few years, however, EU officials have paid at least equal attention to the import of much smaller volumes essentially based on collecting gas from Turkmenistan's offshore fields, and perhaps from some adjoining onshore fields, and transporting something in the order of 10 bcm/y across the Caspian to Azerbaijan for onward shipment to Europe via infrastructure currently being developed as part of the SD2/SGC programmes.

In an era of relatively low gas prices, this is clearly a sensible approach to pursue. Moreover, Azerbaijan's own domestic gas shortage has led to a discussion between Turkmen and Azerbaijani authorities on potential connections between the two countries' offshore gas infrastructure.

Since the spring of 2014, SOCAR President Rovnag Abdullayev has visited Turkmenistan on at least four occasions and high-level talks have also been held between the two countries in Baku. Commercial sources have told the author that some progress has been made in talks to date. This includes a connection from Turkmenistan to the gas-gathering facilities of Azerbaijan's giant Azeri–Chirag–Guneshli offshore oilfield complex, which extends to within 100 km of the offshore infrastructure established by Malaysia's Petronas Carigali to develop gas at its Block One concession in the Turkmen sector of the Caspian Sea.

But although there are strong commercial reasons as to why an agreement for a link-up able to carry around 10 bcm/y should be implemented – not least because Petronas is also a stakeholder in Azerbaijan's giant Shah Deniz project – no agreement has yet been concluded. One outstanding issue appears to be Turkmen insistence that raw gas produced in Turkmen waters should first be brought back onshore for processing at Turkmenbashi before crossing the Caspian again to connect with facilities on the Azerbaijani side. Petronas would clearly like gain from a pipeline to Azerbaijan, since this would enable it to secure commercial prices for its gas output, whereas at present it has to sell the bulk of its current 5 bcm/y output to the Turkmen authorities at extremely low prices, thought to be around $100 per thousand cubic metres, with this gas subsequently exported by the Turkmen authorities to Iran. Petronas is understood to be in a position to raise production to 8–10 bcm/y within a couple of years of being assured of an outlet.

In conclusion, Turkmenistan remains a real prospect for additional supplies into the SGC system currently being developed, not least because it can provide the kind of modest element that would serve to improve the short-term commerciality of such pipelines as TANAP and the SGC without completely filling them up so that they cannot then take the "next wave" of Azerbaijani production in or around the mid-2020s. However, one massive diplomatic obstacle has to be overcome. Turkmenistan, Azerbaijan and Kazakhstan all share the view that pipelines across the Caspian only need the approval of the countries directly concerned and who share maritime boundaries. The other two Caspian states, Russia and Iran, both argue that all five littoral states have to approve Trans-Caspian infrastructure projects. Turkmenistan and Azerbaijan will, therefore, have to persuade Russia that development of a pipeline, particularly a relatively modest pipeline, will not damage Russia's interests. The EU, for its part, will have to stress to Moscow that any action on Russia's part to block the pipeline will in itself be taken as proof that Russia cannot accept reasonable competition and that that, in turn, makes Russia itself an unreliable supplier.

The failure to reach a common agreement on the status of the Caspian is accompanied by a specific failure so far to agree on a common maritime boundary between Turkmenistan and Azerbaijan. At present, this dispute essentially concerns the field known to the Turkmens as Serdar and to the Azerbaijanis as Kyapaz. But this field is thought to contain relatively little hydrocarbons, and since the bulk of it would almost certainly fall on the Turkmen side of any boundary agreed on the basis of median lines, an Azerbaijani suggestion that Baku would be satisfied if any output from Serdar/Kyapaz might be undertaken by Turkmenistan and processed in Azerbaijan might yet offer a solution.

Azerbaijan itself has stated repeatedly that although it is a major producer in its own right, it is also a responsible partner in energy transit, with oil from Turkmenistan and Kazakhstan accounting for close to a quarter of the oil shipped through the Baku–Tbilisi–Ceyhan oil pipeline. In the gas context, the country's deputy minister of energy, Natiq Abbasov, has said: "Azerbaijan should benefit from real opportunity that energy resources should bring to many countries and play a role as crucial bridge between Europe and Asia". He added: "Azerbaijan plays a role as door or gate from Asia to Europe. We know Asian countries will be going to export their resources. Azerbaijan is willing to provide all the facilities for transit to Europe".[12]

Abbasov was speaking after he had delivered a keynote address to an energy conference in which he said: "As an energy security project, the SGC will bring benefit to all of us—producers, transit countries and consumers—for years to come".

14.4.2 Kazakhstan's Priorities

In the very distant future, it is possible to envisage a revival of Kazakh interest in shipping gas across the Caspian. But right now its priority in terms of Caspian field development is to finally ensure sustained development of the much-troubled super-giant Kashagan oil field, on which the consortium developing the field has already spent more than \$55 bn, and which finally entered regular service in October 2016. Kashagan does have a large natural gas cap but almost certainly the immediate priority for Kazakhstan will be to see whether it can harness this for domestic use. Moreover, inasmuch as Kazakhstan is currently looking to develop gas exports, its focus is on China, to which it is connected by means of the Trans-Asian Gas Pipeline system developed by the China National Petroleum Corporation.

14.5 Potential Suppliers: Iran

If only because it possesses the world's largest proven gas reserves, Iran is bound to be considered a prospective supplier of gas to Europe. Indeed, in the early 2000s, the International Energy Agency was seriously contemplating that one day Iran might supply as much as half of all Europe's additional gas imports. This led to Iran being considered a major prospective supplier for the first iterations of the Southern Gas Corridor, when its implementation essentially concerned the development of the Nabucco pipeline proposal. But in 2006, as a result of concerns about Iran's nuclear programme and whether it might have a military component, Nabucco turned away from Iran, and despite some contacts in 2008 with the original developers of the Trans-Adriatic Pipeline (TAP), Iran has not been considered a likely source for SGC input since then.

There are two main reasons for this. The first is the international sanctions imposed on Iran to guard against moves to develop nuclear weapons; the second, in a post-sanctions environment, is Iran's own relative lack of interest in expanding its current gas exports to Turkey or utilising Turkey for transit to secure a niche in European markets beyond Turkey. Moreover, there is a paradox at the heart of current Iranian gas development. The massive multi-stage development of South Pars, the Iranian section of the world's largest offshore gas field, is being completed at an unprecedented rate. In theory, this should mean that something like 105 bcm of raw gas will come on stream between 2014 and 2018, of which perhaps 75 bcm/y would be available for export. However, it is far from clear that Iran will also possess the processing facilities required to enable consumers to use the gas. Moreover, with the notable exceptions of local pipelines being developed to supply gas to Iraq and Oman, Tehran is showing very little interest in additional pipeline exports.

It is against this background that one should assess an assertion made by the Managing Director of the National Iranian Gas Company (NIGC), Hamid Reza Araqi, in October 2015 that "Iran would be able to export 200 mcm/d of gas in four years".[13]

Iran's prime focus is on oil output and oil exports. It is looking to boost gas injection to maintain production levels. It is also looking to use increasing volumes of gas as a substitute for oil in onshore development, notably in automotive transport and to use gas as a feedstock for industry, for power generation and possibly for indirect export in the form of electricity.

Inasmuch as it is focussing on pipeline exports, Iran's prime concerns are the completion of lines to supply Iraq, which is both a commercial and political partner. It has one contract to supply the Baghdad area with 25 mcm/d and, as of mid-2016, was negotiating to supply the southern city of Basra with a further 35 mcm/d. The two contracts cover deliveries of almost 22 bcm/y. Iran is also negotiating to build a 10 bcm/y pipeline to Oman, since it would then be able both to provide some gas for domestic use in the country and some gas for input into Oman's gas liquefaction facility at Sur, and subsequent export as LNG. However, Iranian officials told the author in Tehran in October 2016 that opposition from Saudi Arabia meant that the Oman project, for which onshore planning had already been completed and for which offshore planning was under way, faced a significant problem. The Saudis, the officials said, were putting pressure on the UAE to deny the use of UAE waters for the pipeline. This would mean that instead of laying a line through relatively shallow UAE waters, in order to avoid the UAE, Iran and Oman would have to lay a line across a relatively deep trench, a technology that would almost certainly require the use of specialist foreign pipe-laying expertise and which would increase costs considerably.

A plan to export 22 mcm/d to Pakistan has yet to be implemented since the all-important Pakistani section of the projected Iran–Pakistan–India (IPI) pipeline has yet to be built and it is not clear when, or whether, it will be built.

In the medium to long term, Iran is focussing its gas export efforts on the development of its own LNG export facilities. It should be noted, however, that Iranian efforts to team up with international partners to develop LNG facilities in the early 2000s failed because Iran did not appear to be able to conclude terms with various foreign partners, particularly concerning access to LNG technology.

On various occasions, Turkey has approached Iran for possible increases in supplies to Turkey. These have yet to yield any concrete result and, indeed, when Turkey was engaged in a major search for alternative supplies to Russian gas at a time of high tension between Ankara and Moscow in the wake of the shooting down of a Russian warplane by Turkey over the Turkish–Syrian borderlands on 24 November 2015, Ankara came up blank with regard to anything extra coming from Tehran. From time to time, there have been reports of private Turkish efforts to import vast volumes of Iranian gas, either for use in Turkey or for transiting to Europe, but these have come to nothing.

In the summer of 2016, the possibility of Iran supplying gas into the SGC and specifically into the TANAP pipeline was discussed in tripartitite Azerbaijani, Iranian and Russian talks in Baku. Of rather greater significance, since it stems from corporate rather than governmental interests, were the comments of a senior BP official in Tehran and the reaction from an Iranian gas official in October 2016.

Outlining BP's role in regional energy developments at a major energy conference in Tehran, BP's Dr Jonathan Evans said: "We are involved in developing Azeri-Chirag-Guneshli (oilfield in Azerbaijan), SD2 and the Southern Gas Corridor. We hope that one day Iranian gas could flow through that pipeline to customers in Europe".[14]

When Dr Evans' comment was relayed to Dr Behzad Babazadeh, Director for International Relations at the National Iranian Gas Company (NIGC), the reaction was both positive and practical. Mr Babazadeh immediately asked whether TANAP would be ready to build the necessary connection from the Iranian–Turkey border to the line currently under construction—in effect, a 190-km link from Dogubayezit on the border to Horasan, near Erzurum. Since TANAP shareholders have talked in the past of the possibility of securing gas from Iran—and also from Iraq—it can be safely assumed that should Iran or Iraq be able to secure the necessary sale-and-purchase agreements, TANAP would be ready to supply the requisite infrastructure on Turkish territory.[15]

Mr Babazadeh said that Iran was ready to negotiate for gas sales to the European Union so long as a pipeline connection to Iran could be put in place. But he did indicate that initial supplies would have to be limited as Iran would have to improve its own infrastructure for carrying gas from the processing centre at Assaluyeh to the Iranian–Turkish border. In order to deliver major volumes to Turkey and customers in the European Union, Iran would have to build IGAT-9, a planned 1,850-km, 30–33 bcm/y pipeline from Assaluyeh to the Turkish border, a project currently costed at around $6 bn.

Meanwhile, Babazadeh said that current capacity for delivering supplies to Turkey was limited. "We can increase gas supply to Turkey by 25%, but only in summer, when the domestic consumption is low, but we should construct a new pipeline for any significant amount of gas delivery", he said.[16]

In the light of Iranian concerns that developing an export pipeline to Oman may prove more complicated than expected, a revived Iranian interest in TANAP could yet prove significant. But, as with other prospective input into the SGC, it would need to be underpinned by actual sale-and-purchase agreements. As of late 2016, perhaps the most positive thing that can be said with regard to possible Iranian input into the SGC is that, if a consortium of European gas buyers were to come up with a realistic offer for around 10 bcm of Iranian gas to be supplied via the SGC, and if such gas could be delivered on commercial terms, then this would clearly be of benefit to Tehran in terms of its ambitions to improve its diplomatic as well as its commercial relations with the European Union. But in current market conditions, such an approach seems a very long way off indeed.

14.6 Potential Suppliers: Northern Iraq

In early 2016, the Kurdistan Region of Iraq looked likely to prove the most probable source of new gas in the context of medium-term SGC development—but with an impact that would be essentially indirect. This was because the development of both the region's principal gasfields and the associated export infrastructure was specifically linked to the Turkish market. In principle, the addition of potentially 10 to 20 bcm/y of new supplies to Turkey over the next seven to ten years, as envisaged by both corporate and government leaders in Iraqi Kurdistan, would outstrip any likely increase in Turkish gas consumption, thus freeing up capacity in the SGC for alternatives inputs such as SD1. However, prospects for gas from the Kurdistan Region of Iraq (KRI) utilising the SGC for deliveries to European countries beyond Turkey, although discussed and technically possible, will really depend on whether there is a major upsurge in European gas prices which would serve as an incentive to bolster both gas production and exports over and above current development plans.

Northern Iraq certainly possesses the resource base to supply gas to Turkey, and, in time, to the SGC. The case for this rests on gas that can be made available from specific fields. Although the KRI's Ministry of Natural Resources considers that the region possesses some 165 tcf (about 4.5 tcm) of gas in place, of which some 38 tcf (just over 1 tcm) is recoverable, these estimates are not recognised by the Federal Iraqi Ministry of Oil and it is not clear how they were formulated.

The Government of Turkey has been formally committed to importing gas from northern Iraq since November 2013, when it signed a general sales agreement with the Kurdistan Regional Government which envisaged exports starting in 2017, reaching 4 bcm in 2018, 10 bcm in 2020 and with an option to move to 20 bcm/y thereafter. In November 2015, the KRG Minister of Natural Resources, Ashti Hawrami, and Genel Energy Chairman, Tony Hayward, declared that Iraq Kurdistan would be in a position to deliver 10 bcm to Turkey in 2020 and 20 bcm/y to Turkey in the early 2020s.

In terms of providing gas for Turkey and/or the SGC, the key issue concerns the planned $5.4 bn development of the Miran and Bina Bawi gasfields by the Anglo-Turkish Genel Energy. As and when these fields are connected to Turkey, the possibility would exist for other significant fields, such as Khor Mor and Chemchemal, to be connected as well. Genel estimates that Miran and Bina Bawi between them contain some 11.4 tcf of gas in place, and that this should deliver 8.4 tcf of gas available for commercial sale. Danagas, which has a major stake in operations at Khor Mor and Chemchemal, considers that these two fields contain some 50 tcf (about 1.4 tcm) of gas in place.

In the first half of 2016, two major US companies were working on development plans. Fluor was preparing a draft outline (technically the pre-Front End Engineering Design) for the development of Miran and Bina Bawi, while Baker Hughes was due to deliver a full development plan for the fields around the end of 2016. As of early 2016, Genel Energy was targeting an initial production capacity of 10 bcm/y by the end of 2019. It estimated the cost of initial development of the two fields at $3.5 bn, with $1 bn required for Genel itself to start raw gas production and $2.5 bn for a proposed midstream company to develop the processing facilities required to convert the raw gas into actual sales gas. Additional lifetime costs were put at $1.9 bn. By the end of 2016, however, the status of Genel's plans was far from clear. Genel itself was experiencing financial difficulties, and there was no indication that it was close to bringing in partners to constitute the midstream company.

Nor was it clear whether the Turkish state pipeline company, BOTAŞ, would proceed with plans to build 185-km, 20 bcm/y capacity pipeline from Silopi on the Iraqi–Turkish border to a connection with the Turkish grid at Mardin that would constitute the main component of any system designed to carry gas from Iraqi Kurdistan to Turkish markets. On 24 April 2016, after turning down the responses to an initial tender, BOTAŞ declared that a revised tender would be issued on 26 April. But since then, there has been no word on its outcome.

In terms of a long-term connection to the SGC, it is perhaps worth noting that virtually the last act of the ill-fated Nabucco group was the completion in April 2012 of an environmental impact assessment for the planned 733-km pipeline from the Turkish–Iraqi border at Silopi to the central Turkish city of Sivas. The line was intended to act as a feeder for the main Nabucco pipeline, which until 2011 had hoped to win the competition to carry Azerbaijani gas to European markets. The TANAP line largely follows the route initially plotted by Nabucco (indeed quite a few Nabucco personnel joined the TANAP team) and should Genel Energy decide that it was in a position to supply gas to the SGC, then the route to reach the TANAP line at Sivas has already been mapped out.

Although the strong Turkish involvement in the development of these planned gas sales argues in favour of gas from northern Iraq making an early entry into the Turkish market and thus impacting on SGC balances, regional volatility means that major security issues will have to be addressed. These issues, which appear to be the prime reason for the apparent lack of progress in developing gas in Iraqi Kurdistan in 2016 and during the first half of 2017, are considered below in the section on internal security in Turkey.

14.7 Potential Suppliers: The Eastern Mediterranean

In theory, the Eastern Mediterranean offers some potential for contributing gas into the SGC. In practice, however, regional political problems, notably the wars in Syria and the Cyprus problem, militate against such a contribution. Moreover, even if Eastern Mediterranean gas should find a home in Turkey, which is almost certainly its most attractive commercial market, that gas would essentially serve to help meet increases in Turkish demand and to serve as a substitute for alternative supplies to Turkey, and only after fulfilling those functions might it play a role in terms of direct input into the SGC's pipeline system.

There is plenty of gas in the East Mediterranean to ensure a revival of the region's gas export prospects. Egypt's proven reserves, which stood at 1.8 tcm at the end of 2014, were bolstered by the discovery of the giant Zohr field, which contains an estimated 850 bcm, in mid-2015. Israel, with its Leviathan and Tamar fields, possesses a further 860 bcm, and Cyprus has a modest discovery at Aphrodite with 145 bcm in proven reserves.

Until 2012, Egypt was a significant regional exporter. But gas flows through the Arab Gas Pipeline (AGP) to Syria and Lebanon came to an end with the eruption of civil war in Syria in 2011. This also puts an end—at least for the foreseeable future—to plans to extend the AGP from Aleppo in northern Syria up into Turkey.

Militant unrest in Egypt itself also forced the end of exports to Egypt's biggest pipeline customer, Israel in 2012 when a series of attacks on infrastructure connecting the two countries also led to interruptions in deliveries to Jordan. Moreover, as Egypt dealt with the consequences of the Arab Spring and the natural wish of government, almost regardless of its political orientation, to avoid antagonising the population unnecessarily, massive subsidies on domestic energy use prompted a collapse in the amount of gas available for export in the form of LNG produced at the Idku and Damietta LNG liquefaction plants. Thus when Zohr was discovered by Italy's ENI, the natural inclination was to seek to revive LNG exports, although some of Zohr's gas will also be used to serve the domestic market.

The revival of Idku and Damietta has the potential to serve as a magnet to attract gas from both Israel and Cyprus. The Government of Cyprus has already agreed to this in principle and, given that Zohr and the planned infrastructure that will connect it to the Egyptian mainland, starts just across the boundary line separating Cyprus's exclusive economic zone from that of Egypt, this would seem to represent the most cost-effective way of monetising the Aphrodite field.

As for Israel, a long-drawn-out political dispute over how much of its newly discovered offshore gas resources should be reserved for domestic use on strategic grounds effectively resulted in Israel missing the boat in terms of developing its own export-oriented project before gas prices started their decline in 2014. Egypt remains an export option, and Israeli and Egyptian energy officials have continued to discuss potential Israeli input into Egypt's LNG plants. However, these talks do not look particularly promising as there are continuing tensions between the two countries, not least as a result of the failure of previous Egyptian authorities to cope with anti-regime attacks on the gas pipeline through which Egypt supplied Israel until 2012.

In an ideal world, both the Israeli Government and the companies developing the Leviathan and Tamar fields would like to see much of their output exported to Turkey, not least since Turkey's southern coast is in an area where gas demand is expected to grow rapidly as a result of both industrialisation and development of tourism. In December 2015, Turkish Energy Minister, Berat Albayrak (who is also the son-in-law of Turkish President Recep Tayyip Erdoğan), said that "diplomatic relations between Turkey and Israel had to be normalized in order to transport natural gas from the Leviathan field to Europe through Turkey".[17] Relations between the two countries were restored in June 2016. In September 2016, a senior Israeli official confirmed to the author that Turkey remained the preferred destination for Israeli gas exports.

However, there is a serious question as to whether Israeli gas can be piped to Turkey without a settlement of the Cyprus problem.

In geographical terms, the alternative routes for a pipeline from the Israeli fields to landfall in Turkey are as follows:

- It can go up the coast through Lebanon and Syria towards Turkey's industrial port of Iskenderun. This is not possible because of civil war in Syria and lack of a peace treaty between Israel and Lebanon.
- It can take an inshore route just off the Lebanese and Syrian coasts. But the same issues that block an onshore pipeline also render an inshore line unfeasible.
- It can take a route through Cypriot waters, either to the east or to the west of the island of Cyprus, or even straight across the island.

In practice, the only feasible pipeline routes require the line to traverse the Cypriot Exclusive Economic Zone in order to reach Turkey, passing either to the east of Cyprus or traversing both the Cypriot EEZ and the island of Cyprus itself. The route to the west of the island is both unnecessarily long and also fraught with political problems in view of potentially conflicting claims concerning the extent of the respective exclusive economic zones of Turkey, Greece and Cyprus.

There are some extremely serious efforts to develop a system passing to the east of the island, not least by two business groups in Turkey, the Zorlu Group, which is currently operating three gas-fired power stations in Israel, and the Turcas group of companies. Both are looking to secure gas from Leviathan for the Turkish market (and, in Zorlu's case, also for its Israeli plants).

Turcas officials have said that such a line cannot pass through Cypriot waters or cross Cyprus unless a resolution is in sight to the 43-year-old partition of Cyprus between its Greek and Turkish communities. Whether resolution of the Cypriot issue, or at least significant progress in resolving the issue, is an absolute *sine qua non* for an Israel–Turkey pipeline remains a matter of controversy. So long as the initiative for laying a line through Cypriot EEZ waters comes from a private company or consortium—and both Zorlu and Turcas are privately held groups—it would be able to approach the Cypriot authorities directly concerning Cyprus's views on the environmental impact of a route through the Cypriot EEZ. This is important for two reasons: firstly, it might prove difficult for any government in Cyprus persistently to block the development of such a line, since it can only do so by arguing that a particular route is not appropriate on environmental grounds, which means that eventually it should prove possible to find a reasonable route. Secondly, if an approach comes from a private group this averts the problem that the Turkish Government does not recognise the Cypriot Government, and therefore declines to talk to it, while the Cypriot Government remains averse to talking to the Turkish Government so long as the role of Turkey and the presence of Turkish troops in northern Cyprus remain key factors to be resolved in negotiations aimed at ending the partition of the island.

As of late June 2017, Cypriot President Nicos Anastasiades and Mustapha Akinci, the President of the self-proclaimed Turkish Republic of Northern Cyprus, were engaged in a major round of UN-sponsored talks in Switzerland aimed at resolving the Cyprus problem. If there should be a peace settlement, then the way would almost certainly be open for Israeli gas to flow to Turkey by pipeline. But this is not something one can automatically count on. There is also a further element to be considered. Zohr was discovered in a geological formation that appears to span the EEZ boundary line between Egypt and Cyprus. This has raised Cypriot hopes that further gas discoveries will be made on their side of the line.

When Aphrodite was first discovered in 2011, the Cypriot government immediately started to make ambitious plans for the country to develop its own liquefaction plant at Vassilikos which would serve both Israeli and Cypriot gasfields. Nothing came of this at the time, and the subsequent fall in gas prices, coupled with a reduction in the estimated size of Aphrodite's reserves, effectively meant the LNG project had to be put on ice. But if further discoveries were made, and they proved to be on the scale of Zohr, then no doubt the Cypriot government would once again seek to revive the Vassilikos LNG project. Under such conditions, it is reasonable to suppose that its willingness to cooperate in pipeline developments involving Turkey would be substantially reduced. The prime issue would then be whether gas brought ashore to the island would be harnessed for the benefit of both the Greek and Turkish communities or whether it would solely be used to gasify southern Cyprus.

14.8 The Question of Russia

There is a general consensus that Russia's revived plans for developing a TurkStream pipeline would not fundamentally change the volumes of gas that Russia would seek to export to Turkey. But they do raise at least a theoretical possibility that Gazprom might actually seek to utilise a part of the SGC so that some of the gas shipped through TurkStream to Turkey can then be forwarded to Greece and Italy without the need for physical construction of a major new line within the European Union and operating under EU regulations. This was, of course, the issue that eventually led to the collapse of Russia's earlier South Stream project which was to have involved the construction of a major pipeline from the Bulgarian coastline to Baumgarten in Austria and/or Tarvisio in Italy.

The original TurkStream concept was simply an adaptation of the maritime section of South Stream: a set of four lines—known as "strings"—each capable of transporting 15.75 bcm/y from the Russkaya terminal on Russia's Black Sea coast near Anapa to landfall at Kıyıköy, on the Black Sea coast of Turkish Thrace. Most of the 63 bcm/y of gas to be landed at Kıyıköy would then be carried by a 180-km onshore connection to the Turkish–Greek border where, somehow, it would then enter the European Union.

For a variety of reasons, which almost certainly include low gas prices, limited availability of capital to pay for a four-line system and the vexed question of what to do if prospective European customers declined to pay for a new line from the Turkish–Greek border onwards, Gazprom concluded in the autumn of 2015 that a two-line system would be more appropriate. In the revived TurkStream era, which follows the reconciliation between Moscow and Ankara in the summer of 2016, it is a two-line system on which Russia's sights currently appear to be set. This makes sense since the first line would essentially serve to replace gas delivered to Turkey via the Western pipeline system across Ukraine and the Balkans, which Russia hopes to cease using when its current transit contract with Ukraine comes to an end on 31 December 2019.

Under this scenario, the first string would then essentially be used to substitute for gas delivered to Turkey via the Balkans line. This amounted to just 11.4 bcm in 2015 but more usually has run at around 14 bcm/y. The second line would then partly be used to help cover expected the expected increase in Turkish gas demand while some volumes from both lines would also be used to service existing customers in Bulgaria and Greece, and, if the relevant interconnectors were available, other Balkan customers.

For Gazprom, there is one key advantage in building a two-string system and one key drawback. The advantage is that the physical pipe required for laying a two-string system across the Black Sea has both been purchased and delivered to the Bulgarian port of Varna, since it was originally ordered for South Stream's offshore section. TurkStream is expected to follow the South Stream routeing for around four-fifths of its maritime routeing. The disadvantage is that a two-string system will be capable of carrying 31.5 bcm/y, and that, even allowing for additional deliveries to Turkey and for supplies to customers in south-eastern Europe, it may well prove unable to find any local market for perhaps 8–12 bcm/y of this capacity.

Yet there is one way in which such volumes could reach European customers without the need for developing any additional infrastructure: the Trans-Adriatic Pipeline (TAP). This is because TAP is being built so that it can, with a suitable advance warning, carry at least 20 bcm/y. Moreover, although the European Commission has granted TAP's developers exemption from third-party access for the initial 10 bcm/y to be shipped through the system, there is no such exemption for the 10 bcm/y of expanded capacity. Under EU third-party access rules, Gazprom would be perfectly entitled to ask for an open season at TAP and, since it would almost certainly be the only body actually competing for space on the line, it would gain access. It would have to give proper notice, probably a couple of years in advance, to enable the relevant compression to be put on the line but this should not prove a problem. That is because it will probably take the company longer to bring TurkStream on board by laying both the subsea line and the onshore extension and linkup with TAP at Ipsala/Kipoi.

Overall, accessing TAP is a perfectly reasonable option for Gazprom to pursue, since the whole point of the EU's insistence on open access is to ensure that no single supplier monopolises a pipeline. The irony is that it might be Gazprom itself that seeks to secure such access, rather than a company seeking to compete with Gazprom. The first indication that this might be the case came in January 2017 when Gazprom Deputy Chairman Alexander Medvedev, at a major gas conference in Vienna, said: "We have installed available capacity ready to produce more than 100 bcm of gas today, so we don't need any additional investment to produce more than 100 bcm. But in order to bring this gas to Europe we need additional infrastructure which we are working on with our European partners – NordStream 2 and Turkish Stream. This capacity will not be sufficient to bring all this to Europe. So this is why we are talking to use available capacity on the Poseidon project, the studies for which will be ready soon – or maybe TAP." In mentioning Poseidon, Medvedev was indicating that Russia was still considering the concept of shipping gas to southern Italy by way of the long-proposed Interconnector Turkey-Greece-Italy (ITGI) project and its final subsea element, a 210-km line from Greece to Italy known as Poseidon. But while officials from Russian, Italian and Greek energy companies have long discussed developing the ITGI-Poseidon system, so far they do not appear to have secured the necessary financial commitments required for construction of the system, which would be expected to cost around €5 bn to €6 bn. Meanwhile, the TAP system is not only fully financed but under actual construction. Moreover, TAP is bound by EU regulations to offer its services at commercial rates to any third party user who wishes to use its second-stage expanded capacity. Gazprom would have to give TAP some notice of its intentions, in order to enable TAP to add the necessary extra compression to secure the increase in capacity, but the bottom line is absolutely clear: if Gazprom wants to seek access, TAP has to hold an open season, even if Gazprom is the only bidder. In this context, the key element is Gazprom's ability to bring TurkStream on line in time to book space when TAP opens for business in 2020, whereas the earliest alternative bidders for space on the line, notably the Next Wave of Azerbaijani exports, may not be in a position to ship gas through TAP until 2026 or thereabouts. In October 2016, Russia and Turkey signed an intergovernmental agreement on TurkStream which specified that the line would enter service by the end of 2019. In December 2016, a Gazprom subsidiary signed an agreement with the Swiss-Dutch Allseas Company to lay TurkStream's first string in the second half of 2017 and in February 2017 Allseas signed a second contract to lay the second string.

14.9 A Necessary Caveat: Internal Security in Turkey

The physical security of pipelines is an issue usually addressed in the context of long-standing regional conflicts and disputes, notably concerning the supposedly "frozen conflicts" in the Caucasus and the persistent attacks on Iraqi sections of the Kirkuk–Ceyhan oil pipeline in the aftermath of the overthrow of Saddam Hussein in 2003. These issues still demand attention as was demonstrated by the four-day flare-up of the war between Azerbaijani and Armenian forces over Nagorny-Karabagh in April 2016 and, in Iraq and Syria, by the rise of ISIS/Daesh.

Other security issues that could threaten the delivery of gas through—or to—the Southern Gas Corridor include the still delicate situation in Georgia where Russian forces actually control a 1.5-kilometre stretch of the Baku–Supsa oil pipeline and the feared persistence of extreme Islamist militants within Egypt who continue to keep energy exchanges between Israel and Egypt in their sights.

But there is a prospect that in energy terms, and possibly in human terms as well, may prove more worrisome than any of these threats: general insecurity and the spread of war within Turkey. Until the development of the SGC, Turkey's role in energy transit was dominated by oil. Some 2.09 mb/d is routinely carried by tanker through the Turkish Straits while in the first half of 2016 the BTC line, which is capable of carrying as much as 1.2 mb/d, was carrying 735,000 b/d and the Kirkuk–Ceyhan line was carrying 467,300 b/d.

In gas, what currently counts for Turkey is the volume of gas it receives for its own use. Even when the SGC enters service and starts delivering its initial 10 bcm/y to Europe, the volumes that transit Turkey will be dwarfed by the volumes that Turkey itself imports. In 2015, Turkish gas imports totalled 48.4 bcm, with 6.6 bcm coming from Azerbaijan; 7.8 bcm from Iran; 15.6 bcm from Russia via the Blue Stream pipeline across the Black Sea; and 11.4 bcm from Russia via the Western line through Ukraine and the Balkans. Imported LNG, mainly from Algeria and Nigeria, accounted for the rest. In terms of future gas supplies, as part of the SGC-related sales agreements, Turkey will be importing 6.0 bcm/y of additional gas from Azerbaijan delivered via TANAP from the second half of 2018 onwards and 10 bcm/y of gas for onward delivery to Greece, Bulgaria and Italy—and thence to countries further afield—from 2020 onwards.

The pipelines that bring this gas to Turkey or that carry the transit gas through Turkey are all at risk in the event of prolonged conflict within the country. The problem is twofold. Firstly, there is a prospect of intensified war within Turkey; secondly, rebel militants of the Kurdish PKK, an organisation regarded as terrorist both by the Turkish Government and its NATO allies, marked the start of a resumption of open hostilities between the Turkish State and the PKK in July 2015 by launching a series of attacks on oil and gas pipelines within Turkey. These comprised the following:

- A PKK attack on the Iran–Turkey gasline near Agri in eastern Turkey (27 July 2015).
- A PKK attack on the Kirkuk–Ceyhan oil pipeline in Sirnak province of south-eastern Turkey (29 July 2015).
- A PKK attack on a train carrying pipe for TANAP near Sarıkamış in the north-eastern Turkish province of Kars (30 July 2015).
- A PKK attack on the Baku–Tbilisi–Erzurum (BTE) gasline near Sarıkamış (4 August 2015).
- A further PKK attack on the BTE line near Sarıkamış (24 August 2015).

Between 2003 and 2015, there were also repeated attacks by various anti-Baghdad forces on the Iraqi section of the Kirkuk–Ceyhan oil pipeline.

Do these attacks amount to war, or should they be regarded as terrorist incidents that can be contained by Turkey's security forces? The answer is that ever since the truce between the Turkish state and the PKK came to end in July 2015, after two years of largely indirect negotiations, war has once again descended on the south-east part of the country.

The characterisation of this as war is openly acknowledged by the Turkish military. In October 2015, the commander of the Turkish Air Force, Abidin Ünal, said: "Today the Turkish air forces are actually waging a war, more than just a medium-scale war, it is fighting on two fronts".[18] He was referring not only to the classic war that had been going on previously in south-eastern Turkey but also to the persistence of Turkish military attacks on PKK positions in northern Iraq.

It is a war that appears to have claimed close to 2,500 lives and caused countless other casualties. As of late December 2016, the International Crisis Group calculated that since the truce broke down in July 2015, the confirmed death toll comprised 1021 PKK militants, 858 members of the security forces, 383 civilians and 219 youths of unknown affiliation.[19]

The last group, it said, "cannot be positively identified as civilians or members of plainclothes PKK youth militias due to the blurred line between civilian and militant in an urban conflict setting". Casualty reports by the security forces and by pro-PKK sources both indicate their antagonists have suffered much higher casualties. As of mid-2016, the government was claiming to have killed 4,949 PKK fighters and youth wing militants (with almost half said to have died in air strikes in Iraq) in almost a year of fighting while acknowledging the loss of 483 members of the security forces. The PKK claimed to have killed 6,705 members of the Turkish security forces, while admitting the death of 721 of its own fighters and militants over a similar period. Caught in the middle were the civilians, with the International Crisis Group reporting in March 2016 that at least 400 civilians had been killed since hostilities resumed in July 2015, while no less than 350,000 people had lost their homes.[20]

Pipelines are both current and prospective targets, particularly in south-eastern Turkey. Moreover, the lack of clarity concerning one major incident, the closure of the Kirkuk–Ceyhan oil pipeline from 17 February to 11 March 2016 near the town of Idil in Turkey's Sirnak province, raises the possibility that it is not only the PKK that considers the interruption of energy flows to be a legitimate tactic. The PKK had attacked the line previously, most notably in August 2016 when it caused an outage that cost the Kurdistan Regional Government (KRG) some $250 m in export revenues, and the initial assumption was that the PKK had once again targetted the line. Curiously, however, the first reports of an actual explosion on the line only emerged on the 25th February—a week after the line had been closed. At the time, one senior Iraqi Kurdish source said that the Iraqi Kurdish leadership thought the explosion might have been an accident, occurring during possible Turkish mine-clearing operations. Subsequently, however, the KRG became far more worried as one analyst in the USA—Marina Ottaway, who has many decades of experience covering the region—even went so far as to suggest that the closure was actually orchestrated by the Turks themselves to send a message to the KRG concerning its dependence on Turkey for its all-important energy export revenues. "It's not impossible that the Turks are sending a warning to Iraqi Kurdistan [about their desire for independence] saying, 'you can only go so far before we yank your leash'", Ottaway was quoted as saying.[21]

What makes the unrest truly worrisome is that it may yet presage a far more intense civil war. The reason why this needs to be taken seriously is that the Turkish Government's crackdown on critics of its action, intensified in the wake of the abortive military coup of 15 July 2016, is at least partially aimed at the elected parliamentary leadership of Turkey's pro-Kurdish HDP party. President Erdoğan, who was supposed to be filling an essentially honorary role, consistently acted as Turkey's executive president well before he secured a narrow – and disputed – victory in a referendum on 20 April 2017 held to authorize the introduction of an executive and legally partisan presidency. The government's pre-coup move in May 2016 to lift parliamentary immunity from a number of opposition MPs, including leading members of the HDP, can be viewed as a tactic to help secure this goal.

In August 2016, an Istanbul prosecutor called for the HDP leader to be jailed for five years on a charge of disseminating terrorist propaganda, an apparent reference to Demirtaş's calls in 2014 for a dialogue between the Turkish State and the PKK. On 4 November, Demirtas and the other co-leader of the HDP, Figen Yuksekdag, along with at least nine other HDP parliamentarians, were arrested on charges of spreading pro-PKK propaganda. On 10 December, Erdoğan's ruling AK party introduced a bill to change the constitution in a way that would create an executive presidency with full control, inter alia, of drawing up the national budget. All 316 AK members of the 550-seat Grand National Assembly appended their names to the bill.

The danger is that so long as President Erdoğan continues to crack down on the elected representatives of a pro-Kurdish party, as well as on the militants of the PKK who are truly waging a vicious war against the Turkish state, he risks extending the war from south-eastern Turkey (and the northern reaches of the Kurdistan Region of Iraq) to the major cities of western Turkey, notably Istanbul and Ankara, in which so many millions of Kurds now live. At a time when bomb attacks in Ankara and Istanbul, whether attributed to terrorists backing the PKK or the Daesh Islamist forces in Syria and Iraq, are already impacting on tourism and inward investment as well as on the daily lives of ordinary people, a return to the peace process with Turkey's militant Kurdish forces would seem to be vital for the country's future well-being. However, calls for a return to peace talks are increasingly being interpreted by the Turkish authorities as being tantamount to treason against the Turkish state.

It should be stressed that it is not inevitable that Turkey will descend into civil war. A resumption of the peace process that characterised a two-year period from mid-2013 to mid-2105 would radically improve prospects, even if it took place alongside the current continuing crack-down by security forces in south-east Turkey. But in the absence of any peace process, it would come as little surprise if Kurdish militants at least attempted to carry out their threat, made in February 2016, to attack any future line intended to carry gas from the Kurdistan region of Iraq to Turkey. That could prove sufficient to prevent the construction of the Silopi–Mardin pipeline.

Still worse, as was demonstrated by the pipeline attacks of July and August 2015 and by a further attack on the gasline from Iran in October 2016, the PKK retains the ability to attack pipelines bringing in gas from both Azerbaijan and Iran—and, no doubt, oil from Iraq as well. The corporate and governmental backers of the TANAP line will naturally work to ensure the line's security and, so long as the current conflict is characterised by occasional attacks on pipelines, they should be able to cope. But in the event of more widespread conflict, TANAP's vulnerability, even though the line is buried for almost all its length, may prove too great for effective security. In such circumstances, a substantial supply source for both Turkey and countries beyond Turkey would be at risk. And there would be no prospect whatsoever of further expansion to the SGC system, short of opting for a route that bypassed Turkey altogether.

14.10 Conclusion

In theory, the Southern Gas Corridor can draw on a wide variety of prospective suppliers to fill its expanded capacities of 30–33 bcm to Turkey and 20–22 bcm beyond Turkey. In practice, however, at this stage there is no way of knowing for certain which of them will eventually contribute to SGC input. And yet, such is the variety that it seems extraordinarily unlikely that none of the suppliers detailed above will be able to contribute. The problem is exacerbated by the difficulty of working out a reasonable timeframe for additional input.

Commercially, there may well have to be some radical change in the European energy market, such as the general introduction of carbon taxes, to encourage both increased demand for gas and an increase in gas prices to attract the next wave of gas into the SGC. But politically, there can be no doubt what is the most critical issue: the need to resolve the Kurdish situation in Turkey. If Turkey continues in its current state of unrest, it should still prove possible to safeguard existing pipelines but it might well prove to be an impossible environment for the construction of a new gasline from northern Iraq. If the unrest spreads to the cities, then one is talking of civil war across the country and not only would SGC expansion prove difficult, but safeguarding existing pipelines, including TANAP, might prove impossible.

Perhaps the last word should be given to Vitaly Baylarbayov, SOCAR's SGC Manager, concerning the uncertainty surrounding both the filling of the SGC lines currently under development and the much longer-term expansion of the Corridor through the development of new infrastructure. "It is regrettable when such pipeline projects are not implemented, because they are like bridges, linking people together, linking Asia to Azerbaijan, Azerbaijan to Turkey and Turkey to Europe. It's like a highway to future progress. If they don't happen soon, they will materialise later. I don't care which one will come first, it's important that thanks to these developments, on one side of the pipe people would feel safer, warmer and pay less, and on the other side it will make the lives of people richer, better, nicer. And the transit countries will benefit as well. Everyone would be a winner".[22]

Notes

1. A more recent set of costs was provided in a **prospectus for $1 bn Eurobond offering in March 2017 for the Southern Gas Corridor CJSC**, a company set up in Azerbaijan which will eventually assume the

ownership of the Azerbaijani stakes in the various SGC projects. This prospectus, which detailed SGC costs as of 31 December 2016, was prepared by three banks: Citigroup, J.P. Morgan and UniCredit Bank. It was dated 10 March 2017 and subsequently made available by the Central Bank of Ireland. However, it is possible that even this prospectus, which clearly draws on official company data, may have made a mistake in concluding that total costs amounted to approximately $43 bn.

The prospectus does indeed list the costs for four key elements as follows:

SD2 (upstream, offshore and onshore processing): $24.9 bn.

SCP-X: $4.8 bn.

TANAP: $8.5 bn.

TAP: €4.5 bn, worth c.$4.74 bn as of 31 December 2016.

The costs do indeed total $42.94 bn, which can quite reasonably be described as approximating $43 bn. But the partners in the Shah Deniz project usually cite a single combined figure for the costs of both SD2 and SCP-X, since they are projects being carried out by a single venture. It is thus possible that whoever prepared the prospectus may have accidentally double-counted the cost of SCP-X, and that the true cost, as of 31 December 2016, was around $38.1 bn. This would certainly be in line with reductions in costs reported separately by parties involved in developing the SCP upstream, offshore and onshore processing facilities; the reduced costs of TANAP; and the fact that the estimated cost for TAP is much lower than the figure given by Natiq Aliev in June 2016.

The prospectus can be found online at: https://www.centralbank.ie/docs/default-source/Regulation/prospectus-regulation/2017/03/313136---prospectus-pdf--.pdf?sfvrsn=2.

2. Interview with Bakhtiyar Aslanbeyli, Tbilisi, Georgia, 7 December 2016.
3. https://ec.europa.eu/energy/en/topics/imports-and-secure-supplies/gas-and-oil-supply-routes.
4. Natiq Aliev, interview with The Business Year—Azerbaijan 2014. https://www.thebusinessyear.com/azerbaijan-2014/in-energy-we-trust/interview (The Business Year, Azerbaijan 2014).
5. Baylarbayov, interviewed by EurActiv's Georgi Gotev, 26 May 2016.https://www.euractiv.com/section/europe-s-east/interview/socar-it-is-impossible-to-stop-the-southern-gas-corridor/ (EurActiv.com 2016).
6. The capacity of a pipeline depends on various factors, of which the two most important are usually the diameter of the pipe and the amount of compression used to push the gas through the system. A 48-inch system is

generally reckoned capable of carrying some 20-23 bcm/y, depending on compression power, while a 56-inch system is generally capable of handling 30-33 bcm/y. However, the quality of steel used in pipe fabrication also affects operations with a thicker, tougher pipe boosting performance.

7. Birrell. The Business Year, Azerbaijan 2014. https://www.thebusinessyear.com/azerbaijan-2014/crunch-time/interview (The Business Year, Azerbaijan 2014).
8. Ms. Rzayeva's assessments of Azerbaijani prospects can be found in two documents:
1. **The Outlook for Azerbaijani Gas Supplies to Europe**: Challenges and Perspectives. Gülmira Rzayeva, OIES Research Associate. Oxford Institute for Energy Studies. OIES Paper: NG 97, June 2015. https://www.oxfordenergy.org/wpcms/wp-content/uploads/2015/06/NG-97.pdf.
2. **Materializing mega-gas projects in Azerbaijan in the low price environment.** Gülmira Rzayeva, Senior Research Fellow, Oxford Institute for Energy Studies. Natural Resources Forum, London, 28th June 2016. http://naturalresourcesforum.com/companies/oxfordinstitute2/.
9. Aliev, Fineko/abc.az., 25 April 2016.
10. Yusifzade, op cit.
11. Yusifzade to author, Baku, 29 August 2016.
12. Abbasov remarks. Comments to the author, Baku 29 August 2016.
13. Interview with Dalga Khatinoglu of Azerbaijan's Trend News Agency, conducted in advance of the Gas Exporters Forum Meeting in Tehran and communicated to the author.
14. Dr Jonathan Evans, VP Middle East Exploration & VP Exploration Technical Functions, BP Exploration. Address to the Iran Upstream Congress 2016, Tehran, 17 October 2016. Author's notes.
15. The Nabucco Pipeline Project, which was in many ways the forerunner of TANAP, was initially predicated on receiving supplies from Iran as well as Azerbaijan, and its initial plans thus included an entry point for Iranian gas at Dogubayezit and a junction with the input line from Azerbaijan and Georgia near Horasan. A large number of personnel who formerly worked on Nabucco subsequently joined the TANAP team, and it is a reasonable assumption that TANAP has access to Nabucco's plans.
16. Interview with Mr Babazadeh, Tehran, 18 October 2016. Author's notes. (It was the author who relayed Dr Evans' comments to Mr Babazadeh, eliciting his response).
17. Albayrak is quoted in a paper by Turkish analyst (and former ambassador to Azerbaijan) Ünal Çeviköz. *Could Turkey become a new energy*

trade hub in South East Europe? Çeviköz cites the following reference from Turkey's English-language Daily Sabah newspaper. http://www.dailysabah.com/energy/2015/12/18/no-energy-deals-with-israel-without-normalization-of-relations-turkish-energy-minister. The link to that website has since been broken. But there is no reason to doubt Albayrak's remark. The Çeviköz paper can be found at http://turkishpolicy.com/files/articlepdf/could-turkey-become-a-new-energy-trade-hub-in-south-east-europe_en_5390.pdf.

18. *Hurriyet*, "Top commander says Turkey 'actually waging war' since July", October 6, 2015, http://www.hurriyetdailynews.com/Default.aspx?pageID=238&nID=89490&NewsCatID=338 (Hurriyet 2015).
19. See: http://www.crisisgroup.be/interactives/turkey/.
20. For government casualty claims, see Ana Sayfa, "7 bin 78 PKK'lı etkisiz hale getirildi", Yenicag, 23 May 2016, http://www.yenicaggazetesi.com.tr/7-bin-78-pkkli-etkisiz-hale-getirildi-138066h.htm. For PKK claims, see Hisham Arafat, "PKK claims killing 7,000 Turkish soldiers", kurdistan24.net, 22 June 2016, http://www.kurdistan24.net/en/news/8c60f8cf-02e5-4037-8e6a-dd9dd4970941/PKK-claims-killing-7-000-Turkish-soldiers. For civilian casualties, see International Crisis Group, *The Human Cost of the PKK Conflict in Turkey: The Case of Sur* (Diyarbakir/Istanbul/Brussels, March 2016), https://www.crisisgroup.org/europe-central-asia/western-europemediterranean/turkey/human-cost-pkk-conflict-turkey-case-sur (International Crisis Group 2016; kurdistan24.net. 2016).
21. Ottaway, Quoted in *Foreign Policy* blog by Keith Johnson, 2 March 2016. http://foreignpolicy.com/2016/03/02/a-mysterious-pipeline-closure-is-bankrupting-iraqi-kurds/ (Johnson 2016).
22. Baylarbayov. Op. cit.

Bibliography

EurActiv.com SOCAR. 2016. *It is impossible to stop the Southern Gas Corridor*. Vitaly Baylarbayov is interviewed by Georgi Gotev. 26 May, https://www.euractiv.com/section/europe-s-east/interview/socar-it-is-impossible-to-stop-the-southern-gas-corridor/.

European Commission. *Gas and oil supply routes.* https://ec.europa.eu/energy/en/topics/imports-and-secure-supplies/gas-and-oil-supply-routes. Accessed 30 June 2017.

Foreign Policy Blog by Keith Johnson. 2016. http://foreignpolicy.com/2016/03/02/a-mysterious-pipeline-closure-is-bankrupting-iraqi-kurds/. Accessed 2 Mar 2016.

Hurriyet. 2015. Daily News, *Top commander says Turkey 'actually waging war' since July*. 6 Oct, http://www.hurriyetdailynews.com/Default.aspx?pageID=238&nID=89490&NewsCatID=338.

International Crisis Group. *Turkey's PKK Conflict: The Rising Toll*. http://www.crisisgroup.be/interactives/turkey/. Accessed 30 June 2017.

International Crisis Group. 2016. *The Human Cost of the PKK Conflict in Turkey: The Case of Sur* Diyarbakir/Istanbul/Brussels. March, https://www.crisisgroup.org/europe-central-asia/western-europemediterranean/turkey/human-cost-pkk-conflict-turkey-case-sur. Accessed 30 June 2017.

kurdistan24.net. 2016. *PKK claims killing 7,000 Turkish soldiers*. Hisham Arafat. 22 June, http://www.kurdistan24.net/en/news/8c60f8cf-02e5-4037-8e6a-dd9dd4970941/PKK-claims-killing-7-000-Turkish-soldiers.

The Business Year, Azerbaijan 2014, *In Energy We Trust*, Interview with Natiq Aliev https://www.thebusinessyear.com/azerbaijan-2014/in-energy-we-trust/interview.

The Business Year, Azerbaijan. 2014. *Crunch Time*, Interview with Gordon Birrell. https://www.thebusinessyear.com/azerbaijan-2014/crunch-time/interview.

Author Biography

John M. Roberts is one of Europe's leading energy security specialists. He has testified to UK parliamentary committees on Caspian, Russian, Mideast, Turkish, and Kurdish energy security issues. He has written regularly on Eurasian energy connections for more than 30 years. A Senior Partner with Methinks Ltd and Senior Fellow at the Atlantic Council of the US, he was previously a Managing Editor with Platts and Financial Times Energy. He is the author of Visions & Mirages: The Middle East in a New Era (Mainstream, Edinburgh, 1995), Caspian Pipelines (Chatham House, London, 1996) and Beyrouth: L'Été '82 (Beirut, Summer of '82) (1983).

15

The Evolution of European Gas Pricing Mechanisms

Jonathan Stern and Howard Rogers

15.1 Introduction[1]

Historically, gas was traded in Continental Europe on long-term (15–25 years) contracts with prices based on, and indexed to, those of oil products. In the six major Continental European national gas markets, the rationale of linkage of long-term contract gas prices to those of oil products began to weaken during the 1990s, a process which continued during the 2000s (Stern 2007, 2009).[2] Its original rationale—that end-users had a real choice between burning gas and oil products, and would switch to the latter if given a price incentive to do so—was robust when the netback market pricing mechanism (largely) based on oil product linkage was established in the 1970s (and earlier in some countries).[3] But a combination of:

J. Stern (✉) · H. Rogers
Oxford Institute for Energy Studies, Oxford, UK
e-mail: jonathan.stern@oxfordenergy.org

H. Rogers
e-mail: howard.v.rogers@gmail.com

M. Hafner and S. Tagliapietra (eds.), *The European Gas Markets*,
DOI 10.1007/978-3-319-55801-1_15

- the virtual elimination of oil products from many stationary energy sectors in these markets;
- the cost and inconvenience of maintaining oil-burning equipment and substantial stocks of oil products;
- the emergence of modern gas-burning equipment in which the use of oil products means a substantial loss of efficiency;
- tightening environmental standards in relation to emissions, particularly sulphur content and nitrogen oxide;

rendered the original rationale increasingly dubious, particularly in north-west Europe.[4] There is no commercial scenario in which users installing *new* fuel-burning equipment will choose to use oil products rather than gas in stationary uses, unless they have no access to a gas supply.[5]

Nevertheless, traditional oil product-linked pricing in long-term gas contracts remained largely unchallenged until the dramatic events of the late 2000s and early 2010s, which set in motion a process which we have argued in previous research and set out again in this chapter, have led inexorably towards hub-based gas pricing.

This chapter is divided into six sections. Following this introduction, we examine the transition to hub-based gas pricing, followed by an assessment of the development of trading at gas hubs and exchanges over the past decade. We then look at the correlation between the prices at the different hubs which is the test for whether these can be regarded as 'market prices'. The following section looks at the growing impact of external influences from Asia and North America on European hub prices, principally via the increasing influence of the global LNG market. Finally, we look at some of the contractual problems which have been encountered in moving from oil-linked to hub-based prices, before drawing conclusions.

15.2 The Transition to Hub-Based Pricing in Europe

Continental European Gas Supply, Demand and Pricing—The 'Perfect Storm'

As the global financial crisis hit economic activity towards the end of 2008, European gas demand fell. Demand in 2009 was 5.7% below its 2008 level and although it recovered in 2010, this was in large part due to abnormally cold weather and it subsequently continued on a downward path through the first half of the 2010s; in 2015, it was nearly 20% below the 2010 level.[6] New LNG supplies from Qatar, Yemen, Russia, Peru and Indonesia came on stream between 2009 and 2012 adding some 100 Bcma to global supply. Some of these new supplies had originally been intended for the US market, but the remarkable and unforeseen growth in shale gas production resulted in it becoming surplus to US requirements with much of it ending up in Europe.

Not only had demand fallen and available supply increased, but long-term contract gas prices were rising rapidly, driven by oil prices increasing to more than $100/bbl. These changes to European gas fundamentals were especially unwelcome to the midstream utilities in north-west Europe who were caught in the unenviable position of being obliged to buy high-priced, oil-indexed gas under their long-term contracts, but increasingly forced to sell at hub-based prices demanded by their customers. The rise of the north-west European hubs (discussed below) with transparent prices available on the Internet, and legal rulings which freed customers from multi-year purchase agreements for gas at oil-indexed prices, heralded a fundamental challenge to the midstream utility business model.[7] With the progressive merger of gas and power utilities, management of the newly combined entities became increasingly influenced by concepts such as 'mark to market' and, to this mindset, long-term oil-indexed contracts represented a potentially unbounded future liability. With some utilities losing around €1 billion/year in gas trading operations in the early 2010s, their commercial position rapidly became unsustainable (Stern and Rogers 2014).[8]

At the Offshore Northern Seas conference of 2010, the new CEO of E.ON Ruhrgas, Klaus Schafer, announced that: 'Hubs are the reference point when customers talk to us ...[long term contracts] LTCs in their current form no longer reflect the market.... We have to re-engineer the LTCs to anticipate the future needs of the market: price levels, indexation and review mechanism'.[9] This was the first time that a major European gas buyer had publicly endorsed hub pricing as the future price formation mechanism and marked the starting point of a series of renegotiations of long-term contract prices—many of which required international arbitration proceedings to resolve (see below).

During 2008–2012, as the gap between hub-based and oil-linked prices fluctuated, the rationale for retaining oil-linked gas pricing in Continental European long-term gas contracts was replaced by arguments suggesting that the supply/demand conditions encountered post-2008 were of a temporary nature, principally due to economic recession, and that hub and oil-linked prices would 'recouple' by 2012. Furthermore, it was argued that no other appropriate pricing mechanism was available, and specifically that European gas hubs remained both insufficiently liquid and prone to manipulation by local market players to be considered an appropriate price reference. However, as the 2010s unfolded it became clear that, driven by both commercial and regulatory pressures, hub pricing was progressively taking over from oil product linkage as the main wholesale price formation mechanism in Europe.

Figure 15.1 shows that not only was there no recoupling between oil-linked and hub prices (represented by TTF) and representative oil-linked long-term contract prices in north-west Europe (NWE oil-indexed LTC), but the problem worsened in the 2010s with the gap

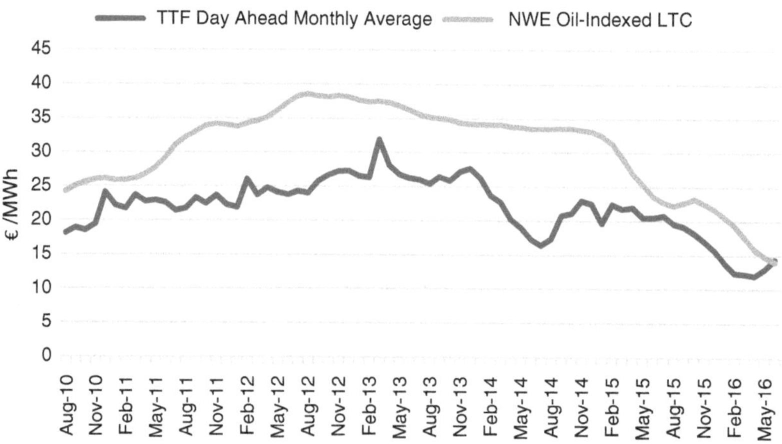

Fig. 15.1 A comparison of European hub and long-term contract prices 2010–2016. *Source* European gas daily monthly averages (respective months)

widening significantly. In 2014 and 2015, TTF prices averaged 33 and 25% (respectively) below long-term contract prices; only in mid-2016 did the two curves finally come together for the first time. This meant that European buyers required to take (or pay for) gas at oil-linked prices which they could only sell to their customers at hub-based prices were exposed to multi-billion euro losses.

Figure 15.2, taken from the International Gas Union (IGU)'s survey of pricing, shows that by 2015, nearly two-thirds of gas sold in Europe was priced at hub (GOG) compared with around 15% a decade earlier. Correspondingly, the share of gas price in relation to oil and oil products (OPE) had shrunk from nearly 80% to less than 30%.

However, as in so many other respects, as far as gas pricing is concerned Europe is not a single entity. Figure 15.3 shows IGU data split into different regions from which it is clear that the major European markets in the north-west, which account for half of European demand, are almost completely hub-priced. More than half of gas sold in Central Europe (10% of European demand) was also hub-priced, but smaller markets further south and east—with the exception of Italy—remained dominated by oil-linked and regulated prices.

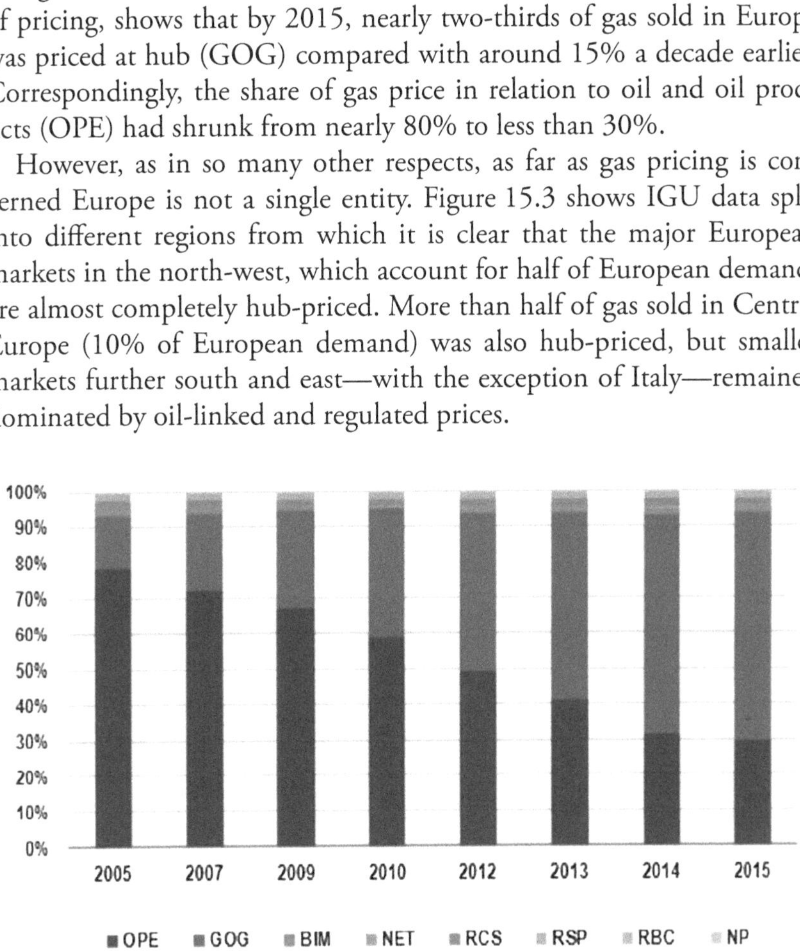

Fig. 15.2 European gas price formation mechanisms 2005–2015. *Note* For precise definitions of categories, see source. *Source* IGU (2016)

REGION AND APPROX % OF DEMAND	Oil-Related Pricing OPE	Gas on Gas Competition GOG	Bilateral Monopoly BIM	Regulated	
				Cost of Service RCS	Social Pricing RSP
North West 50%	8	92			
Central 10%	29	56			15**
Mediterranean 30%	63	32***	5		
South East 8%	38	4****		52	4
Scandinavia and Baltics 2%	48	15		Netback from final product 3%; No price 35%	

Fig. 15.3 Gas price formation mechanisms by European region (%)*, 2015. *Note* *North West Europe: Belgium, Denmark, France, Germany Ireland, Netherlands, UK; Central Europe: Austria, Czech Republic, Hungary, Poland, Slovakia, Switzerland; Mediterranean Europe: Greece, Italy, Portugal, Spain, Turkey; South East Europe: Bosnia, Bulgaria, Croatia, FYROM, Romania, Serbia, Slovenia Scandinavia/Baltics: Estonia, Latvia, Lithuania, Norway, Sweden**Hungary and Poland ***mainly Italy ****Croatia. *Source* IGU (2016)

15.3 The Development of Hubs and Exchanges

15.3.1 Background and Regulatory Context

The first European gas hub to be created was the UK's National Balancing Point (NBP) in 1996 (Heather 2010), which was followed by a number of 'false starts' for Continental European hubs: the Zeebrugge hub in 2000 was followed by EuroHub and NWE-Hubco in 2002. The latter shows how gas trading could be frustrated rather than promoted. It was not until the creation of the Title Transfer Facility (TTF) by Gasunie in 2003, followed by Central European Gas Hub in 2005 and the two German hubs—Net Connect Germany (NCG in south Germany) and Gaspool (north Germany) in 2009—that the incumbent European companies became serious about facilitating gas trading.

The 'perfect storm' in European gas markets was described above. However, in addition to these market developments, in 2010, the Madrid (EU Gas Regulatory) Forum invited the EC and its regulators to 'explore...the interaction and interdependence of all relevant areas for network codes and to initiate a process establishing a gas target model' (GTM). The Council of European Energy Regulators (CEER) *Vision for a European GTM* was endorsed by the Madrid Forum in March 2012. The GTM provided a vision of a single liberalised EU gas market. It defined an end-point of the liberalisation process of the EU gas sector as establishing functioning wholesale markets and connecting them with one another, as well as ensuring secure supply to and economic investment in these markets, with these ends to be achieved through a series of network codes. The GTM envisaged functioning wholesale markets structured as 'entry-exit zones, with entry capacity allocated separately from exit capacity so that any gas that enters the zone can be delivered, at least commercially, to any exit point in that zone', with each zone having its own hub (or virtual trading point, VTP).[10]

With trading as a central element of the GTM vision, and wholesale contracts needing to be priced using a gas price reference, it became essential that newly formed gas hubs across Europe should perform a balancing role, and that at least one hub in the region performs a risk management function.[11] The regulatory context in Continental Europe therefore changed in the direction of hubs not just being required to exist, but also to function as trading locations with market-reflective prices.

Figure 15.4 ranks European gas hubs which are operating, or have been announced, by the level of activity and the stage of maturity. The development of a liquid hub takes time, commitment and, as history has shown us in North America, Britain and now north-west Europe, can result in disruption and financial cost to (particularly) the incumbent players which dominated the pre-liberalisation landscape. Based on the transition experience of North American and British markets, the process takes 5–10 years, and this is proving to be the case in continental Europe. It also requires the commitment of governments, suppliers and system operators to achieve a smooth transition. However, a market that has indigenous production and/or is well supplied by competing sources of gas is likely to achieve the goal of a liquid trading hub more quickly.

The process starts with a move to third-party access (TPA) to network infrastructure, often requiring legislative changes to force incumbents to release infrastructure capacity and gas supply volumes, thus incentivising independents to enter the market. There is a requirement

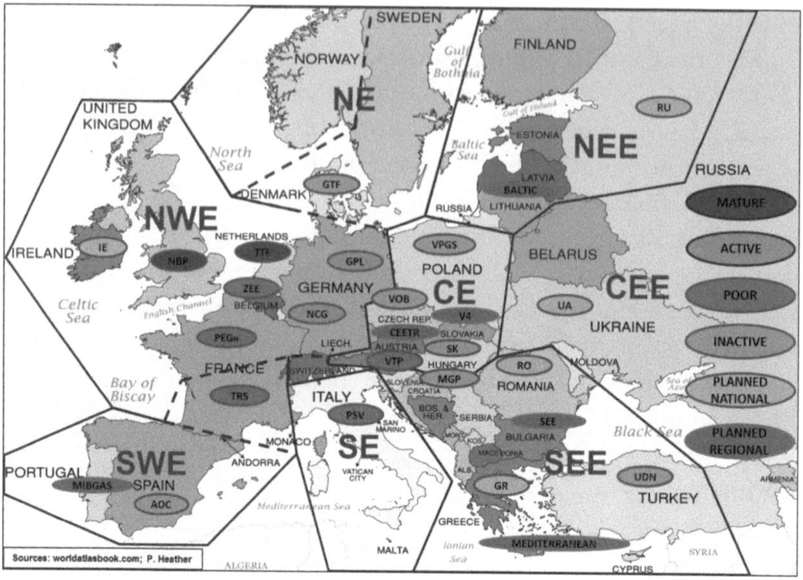

Fig. 15.4 European gas regions, markets and hubs. *Source* Heather (2016)

for the adoption of rules and regulations governing the physical side of the business, while the emergence of standardised contracts will favour commercial development. This will then be followed by bilateral trading, often aided by the first brokers helping to create trading opportunities between counterparties. These trades start to be reported in the trade press, thus creating the beginnings of a transparent market. With price disclosure, comes price discovery which, in turn, attracts more players to the market, often accompanied by smaller physical traders and the first tentative moves by financial players. The creation of exchange products (futures), based on the underlying physical contracts, offers greater access to the market, especially by non-physical players (who will always close out their trading positions before maturity).

Gradually, as increasing numbers of varied participants come to trade in a particular market, a forward curve will develop, and this will be used for risk management purposes. The final stage of maturity is when the hub develops sufficient liquidity for traders to use specific traded products (such as day-ahead or month-ahead quotes) as indices on which to price their physical transactions.

Serious development of continental European hubs generally dates from 2009, with the wave of LNG supply reaching Europe at a time of low demand serving to catalyse hub liquidity and development. By 2012, virtually all Britain's physical gas supplies were market (i.e. hub)-priced and very few traditional long-term contracts remained. A survey in that year concluded that in Europe as a whole, the NBP hub was at the forefront of gas market development, with a liberalised, fully mature traded market, offering reliable marker prices; the Dutch TTF had emerged as the pre-eminent hub in Continental Europe (Heather 2012).

Table 15.1 shows Heather's 2016 analysis (based on 2015 data) of European gas hubs along three key parameters: number of active market participants, traded volumes and churn rate (which is a standard measure of liquidity), confirming that NBP and TTF were far ahead of all competitors. At the end of 2015 (based on 2014 data), Heather concluded:

> The reality at the end of 2014 is that Europe does have two leading, mature, benchmark hubs, a few 'active' hubs and several mid-market hubs that do trade, especially in the spot/prompt and near curve and that are primarily 'balancing' hubs…The model of concentrating liquidity on one benchmark hub has served the North America market well, where trading and liquidity is focussed on a single hub (Henry Hub) and each of the other 32 hubs trades at a 'basis' to the benchmark. This is a system that is beginning to develop in Europe and one that should provide the required physical flexibility as well as the financial risk management tools. The emerging hubs..should..be in a position to become efficient balancing hubs; but, as we have seen, this process has already taken a long time and the end point is not clearly determined yet.[12]

While these observations relate to 2014, more recent data from 2015 (Table 15.1) reinforce those conclusions. During that year, NBP, TTF, NCG, PSV and VTP increased participant numbers. In terms of products and volumes traded, NBP and TTF increased exchange trading and options trading. TTF was the 'star performer' among the hubs in 2015. Other hubs also making progress included PSV (with a large increase in trading including exchange on the Pegas platform) and the Czech VOB hub (Heather 2016).

Table 15.1 European gas hubs in 2015: Participants, volumes and churn rate

	Active market participants	Traded volumes (TW h)	Churn rate
NBP	>50	20955	26.2
TTF	>50	17080	45.9
NCG	30	1790	3.9
GPL	30	950	2.5
ZEE	15	805	4.3
PSV	15	720	1.0
PEG Nord	10	500	1.7
PEG TRS	5	90	0.6
CEGH/VTP	15	240	3.7
VOB	<10	80	1.0
AOC	<10	10	<0.1

Source Heather (2016)

In this context, the aspirations of the European Commission to create a 'Western Balkan hub' based in Bulgaria to act as a regional hub for south-east Europe appear extremely challenging. The small size of south-east European markets places limits on the potential for diversified supply sources, and the lack of significant numbers of privately owned market players are additional obstacles to hub creation.[13]

15.3.2 The Development of OTC and Exchange Trading

Figures 15.5 and 15.6 show the evolution of OTC trading at the hubs in the period 2007–2015, for day-ahead[14] and all products. It is immediately clear that in terms of volumes traded, NBP and TTF are an order of magnitude larger than the other hubs, many of which cannot easily be shown on the same chart. Day-ahead volumes at many hubs peaked in the early 2010s, but the volume of all products continued to increase with the exception of NBP where a substantial proportion of trading moved from OTC to exchange (Fig. 15.7). This means that, in terms of both day-ahead and total OTC trading, TTF overtook NBP in 2014. It is a little surprising that trading at the German hubs remains relatively small (despite the size of the market and its geographic centrality). OTC trading at the rest of the hubs (mostly) continues to make progress, but it is hard to see any of them challenging for a leading position in Europe.

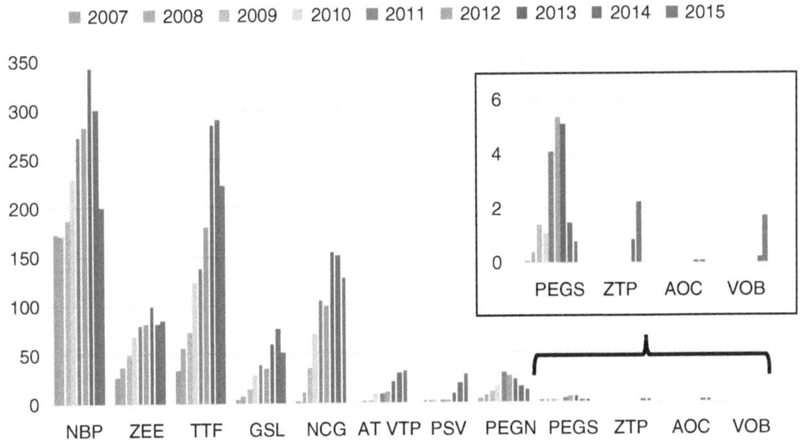

Fig. 15.5 Gas volumes traded, OTC day-ahead 2007–2015 (TWh). *Note* OIES estimates that the Tankard Parties (ICAP, Tullett Prebon, Marex Spectron) market represent about 70-80% over total OTC traded volumes as published by LEBA. PEGS volumes since April 2015 refers to TRS. *Source* Petrovich (2016) based on Tankard parties data

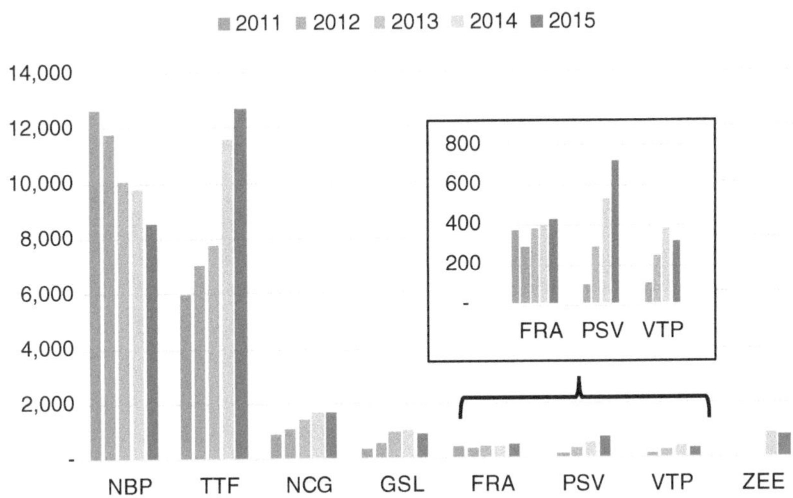

Fig. 15.6 Gas volumes traded, OTC all products 2011–2015 (TW h). *Notes* FRA = all PEGS; GSL = Gaspool; LEBA volumes available since January 2011 for NBP, TTF, German hubs, French hubs, since May 2012 for PSV and Austrian VTP, since January 2014 for Zeebrugge. Other hubs are not included. The data do not provide a split between PEG Nord, PEG Sud and PEG TIGF. *Source* LEBA volume reports

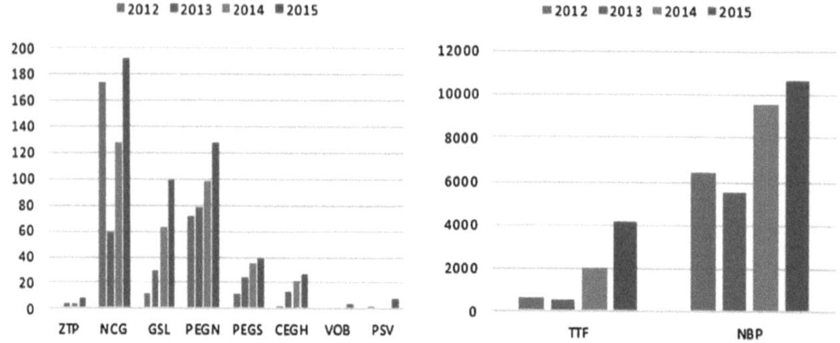

Fig. 15.7 Exchange traded volumes, 2012–2015 (TWh). *Notes* The euro-quoted virtual hub ZTP, located in Belgium, was created in 2012; trading at Zeebrugge was all OTC in 2013 and 2014; PSV volumes do not include the centralized balancing platform (PBGAS) for trading gas in storage. At Czech VOB, spot gas exchange trading launched in May 2015. *Sources* Heather (2016, 2015)

Exchange-traded volumes are shown in Fig. 15.7, demonstrating that although TTF has made significant progress since 2013, NBP remains far ahead with exchange volumes exceeding those of OTC.

15.4 The Evolution of Hub Price Correlation and the Development of Market Pricing

Increases in trading volume and liquidity (churn rate) are the important indicators of hub maturity. But the key demonstration of progress towards market pricing is the degree of price correlation between European hubs. If we can demonstrate that good correlation between prices at different hubs exists, this should resolve any suspicion of manipulation at the local level and indicate the possibility to freely trade the commodity across borders, as envisaged by the GTM. Where price correlation is poor, we need to see whether this can be explained by physical or contractual reasons.

The OIES has been conducting a research project on hub price correlation using a proprietary dataset of anonymised trade data for European hubs since 2007.[15] The papers published in this series

(Petrovich 2013, 2014, 2015, 2016) have established that the correlation between European hub prices in the 2007–2015 period was generally good and improving for those hubs with significant OTC and exchange trading, and express a reliable price signal. Figure 15.8 shows a summary of the correlation data for selected groups of hubs and pairs of hubs for the period 2012–2015. Exceptions to good correlation occur in periods where pipeline bottlenecks (physical or contractual) or closure (for maintenance) cause a 'de-linkage' between prices at the main north-western European hubs and typically those of Italy (PSV), Austria (CEGH), southern France (PEG Sud) and at times UK (NBP). ACER reports the same finding in its 2014 Market Monitoring Report[16] and comments that one of the main reasons for this was the renegotiation of long-term contract conditions where hub prices have been increasingly used as a reference, or discounts have been granted, placing downward pressure on prices in higher priced markets.

The north-west Europe core group (Zeebrugge, TTF, the German hubs and PEGN) stands out as almost perfectly correlated over the

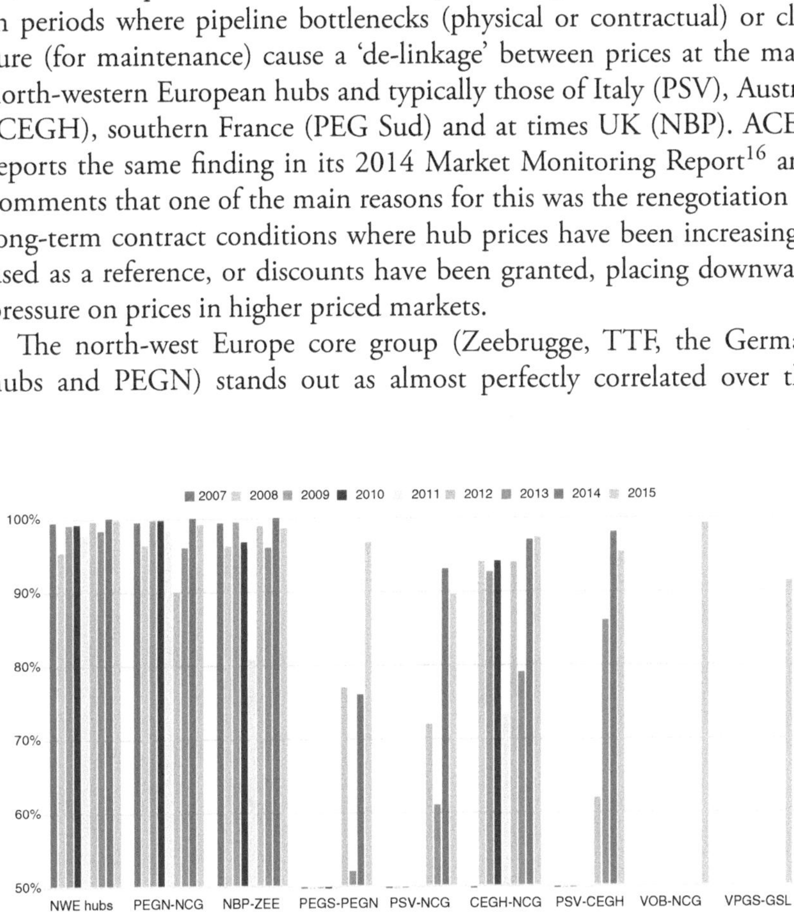

Fig. 15.8 Correlations between OTC day-ahead daily prices for selected groups of hubs (Pearson coefficients %). *Notes* NWE hubs= NCG, GSL, TTF, ZEE. GSL = Gaspool, CEGH = Austrian VTP. Exchange day ahead prices used for pairs including Polish VPGS hub, where OTC trading had not developed as of 2015. *Source* Petrovich (2016), Polpx, EEX, CEGH

whole period with these hubs behaving as a single integrated market area. Periodic de-linkage, however, occurs at the more peripheral hubs: at PEGS, and also, although to a lesser extent, at the PSV (Italian) and VTP/CEGH (Austrian) hubs. Reductions in correlation scores signal that there are barriers that prevented a full integration between the main gas hubs in Europe. The analysis shows that gas flows do not always fully exploit the arbitrage opportunities emerging between these peripheral markets and the core group, so impeding full price alignment. The nature of these barriers was found to be mainly physical: de-linkages occur when there is physical congestion of the interconnecting infrastructure. On the other hand, with the exception of Italy and Poland, there is no evidence of widespread non-physical barriers to trade.

Petrovich (2016) also found that day-ahead OTC and exchange prices at the same hub also showed a strong correlation (Fig. 15.9).

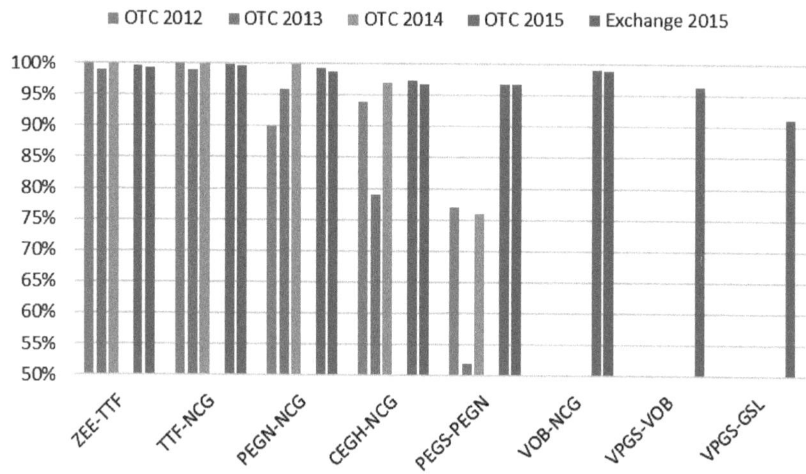

Fig. 15.9 Cross correlations between OTC day-ahead daily prices and exchange day-ahead prices (January–December 2015). *Note* Exchange data for Zeebrugge (ZEE) in 2015 refers to ZTP exchange price which are available starting from June 2015 up to November 2015; OTC data for VPGS not available. *Source* Petrovich (2016)

Periodic price de-linkages were further investigated for the French, Austrian and Italian hubs. For *southern France*, the analysis shows that PEGS de-links when it is physically separated from PEGN due to LNG supply being diverted, requiring consumption to be met by higher flows from northern France, which, in turn, congests the interconnecting infrastructure between the two French market zones. As soon as LNG supply increases in the south of France (which occurred at the end of 2014), the spare transmission capacity between the two adjacent French zones restored price alignment within France. Petrovich (2015) found that the pilot project of coupling the PEG Nord and PEG Sud markets through an implicit allocation scheme, which has been in place since 2011 for a limited amount of the full interconnection capacity, was not enough to solve the bottleneck between PEGS and PEGN at times of LNG scarcity.

Austrian (CEGH) hub de-linkages are related to physical congestion at Oberkappel (IP between NCG and CEGH), which tend to occur due to heavy exports from Germany to Austria during the summer which saturate transmission capacity, especially due to physical constraints on the German side (disparity between entry and exit capacities, plus pressure constraints in the MEGAL system). Petrovich (2014) found that in 2014 the offer of interruptible capacity was not enough to solve the bottleneck between NCG and CEGH. Moreover, requests to flow gas from Germany to Austria were supported by increasing volumes being shipped towards Ukraine, or possibly to other European markets, in 2014 as a consequence of Russian gas supplies not meeting nominations that summer. The trigger factors which created physical bottlenecks were related to changes in supply patterns to Europe: LNG being diverted from Europe in response to demand in the Asian market,[17] and the start of the reverse flow to Ukraine, after Russian direct sales to Ukraine were suspended in March 2014 (Stern et al. 2014).

The *PSV (Italy)* story is somewhat different. Although the PSV premium increased significantly in the second half of 2014, most of the time the route from the lower-priced NW European hubs to the Italian hub was not physically congested. In 2014, the PSV–NCG price spread stayed at a level roughly equal to double the estimated transmission tariff at a time when more than 10% of the capacity connecting NCG to PSV (via the Transitgas pipeline) was unused. The persistence of the price differential and the under-utilisation of cross-border transmission capacity on the NCG–Switzerland–PSV pipeline system suggests that neither resales of pre-booked capacity on an interruptible basis carried out by TSOs, nor the release of long-term-booked capacity undertaken by ENI through periodic auctions, were enough to fully exploit arbitrage opportunities between the Italian market and other north-west European markets. Analysis also suggests that traders may not be in a position to procure transmission capacity from NCG/TTF to PSV to exploit the price differential, possibly because allocation procedures for transmission capacity on the Swiss side are not harmonised or not transparent.

Petrovich (2015) found that the costs of these periodic price de-linkages were non-trivial. This is due to the fact that more than 60% of consumption in the countries with the less aligned hubs (France, Italy and Austria) is priced based on gas-on-gas competition.[18] Physical congestion between Germany and Austria resulted in an estimated additional gas procurement cost in 2014 of about €60 million, most of which was accounted for by CEGH prices being higher than NCG in September and October 2014.

15.5 The Growing Impact of Global Developments on European Prices: Four Pertinent Issues

In former times, gas developments and prices elsewhere in the world were not of immediate relevance to those operating in Europe due to physical and contractual disconnects between regional gas markets. Starting in the late 2000s that situation changed, as developments first in North America, with shale gas development eliminating the need for US LNG imports; and then in Asia with the surge of LNG demand due to the Fukushima disaster in Japan and double-digit demand growth in China. These regional developments led first to LNG flowing into Europe, and then flowing out again, significantly impacting price levels and competition between LNG and pipeline gas.

The aftermath of warmer than normal 2013/2014 winters in Europe and Asia, evidence of slowing Asian LNG demand growth through 2014 and the collapse of the oil price in late 2014 has resulted in a painful 'new normal' for key players in the global gas system, specifically LNG project investors and Russia/Gazprom. While the slowdown in Asian LNG demand and stagnant European gas demand has a direct causal impact on European hub and LNG spot prices these developments, together with the fall in oil prices, brought oil-indexed gas and LNG contract price levels down to levels in 2016 which would have been unimaginable 3 years previously. With project economics challenged and cashflows crimped, investors in new gas (especially LNG) supply projects are inevitably holding back, cutting costs and awaiting a more positive market outlook.

At a more fundamental level however, we may be about to witness a significant disruption to regional gas equilibria as a wave of new (principally Australian) LNG supply meets a slowing Asian market and a significant regional component (US/North America) reconnects with the global system in the form of 80 bcma (and counting) of new LNG export projects. Europe will be a passive recipient of excess LNG supply at a time when its gas demand growth is at best tepid, but with an import requirement which may be rising due to declining domestic production.

In this environment, we believe it is important to address the following four pertinent issues:

The Impact of Lower Oil and Lower Gas Prices on Existing and Future Gas and LNG Projects

As the majority of upstream gas projects are undertaken by oil and gas companies (whether IOCs or NOCs), lower oil and gas prices feed directly through to lower discretionary cashflows, and hence funds available for capital investment. Capital allocation is prioritised on projects with the most robust returns, although such decisions are based on a view of long-term price developments, given that these projects will have productive lives beginning 4–8 years in the future and generating revenues for some 20–30 years thereafter. The problem faced by oil and gas companies at present is accentuated by uncertainties related to Asian economic growth (particularly China) on the demand side for both oil and gas, and the future potential of shale (tight) oil and gas production in the USA.

The oil price outlook is, as always, prey to geopolitical events, mainly relating to the MENA region which could reduce or increase oil supply and hence influence price. For gas, the oil price collapse has undermined the rationale for relying on this pricing basis as the 'gold standard' for underpinning the economics of high cost-base LNG projects (and Russian pipeline export contracts). The mid-2016 level of hub prices and oil-linked contract reference prices would render uneconomic virtually all tranches of incremental gas supply—apart from the 100-plus bcma of developed, but 'shut-in', Russian gas which could flow to the European border at a price of around $4/MMbtu (€12.4/MWh) (Henderson 2016a).

The outlook for the period to 2030 for markets connected by flexible LNG supplies given the uncertainty in regional demand outlooks in the light of new LNG supply currently under construction.

This involves the considerable range of uncertainties regarding supply and demand trends in the global system connected by flexible LNG. Conventional wisdom suggests that the creation of a network of physical commodity flows might be expected to lower general volatility of supply and demand (the portfolio risk diversification effect). With gas being a low energy density fuel (in terms of heat content per unit of volume), this is not necessarily the case. In order to recover fixed costs, all gas producers tend to produce at maximum sustainable rates, storage facilities are rarely sized to compensate for multi-year demand fluctuations, and hence demand and supply events will tend to have an impact on all markets connected by gas trade flows—whether by pipeline or LNG. By contrast, the situation of Russia is unique. Gazprom has built up excess productive capacity of more than 100 bcma—largely due to investments in the Yamal Peninsula fields in anticipation of higher European demand than has transpired, but also because of reduced domestic and CIS demand for Gazprom production through the erosion of its domestic market share by competing producers (Rosneft, Lukoil and Novatek).

The three key uncertainties on the demand side in this system are as follows:

- The future growth of Asian (and particularly Chinese) demand for gas and LNG (Rogers 2016).
- The emergence of 'new' markets for LNG in the 2020s, particularly marine transportation fuel ('bunkers') (Le Fevre 2014, 2016).
- The uncertainty around European future demand recovery for gas, which is subject to the rate of nuclear and coal plant closures and the future pace of renewable investment and capacity build.

The three key uncertainties on the supply side in this system are as follows:

- The scale and timing of US LNG exports, given the likely affirmation of the resource base, depends on the future pace of project FIDs beyond the 80 bcma of projects under construction (Jensen 2016).
- The scale and pace of non-US LNG projects from Australia (85 bcma under construction but additional brownfield/expansion potential), Canada, East Africa and Russia (other than Yamal, which are likely to be delayed by the current imposition of sanctions) (Henderson 2016b and Corbeau 2016).
- Russian pipeline export volumes. These relate both to China—where its two pipeline export deals are by no means immune to future renegotiation and delay, but also to Europe where Russia will likely be forced to make a choice between a 'high volume, low price' or a 'low volume high price' strategy, as the threat of losing market share to LNG volumes from existing (including under construction) and (in the longer run) new projects.

We might add an additional element of uncertainty to the supply side—specifically the rate of decline of domestic production in the European region. The three main producing centres—UK, the Netherlands and Norway—have reached maturity and in the case of the UK and the Netherlands are declining (Stern et al. 2014). Future production levels depend on the successful development of new (but typically smaller and price sensitive) discoveries and the rate of decline of older fields approaching abandonment (notoriously difficult to predict). Given cost levels in the North Sea, the 2016 price environment—should it continue for several years—may have the effect of both delaying new developments and accelerating the decommissioning of existing fields. The ongoing uncertainty surrounding Groningen production capacity (due to environmental restrictions imposed by government) adds to concerns about declining European domestic production.

These demand and supply-side uncertainties are to some extent price-related; however, it is important to be aware of the real-world constraints to such price responses. On the demand side, the response to lower wholesale gas prices will generally operate on a multi-year timescale. In a liberalised market, midstream utilities and local distribution companies will purchase supplies up to 2 years in advance on forward markets. Lower prices will take a while to feed through into domestic tariffs and even then it is questionable whether consumers will consciously increase gas usage (by turning thermostats up).

Gas consumption in the industrial sector is a function of GDP growth and long-term trends in the preponderance of energy-intensive industry in national economies—rather than a short-term response to gas prices. In the power sector, gas prices would generally have to drop below €12.4/MWh ($4/MMbtu) to materially displace coal in power generation in most European countries (probably €15.5/MWh or $5/MMbtu in the UK which has a higher carbon price). The demand response to lower gas prices therefore is generally longer term and subsumed within other economic trends, except in periods of gas prices below €13/MWh, where it may be shorter term in markets with a responsive power sector.

On the supply side, we need to distinguish between gas from fields (and LNG facilities) already in production which will tend to continue at the maximum rate (with the exception of Russian supply), future supply from projects already under construction (which will be produced at design rates when commissioned), and future projects for which FID has not yet been taken. Here it is likely that such investment decisions will be delayed until market fundamentals support a view on the part of project investors that future prices (in the case of LNG projects those from 5 years into the future and beyond) will be sufficient to adequately remunerate investment. In the case of future non-US

LNG projects, this uncertainty is compounded by the stated desire of Asian buyers to seek 'hybrid' prices (some combination of Henry Hub 'plus costs', oil, European hubs and potentially other elements) with the possibility of a move to Asian hub pricing in the future (Stern 2016). North-west European LNG buyers will be unwilling to buy any new (LNG or pipeline) gas on anything other than a European hub price, and may be unwilling to sign contracts for more than 10 years.[19]

The impact of the probable delay on new LNG project FIDs given demand uncertainties and the apparent need to move from oil indexation to new contract price formation structures; and *the extent to which Russia can use its market power in Europe to 'control' hub prices to influence outcomes.*

These last two issues are somewhat inter-related. Gas hub or spot prices in 2016 were clearly signalling that the market at present was adequately supplied. This is not particularly helpful for a project which, if it takes FID in 2016, will come on stream around 2021. Neither were 2016 oil prices a particularly valid signal as to the need (or otherwise) for new LNG project FIDs. Equally important is the future strategy which Russia might adopt. Given its comparative advantage of 100 plus bcma of developed (currently 'shut-in') production at low variable cost, Russia might decide to manage physical exports to Europe to keep European hub prices (and by arbitrage Asian LNG spot prices) too low to support new LNG projects which have not secured contracts with an Asian buyer on an acceptable 'hybrid' price basis. This would see Russia's gas market share increase substantially in Europe through the 2020s, but at some point the temptation to use such market power to increase price levels would probably prove hard to resist—resulting in a subsequent renewed surge of competing new LNG supply.

While this 'new great game' dynamic is certainly possible, and has a compelling logic, there was scant evidence in mid-2016 that Russia was immediately contemplating such a course of action. But its hand may be forced should:

- its buyers continue to nominate high volumes in order to sell a share on the hubs and hence stimulate gas demand (in the belief that concessions and rebates from Gazprom would keep them financially whole) or…

- a surge of European LNG imports (such volumes of LNG being not required elsewhere) take hub prices down to levels which either Gazprom deems 'too low' or where its midstream buyers, unable to meet take or pay requirements demand further substantial contractual concessions.

From a more positive perspective, more flexible use of 'spare' West Siberian production capacity would moderate a potential early 2020s 'tight' LNG market situation (perhaps caused by a sudden acceleration in Asian LNG demand). From a less positive perspective, Russia's market power would, in this system, extend beyond Europe. The impact of higher or lower Russian physical flows would certainly impact European hubs and also (by arbitrage) Asian LNG spot prices. In certain circumstances (e.g. overbuilt US LNG export capacity not fully utilised) European hub price levels could also, through arbitrage, influence the Henry Hub price, especially if US LNG exports continue on the basis of only covering variable shipping and regasification costs. The moderating factor, however, is that if Russia maintains European hub prices at levels high enough to trigger new LNG FIDs, this would create competing supply which, once built, will have low variable costs and so will tend not to respond to subsequent lower prices.

Lack of clarity on Russia's future preferred commercial behaviour adds a level of complexity most market participants would prefer to ignore. Gazprom is occupied on many fronts in both political and commercial spheres. At some point, however, the need to adopt a more market-oriented strategy is likely to rise on its list of priorities. While the timing of this is at present uncertain, we would strongly suggest that this is a development that players in both Europe and the wider LNG-connected global system should be closely monitoring.

15.6 Contractual Problems of Price Evolution

Price Review and Arbitration Traditional Continental European long-term gas contracts contain clauses which provide for negotiations (usually) every 3 years to change certain elements of the price formula—indexation and base price. In some contracts, it may also be possible to change volumes and volume flexibility arrangements (take or pay levels and treatment of volumes not taken within a contract year).[20] A key parameter for a price review is the phrase 'changed economic circumstances beyond the control of both the buyer and the seller', which refers to how the energy market—and in particular the fuels with which gas competes in end-user markets—has changed since the previous review.[21] Parties need to agree whether and how these circumstances have changed and therefore whether and how the price needs to be adapted. If they are unable to agree on the resulting price for the upcoming 3-year contractual period, their only option (apart from to continue negotiating and hope that their differences can be resolved) is to go to arbitration.

All European long-term gas contracts contain an arbitration clause which typically states that, in the event of unresolved disagreement about price terms, the parties will request a decision from an appointed expert or an arbitral tribunal. The expert or arbitral tribunal will hear the case and reach a decision which will be binding on both parties. The decision may include the resetting of the price formula by the (expert or) tribunal, the application of which will be backdated to when legal proceedings commenced. From the point of view of the parties to the contract, this represents a process where the outcome is highly unpredictable and therefore risky.

Over the approximately 40-year history of long-term European gas contracts, arbitrations had been extremely rare events. However, starting in the second half of the 2000s, reports of arbitrations became more frequent and continued up to the present.[22] Summarising a very complex (and still unfolding) picture, by 2016 Statoil and Gasterra had both appeared to embrace the new gas market paradigm in north-west Europe by agreeing to move long-term contracts to hub prices where competitive markets existed.[23] But the cases of Gazprom and Sonatrach were fundamentally different.

Sonatrach, having won an earlier arbitral case against Gas Natural in 2010, felt justified in continuing to insist on the retention of oil indexation for customers in Spain and Italy.[24] However, failure to maintain gas production to keep pace with (heavily subsidised domestic downstream pricing and hence) burgeoning domestic demand has left the company short of gas supplies. As a result, it has shown willingness to relax take or pay obligations, but reluctance to compromise on oil-linked prices in long-term contracts, although it was finally forced to do so. Algeria's price maximisation strategy in Europe reflected its lack of available gas to export, lack of pressing need for additional revenues and (during the period 2011–2014) ability to divert its LNG exports from Europe to higher priced markets in Asia. Following the collapse in (oil and) gas prices, the country has found itself in a less comfortable position (Aissaoui 2016).

By contrast, as described above, Gazprom's price-volume strategy is one of the key determinants of the 'new world order' in global gas dynamics and is not straightforward. The company has been in negotiations, often leading to arbitrations, with its European pipeline gas customers under long-term oil-indexed contracts from 2010. Three stages of its negotiations can be identified[25]:

- 2010–2012 where customers were obliged to pay for minimum contract quantities at oil-linked price formulae but could purchase additional volumes at hub prices;
- 2012–2015 where, although oil indexation remained in the price formula, the company agreed with individual buyers a complicated mix of base price reductions and rebates on prices paid under the contract formula relative to hub prices. In many contracts, TOP levels were reduced from 85 to 70%.
- 2016 where rebates were institutionalised for the remainder of the contract and operate for both buyer and seller (i.e. if the oil-linked contract price is lower than the hub price then Gazprom receives the rebate).

At the time of writing in mid-2016, Gazprom had at least six ongoing arbitrations.

Arbitral decisions which have been made public since 2012 have found that the spot price (either at hubs or for LNG cargoes) should be considered at least part of the market price in existing long-term contracts.[26] It may be a reasonable generalisation to say that, as the 2010s have progressed, arbitral tribunals have ruled that an increasing share of hub pricing should be included in the price formula.

Despite all of the problems in long-term contract pricing, there are virtually no examples of European long-term contracts which have been terminated as a result of price disputes. The closest is the contract between Gazprom and RWE Transgaz (in the Czech Republic) which was 'suspended' in 2014 following the arbitration between the parties and only recommenced in 2016.[27] However, it is likely that a significant number of long-term contracts will not be renewed or extended when they reach the end of their lives, either because of insufficient gas availability (in the case of the Netherlands and Algeria) or because buyers are much more comfortable signing short-term contracts with hub indexation and retaining flexibility to deal with unexpected fluctuations in demand or availability of supply.[28]

Many European long-term contracts expire in the late 2010s and early 2020s, and it is therefore likely that for these contracts the 10-year saga of price arbitrations will draw to a close within the next decade. However, for Russian contracts which continue into the 2030s, parties have to decide whether they are prepared to battle on for another 20 years, unless the obvious compromise of moving these contracts to hub prices can be reached earlier.

15.7 Summary and Conclusions

Progressive transition to hub-based pricing: over the past decade, there has been a progressive transition—equivalent to a revolution—in European gas price formation, from dominance of oil-related pricing to a situation in 2015 where nearly two-thirds of gas in European wholesale markets was sold at hub prices. In north-west Europe, this transition is almost complete and it is well under way in central Europe and Italy. However, in Mediterranean and south-east Europe oil and regulated prices (respectively) remain dominant, principally because large-scale physical links to the rest of the Continent remain to be established.

Hubs and hub development: following the establishment of Britain's NBP in 1996, Continental European gas hubs were established in the early to mid-2000s, but serious growth did not take place until the 2010s. By the mid-2010s, the NBP and (the Dutch) TTF were established as 'benchmark hubs' and used as price references for risk management. Hubs located in large gas markets such as NCG (Germany), PEG Nord (France) and PSV (Italy) could also develop into price references, but it seems doubtful this will happen or that they could grow to rival TTF as the 'Euro-hub', while the NBP has no competitors as the 'Sterling-hub'. Although every EU member state (and perhaps also non-member state) will in time have its own gas hub into which and from which physical volumes of gas will be traded, most of these will remain 'balancing hubs' with only national (local) relevance.

The importance of *hub price correlation* cannot be overstated as this demonstrates that hubs do indeed represent market prices and are not the result of manipulation by national dominant players. The results of OIES' (ongoing) research in this area have concluded that the main gas hubs in Europe are already well integrated, and that in general correlation is high and continues to improve. However, some bottlenecks remain, and in order to resolve the related price misalignment in some parts of Europe, investment and new rules on capacity optimisation are needed so that transmission capacity is sufficient to allow for the free flow of gas in response to price signals. Changes in gas flow patterns may mean that, even in a mature and well-integrated European gas market, these problems may recur for some periods. The price at one hub may diverge from the others for a period and display a dynamic which is completely different, and which may result in significantly higher costs in that country.

Impact of global developments: Since the late 2000s, European prices became increasingly impacted by developments in other regional markets—principally North America and Asia—via LNG trade. During 2011–2014, European gas prices were in the range of €30–40/MWh due to very strong demand in Asia (principally Japan and China), a lack of LNG due to delays in bringing (principally) Australian mega-projects on stream, and oil prices in excess of $100/bbl. By contrast, 2014–2016 saw the start of the long-anticipated LNG supply wave, much lower than expected LNG demand in Asia, and an oil price collapse to $30–50/bbl, a combination of which resulted in European gas prices falling below €15/MWh in 2016. As Australian LNG projects come on stream, to be joined by additional US projects later this decade, prices look set to remain low for some years with the main countervailing force being the possibility that lower prices will create additional demand, particularly in Asia.

Approaching 2020, much may depend on the degree of demand recovery, combined with the extent of domestic European production decline and lack of new LNG and pipeline gas projects, both of which are likely to be negatively impacted by the current low-price period. Further uncertainty involves the pace of a transition in Asia from oil-linked to hybrid and eventually hub-based LNG prices, although the latter are unlikely to materialise until the early 2020s. For Europe, a major uncertainty is the strategy of Gazprom which, with a surplus of available gas exceeding 100 bcma, could decide to sell sufficient gas to maintain prices low enough to keep LNG out of Europe and discourage new projects. Lack of clarity on Russia's future preferred commercial behaviour adds a level of complexity most market participants would prefer to ignore, but that European gas players should be closely monitoring.

Contractual problems and price evolution: Starting in late 2008, European utilities began to encounter increasing commercial difficulty in managing their long-term gas contracts, as oil-linked purchase prices rose significantly above hub-based prices, and they were forced to reduce their sales prices in order to retain their customers and meet their take or pay commitments. Since then, the gap between oil-linked and hub-based prices fluctuated, but for much of the 2011–2014 period hub prices were 25–50% lower than long-term oil-linked contract prices.

Exposure to European utilities to multi-billion euro losses led to an upsurge of renegotiation, often followed by litigation, which mostly moved long-term contract prices closer to hub levels. This was less painful and protracted for Norwegian and Dutch contracts and more painful and protracted for Algerian and Russian contracts. In the case of Gazprom, a series of arbitrations have led to agreements to determine a relationship between contract prices and hub prices whereby at the end of the price review period, one side or the other will receive a rebate if too much or too little money has been paid or received.

The situation in 2015 whereby nearly two-thirds of wholesale gas sold in Europe was priced at hubs, with the likelihood that percentage will increase, means that any buyer signing a new, or renewing an existing, contract—of whatever length—must base the price in that contract on some form of hub price. Exactly which hub, or a combination of hubs, and whether an average of day-ahead or month-ahead prices and over what period will be matters for negotiation, but the principle of hub price formation is established. Even countries in south-east Europe—which were not within reach of a hub in 2016—must be anticipating that interconnections being implemented as part of the EU's Security Package will, within a few years, provide them with a market price reference.

Notes

1. The authors would like to thank Patrick Heather and Beatrice Petrovich for their assistance in preparing this chapter.
2. The six markets are Germany, France, the Netherlands, Belgium, Italy and Spain.
3. For the history of this mechanism and its logic, see Stern (2012), pp. 54–59.
4. The position of some smaller markets, especially in south-eastern Europe, is different as they are still burning significant quantities of oil in stationary sectors and have retained greater switchable capacity, see Kovacevic (2007) and Giamouridis (2009).
5. However, they may choose to use other alternatives to gas with the main battleground being in power generation between gas, coal and low-carbon (renewable and nuclear) sources. But none of these sources will have prices set in relation to oil products.
6. OECD Europe gas demand in 2015 was 469.3 Bcm. For a detailed account of gas demand in the post-recession period, see Honoré (2014).
7. See, for example, in the case of Germany: German Energy Blog, 24th March 2010: 'BGH Declares Oil Price Linkage in Gas Contracts Void': see article at: http://www.germanenergyblog.de/?p-2278.

8. E.ON claimed that its long-term contracts were responsible for its gas trading losing €1 billion in 2011. http://millicentmedia.com/2011/08/12/germanys-giant-utilities-are-posting-losses-and-slashing-jobs-what.%E2%80%99s-going-on/
9. The video of the presentation is unfortunately no longer on the ONS website, but see: 'EON Ruhrgas seeks gas contract reform', European Gas Daily, August 26, 2010, pp. 1–2.
10. For many more details of this process, see Yafimava (2013), pp. 2–8.
11. Balancing' refers to the need for TSOs to keep the network in balance for safety reasons, which requires shippers to inject into the system a similar amount of gas to that which they withdraw on a daily basis as required in the balancing network code.
12. Heather (2015), p. 96.
13. 'Balkan hub could reduce NS2 concerns: EC', Platts, European Gas Daily, 30 June 2016, p. 4.
14. The focus on day-ahead is because it is by far the most frequently traded product across all the hubs, and one of the first to develop as it is used for balancing purposes.
15. The data set is provided by the three major brokers, ICAP, Marex Spectron and Tullett Prebon, collectively known as the Tankard Parties and contains more than 4 million trades conducted at all European hubs for the period 2007–2015. This data set is believed to capture 70–80% of all OTC trades. A full account of the methodology and data sources can be found in Petrovich (2013), pp. 5–12.
16. ACER/CEER (2014), pp. 172–174.
17. Global gas scenarios and consequence for the European gas market have been presented in Rogers (2015).
18. As reported by Nexant which is responsible for the annual IGU price surveys.
19. This was made clear the long-term contracts for supplies from Azerbaijan's Shah Deniz 2 development signed in 2014, these are 20-year contracts but priced at hubs.
20. Levy (2014). contains extensive commentary on many aspects of gas price reviews.
21. See Frisch (2010), pp. 15–18 for the various tests which must be carried out to determine whether changes in price provisions are justified under the contract.

22. During the 2010s, public reports of ongoing arbitrations and some details of judgements have become available.
23. But some cases remain unresolved as the 2016 Gasterra/ENI arbitration demonstrates. Eni Press Release: Statement on the arbitration with GasTerra and following measures were taken by the company. 20 July 2016. More examples of arbitral decisions can be found in Stern and Rogers (2011, 2012, 2014).
24. 'Gas Natural loses Sonatrach dispute', European Gas Daily, August 18, 2010, p. 2.
25. The first two stages are explained in greater detail in Stern (2014), pp. 58–66.
26. Arbitral decisions are confidential, so we cannot be certain. This information is from press reports which cannot be verified. But for example: Edison received €450 million from Rasgas (in relation to its long-term LNG contract during 2010/11); in October 2012, Edison received €250 million from ENI (in relation to gas from ENI's long-term contract with Libya); in May 2013, Edison received €200 million from Sonatrach; in July 2013, a tribunal awarded RWE a payment of €1.6 billion from Gazprom and a share of hub pricing in the long-term contract formula.
27. 'RWE removes price risks with new contract', European Gas Daily, 12 August 2016, pp. 3–4.
28. The Polish company PGNiG has said that it will not renew its long-term contract with Gazprom when this expires in 2022 (although there is enough time for it to reconsider this decision).

Bibliography

ACER/CEER. 2014. Agency for the Cooperation of Energy Regulators/ Council of European Energy Regulators, *Annual Report on the Results of Monitoring the International Electricity and Gas Markets in 2013.*

Aissaoui, A. 2016. *Algerian Gas: Troubling Trends, Troubled Policies.* OIES Paper NG108.

Corbeau, A.-S. 2016. East Africa: Beyond 2020. In *LNG Markets in Transition: The Great Reconfiguration*, eds. A.-S. Corbeau and D. Ledesma, 273–289. (2016). Oxford: OIES/OUP.

European Gas Daily and European Gas Daily Monthly Averages: Platts European Gas Daily (various issues).

Frisch, M. 2010. Current European Pricing Problems: Solutions Based on the Price Review and Price Re-Opener Provisions. Centre for Energy, Petroleum and Mineral Law and Policy, University of Dundee, Working Research Paper Series No. 2010:03.

Giamouridis, A. 2009. *Natural Gas in Greece and Albania: Supply and Demand Prospects to 2015*. OIES Paper NG 37.

Heather, P. 2010. *The Evolution and Functioning of the Traded Gas Market in Britain*. OIES Paper NG44.

Heather, P. 2012. *Continental European Gas Hubs: Are They Fit for Purpose?* OIES Paper NG63.

Heather, P. 2015. *The Evolution of European Traded Gas Hubs*. OIES Paper NG104.

Heather, P. 2016. *The Evolution of European Traded Gas Hubs: A Review of 2015*. OIES Energy Comment.

Henderson, J. 2016a. *Gazprom: Is 2016 the Year for a Change in Russian Pricing Strategy in Europe?* Oxford Energy Comment.

Henderson, J. 2016b. 'Outlook for Australian LNG: The Limits of Expansion'; 'The Outlook for Russian LNG, Current Delays, Long Term Upside'; 'Other Potential Suppliers'. In *LNG Markets in Transition: The Great Reconfiguration*, eds. A.-S. Corbeau and D. Ledesma (2016), 255–272 and. 290–306. Oxford: OIES/OUP.

Honoré, A. 2014. *The Outlook For Gas Demand in Europe*. OIES Paper NG87.

IGU. 2016. International Gas Union. *Wholesale Gas Price Survey—2016 Edition: A Global Review of Price Formation Mechanisms 2005 to 2015*.

Jensen, J.T. 2016. North America Enters the LNG Export Race. In *LNG Markets in Transition: The Great Reconfiguration*, eds. A.-S. Corbeau and D. Ledesma (2016), 160–180. Oxford: OIES/OUP.

Le Fevre, C. 2014. *The Prospects for Natural Gas as a Transportation Fuel in Europe*. OIES Paper NG84.

Le Fevre, C. 2016. LNG as a Transport Fuel. In *LNG Markets in Transition: The Great Reconfiguration*, eds. A.-S. Corbeau and D. Ledesma (2016), 442–467. Oxford: OIES/OUP.

Kovacevic, A. 2007. *The Potential Contribution of Natural Gas to Sustainable Development in South Eastern Europe*. OIES Paper NG 17.

LEBA: London Energy Brokers' Association OTC Energy Volume Reports (various issues).

Levy, M. 2014. *Gas Price Arbitrations: A Practical Handbook*. London: Global Business Publishing).
Petrovich, B. 2013. *European Gas Hubs: How Strong is the Price Correlation?* OIES Paper NG79.
Petrovich, B. 2014. *European Gas Price Hubs Correlation: Barriers to Convergence*. OIES Paper NG91.
Petrovich, B. 2015. *The Cost of Price De-Linkages Between European Gas Hubs*. OIES Paper NG101.
Petrovich, B. 2016. *Do We Have Reliable and Aligned Gas Exchange Prices in Europe?* Oxford Energy Comment.
Rogers, H.V. 2015. *The Impact of Lower Gas and Oil Prices on Global Gas and LNG Markets*. OIES Paper NG99.
Rogers, H.V. 2016. *Asian LNG Demand: Key Drivers and Outlook*. OIES Paper NG106.
Stern, J., and H.V. Rogers. 2011. *The Transition to Hub-Based Gas Pricing in Continental Europe*. OIES Paper NG 49.
Stern, J., and H.V. Rogers. 2014. *The Dynamics of a Liberalised European Gas Market: Key Determinants of Hub Prices, Roles and Risks of Major Players*. OIES Paper NG94.
Stern, J. 2007. *Is There a Rationale for the Continuing Link to Oil Product Prices in Continental European Long Term Gas Contracts?* OIES Paper NG 19.
Stern, J. 2009. *Continental European Long-Term Gas Contracts: Is a Transition Away from Oil Product-Linked Pricing Inevitable and Imminent?* OIES Paper NG 34.
Stern, J. 2014. Russian Responses to Commercial Change in European Gas Markets. In *The Russian Gas Matrix: How Markets are Driving Change*, eds. J. Henderson and S. Pirani. Oxford: OUP/OIES.
Stern, J. 2012. The Pricing of Gas in International Trade: An Historical Survey. In *The Pricing of Internationally Traded Gas*, ed. J. Stern. Oxford: OUP/OIES.
Stern, J. 2016. LNG Pricing: Challenges in the Late 2010s. In *LNG Markets in Transition: The Great Reconfiguration*, eds. A.-S. Corbeau and D. Ledesma (2016), 468–501. Oxford: OIES/OUP.
Stern, J. et al. 2014. *Reducing Dependence on Russian Gas—Distinguishing Natural Gas Security from Geopolitics*. OIES Paper NG92.
Yafimava, K. 2013. *The EU Third Package for Gas and the Gas Target Model: Major Contentious Issues Inside and Outside the EU*. OIES Paper NG75.

Authors' Biography

Jonathan Stern founded the OIES Natural Gas Research Programme in 2003 and was its Director until October 2011 when he became its Chairman and a Senior Research Fellow, he became Distinguished Fellow in October 2016. He is honorary professor at the CEPMLP, University of Dundee; visiting professor at the Centre for Environmental Policy, Imperial College London; fellow of the Energy Delta Institute and a Distinguished Research Fellow of the Institute of Energy Economics, Japan (in Tokyo). He is the author and editor of several books and numerous shorter works on gas published over the past several decades.

Howard V Rogers is Chairman and Senior Research Fellow of the Natural Gas Research Programme at the Oxford Institute for Energy Studies. Howard was the OIES Director of Gas Research from 2011 to 2016. He has published papers and book chapters on LNG arbitrage, pricing evolution of European gas and Asian LNG (co-authored with Jonathan Stern), and other gas-related issues. Howard Rogers joined OIES after 29 years with BP, in upstream oil and gas business development and strategy. This included postings to the US, Kuwait and Azerbaijan.

16

The Future of Gas in the Energy Union: Managing Its Decline?

Jean-Arnold Vinois

16.1 Introduction

The story of gas in the European Union is a success story: starting from scratch in the sixties, it is representing today about 25% of the primary energy used in the European Union, before Brexit, representing 28 Member States and more than 500 millions consumers. There is a gas market in 26 Member States (Malta and Cyprus have no gas market, but Cyprus is likely to have one once gas discoveries in its waters will reach it), and gas might be further introduced in some neighbouring countries belonging to the Energy Community, such as Albania. In spite of a wide use in the EU and in the Energy Community countries and in spite of being the cleanest fossil fuel, gas seems to be a "mal aimé" by the national and European decision makers, at least according to the gas industry. The latter is indeed worried to see coal and lignite remaining favoured for various questionable reasons by several countries, such as Germany, Poland, Greece, Spain and beyond, Ukraine and in the Western Balkans,

J.-A. Vinois (✉)
Notre Europe Jacques Delors Institute, Paris, France
e-mail: vinois.ducs@gmail.com

M. Hafner and S. Tagliapietra (eds.), *The European Gas Markets*,
DOI 10.1007/978-3-319-55801-1_16

like Serbia. Others may say that gas industry has missed opportunities, being overconfident about the strengths of its product.

This chapter will examine more in-depth the position of gas in the European energy mix today and in the future, the functioning of the internal gas market and finally the key challenges facing gas in the Energy Union which should lead the European Union towards a low carbon economy, in line with the Paris Agreement of December 2015.

16.2 Gas in the European Energy Mix

16.2.1 Evolution of the Gas Production, Importations and Consumption

It is not the place here to deliver a detailed statistical analysis of the evolution of gas in the European energy mix. Eurostat publishes every year all the data, unfortunately with 2 years delay.[1] First of all, there is no European energy mix because the energy mix remains the result of the decisions taken by the Member States, as part of their sovereignty and by a myriad of actors, such as regions, cities, companies and individuals. In other words, as part of an explicit energy policy when it exists, gas may be promoted more or less by each Member State, according to its own situation: some Member States, such as UK, NL and DK, but also DE, IT, ROM, enjoyed for a long time good gas resources in their territory, leading them to favour an extensive use of gas in industry, agriculture, power generation, heating and cooking. Others were located close to a producing country and developed rapidly imports of gas, such as Belgium (from the Netherlands) and Spain (from Algeria). The development of the continental shelf of Norway offered another opportunity for gas in Europe. Finally, most of the eastern European countries which were part of the Soviet Union before their independence in the 1990s had developed gas use for heating and cooking and were irrigated by numerous pipelines bringing them Russian or even central Asian gas. In the meantime, gas benefitted particularly from the organised reduction of coal production everywhere in Europe to replace it for heating and power generation. Gas also benefitted from the

Tchernobyl accident in 1986 leading to a halt of nuclear investments, enabling it to enter massively into the power generation systems, something strangely prohibited by an EU directive of 1977 in the EU, until it became usual. As a result, by 1995, gas represented already 20% of the gross inland consumption and reached 23% in 2013, while during the same period, coal went down from 22 to 17% and oil from 39 to 33%, and renewables went up from 5 to 12%, while nuclear remained stable at 14%. Interestingly, the reduction of coal and oil was compensated by an increase of gas and renewables in the proportion of one-third/two-thirds. In other words, gas was particularly resilient and was able to increase its market share, in a period where the overall consumption of energy has decreased. And with a decreasing domestic production inside EU (from more than 200 BCM in 1995 to a bit more than 100 BCM in 2015), it meant an increase in imports and a subsequent greater dependency on external suppliers.

Looking at domestic production, EU has little to nothing to say about the production of natural resources which remains a sovereign decision to be taken by each Member State (Article 194 of the EU Treaty[2]). There is today no more or little debate about shale gas at EU level. Some countries such as France and Bulgaria have banned shale gas exploration and exploitation while UK and Poland have taken measures to facilitate such exploration and exploitation. EU may not influence this, and its main possibility of action lies in the environmental rules that could affect such operations. After the Polish disillusion of the early 2010 about its shale gas potential resources, nobody today believes in a major contribution of shale gas to the EU domestic production. In the meantime, the Netherlands is facing a dramatic situation in its Groningen gas fields leading it to reduce drastically its production (up to 30 BCMa). Whether this loss of production will be compensated by the discovery of new gas fields in the Black Sea and in East Mediterranean is still an open question. In any case, the decline of the domestic EU production and its likely stagnation around 100 BCMa in the next 10 years seems to be a reality and has to be matched by a reduction of consumption and/or an increase of imports. However, in our opinion, there are many reasons to consider the Norwegian production as part of the domestic production, as this production is almost completely destined to the EU, before Brexit of course. Brexit is expected to alter significantly this situation as it will remove the UK production and imports from the EU statistics.

Concerning the consumption, three equally important segments are usually identified: industry, heating and cooling or residential and services and power generation. Gas demand was supposed to increase in the power generation segment, while it was expected to decrease in industry because of competition with low gas prices countries (US, Gulf States) and in heating and cooling as buildings are becoming more and more energy efficient. But it was finally in the power generation that gas lost ground because of the low price of carbon (the price of carbon went from €30 per ton in 2012 to €4 in 2016, coppled with a low price of coal (displaced from the US market by shale gas and combined with a lower consumption in China)) and a too high price of gas. Simultaneously, renewables enjoyed (generally too) generous subsidies and a priority of access to the grid without having to contribute to the stability of the grid and to the backup needed in case of absence of wind or sun. The rigidity of gas prices, due to the prevailing oil indexation formula, did not help maintaining gas in the power mix since 2010. In addition, electricity consumption was declining since 2005 as a result of improved efficiency and of the economic crisis. No surprise thus that the EU consumption of gas went down from 550 BCMa in 2010 to 400 in 2014 and 425 in 2015.

Finally, there is the likely impact of the major December 2015 Paris agreement on gas demand in the future. Can we now expect that coal will be removed faster from the mix? UK started this process already but Germany and Poland as well as some others are still investing heavily in coal and lignite to the detriment of their emissions levels and their public budgets. Will nuclear come back and replace coal to the detriment of gas as envisaged in UK,[3] Poland, Hungary and Bulgaria? All these questions might be answered, at least partially, in the two coming years with the implementation of the governance scheme of the Energy Union.[4] Each Member State is required to spell out, by 2018, its energy and climate strategy up to 2035, in line with the Paris agreement and the 2030 objectives agreed in October 2014 by the European Council, of which the target of a reduction of 40% of the GHG emissions is the cornerstone. The last question is whether there will be a more voluntary approach to renewables and energy efficiency, once UK is leaving the European Union. UK was indeed a well-known opponent to binding targets in these two areas, in contrast to France and Germany which came in July 2016 with a joint paper advocating a stronger commitment of the Union in favour of renewables and energy efficiency.

Reverting to the question of gas import dependency, which is likely to increase given the decline of the domestic production and in spite of the reduction of the consumption, it is also useful to analyse the actual situation.

First of all, **Russia**, with 100 to 150 BCMa to the European Union and a large portfolio of long-term contracts up to 2030, is the main external supplier of gas and will remain so in the future as it owns, just after Iran, the largest proven resources, and as it is able to offer the lowest price of production and transport, up to now. Like Saudi Arabia in the oil sector, Russia is the swing producer of gas and may easily dictate the price. This is obvious today at a time US LNG tries to reach the EU coasts and other LNG producers are looking at the European market. Gazprom is probably able to lower its prices to the point to eliminate any competition. It may represent the biggest threat to the gas-to-gas competition in Europe, a situation which should be closely monitored by the DG Competition of the European Commission. The discussion about new import pipelines from Russia to the EU, such as NordStream2,[5] SouthStream or Turk(ish)Stream, is raising many questions beyond removing Ukraine from the transit routes and beyond undermining the southern gas corridor. At stake is the dominance and control of Gazprom of the EU gas market involving even the ability to set the prices at the main hubs. These risks have to be properly understood and monitored. The longstanding EU–Russian energy dialogue, which has been developed since 2000, has been deeply impacted by the Ukrainian conflict and the Crimea annexation of 2014. As long as sanctions are imposed and as the Ukrainian–Russian conflict is not solved, there is no chance of a wide agreement between Russia and EU covering the energy exchanges and infrastructures, which would be the most logical way to solve all the bilateral problems in this area.

Norway, with about 110 BCMa, is the second external supplier and in reality should not be considered as an external supplier. Norway belongs to the European Economic Area which is implementing most of the internal market rules and is part of a robust institutional framework including a Court to judge any possible infringement of these rules including the competition rules. To measure the EU gas import dependency, it could be wise to consolidate the EU and Norwegian gas production as it would definitely reveal a level of import dependency which is well below 50%. An agreement between EU and Norway on this issue and on measures to be taken in case of emergency should be envisaged as part of a strategic gas partnership.

The third major external supplier is **Algeria**, with some 30/35 BCMa. Its importance has been seriously reduced in the recent years, because of production problems and a lack of investments in production, combined with an increase of the domestic consumption reducing the available volumes for exports. EU resumed in 2015 a new dialogue with Algeria which should be seen as a strategic partner.

Last but not least and probably, the most attractive solution in a very flexible market is **LNG**, bringing gas by ship from everywhere in the world, today mainly from Qatar, Nigeria, Libya and now from USA and many other new sources such as Iran and Australia in the future. Key question here is the price of gas. Will the gas prices in Europe be attractive (read high) enough to bring LNG cargoes to EU terminals which are now numerous and representing about a 200 BCMa capacity? This question cannot be separated from the price of competing piped gas from Russia, Norway, Algeria, Central Asia and the well functioning of the European gas hubs.

To conclude this overview, it may be said that (i) EU gas demand has declined sharply since 2010 and is now stagnating in a context of general decrease of energy consumption, (ii) new uses of gas are needed if it is to grow or even remain at the present level, (iii) the need to reduce GHG emissions affects deeply the future of gas, (iv) the abundance of gas as well as diversified sources of gas combined with a lower price is giving it a new opportunity at least to maintain its important share in the EU energy mix.

16.2.2 The Evolution of the Structure of Consumption of Gas

Three main segments of gas consumers are usually identified: industry, residential and services or buildings, power generation.

Industry has always managed in a more effective way its consumption of gas, as it is vital for its competitiveness. Thus, it may be expected to see industry constantly optimising its gas use. It is fair to say that industry appreciated most the opening of the gas market, especially since the third package, giving it the full choice of suppliers across the EU. It is mainly concerned by the GHG emissions because industry has to pay for their emissions and while gas is a fossil fuel, there may be an incentive to replace gas by renewables including biogas, wherever technically possible.

For the residential segment, one of the main features of the creation of the internal market is the choice given to the consumer to choose its supplier and thus to switch easily from one supplier to another.

This is more or less the reality since 2010, with some Member States being much more dynamic than others, where the incumbents and associated regulated prices have been able to slow down, if not to prevent, this process. In some countries, like Belgium, consumers associations or communities started to act on behalf of their members to negotiate better prices or to auction for the gas supply services. Simultaneously, energy efficiency obligations like those set by the 2012 Energy efficiency directive, combined with favourable financing schemes, produced the desired impact of a reduction of consumption for heating and cooling in buildings. In addition, new tools are now available to allow the consumer to manage more closely and economically his consumption, like intelligent thermostats (Google-Nest) and other formulas like those based on comparative behaviour of consumers (Opower). Demand side management, long ignored, is becoming a new fuel per se, as the digitisation and the Internet of things are offering new solutions to better match supply and demand in the field of energy. In addition, the new sensitivity of consumers to energy prices and sometimes their interest for green energy lead them to invest in renewable sources of energy and particularly solar PV for heating and for electricity.

What it means for gas demand is still difficult to assess, but it might have a very significant impact.

For the third segment, that is power generation, the conditions for a come back of gas to this segment may be of two orders: first, the price of carbon should be much more significant to remove coal from the mix, and this should come from the new ETS to be in place from 2020, and second, the new power market design to be proposed by the European Commission before the end of 2016 should create a better level playing field for gas and renewables. It is also true that a number of countries have decided to remove coal from their power generation mix (like UK) and that banks like EIB are not financing any more new coal-fired power plants. This move is likely to favour gas, but at the same time, the price of renewables is continuing to come down.

Given the difficulties encountered by gas in its traditional segments, finding new segment may be the way out for securing a better future for gas. And there is one emerging: the transport segment. It starts with LNG for shipping: new environmental rules promulgated by the International Maritime Organisation oblige ships to abandon heavy fuel for bunkering. LNG offers less harmful environmental consequences and may replace heavy fuel provided the engines of the ship are adapted and that ports offer the appropriate infrastructure for bunkering. These conditions seem to be present in many places in Europe, particularly in the Baltic and North Seas, where LNG for shipping is now flourishing. Another major development appears to be the use of gas for heavy truck transport, where ports may also offer micro-LNG stations to fill their reservoirs. And finally for light vehicles, big car manufacturers include now in their catalogue Natural Gas Vehicles. Some countries like Italy developed this segment long ago already, and it is not sure yet that others will follow, since it may require some tax incentives. In contrast, electric vehicles seem to attract more interest from the public authorities which tend to give, like in Norway, major incentives to purchase such vehicles.

Recognising the need to stimulate alternative fuels on the market, the EU has created already a few years ago a regulatory framework to encourage the construction of the needed infrastructures for alternative fuels in transport and Member States have to establish by November

2016 investment plans to make this a reality. Several EU funds are also helping the financing of these new infrastructures (Connecting Europe Facility, EFSI, etc…).

16.2.2.1 Prospects of the Gas Market

As explained, gas is and will remain an important fuel for the energy sector. After a continuous increase between 1970 and 2010, gas market in Europe reached a certain level of maturity and needs now to adapt to a new environment which is governed by the Paris Agreement of December 2015, the 2030 objectives of the European Union and the subsequent transition towards a low carbon economy.

Against this background, the key words of the European Union will be decarbonisation, energy efficiency and renewables which all affect the future of gas.

Decarbonisation of gas has to be sought by the gas industry as, by 2030, this will become a key requirement for keeping gas in the mix. In spite of the efforts of the European Union in 2010–2012 and several billion euros committed, not a single project aiming at demonstrating the viability of carbon capture and storage technologies has been completed, leaving the impression that decarbonisation of coal and gas will not happen. In 2016, the same impression prevails and gas industry should at least demonstrate concretely its willingness to address the issue. This is going beyond the need of a carbon price making such investments more economic.

As already said, investments in energy efficiency and more particularly the improvement of the energy performance of existing buildings will have a great impact on the gas demand. Depending on the speed of the rehabilitation of the building fleet, the reduction may be spectacular. As may be the impact of modernisation of the district heating systems in many countries, especially in Central and eastern Europe.

Renewables and decentralised power generation will continue to attract major investments by individuals and companies, reducing again the need for gas in large combustion plants. Instead, small and fast units powered by gas should be used, but this will never represent huge volumes of gas in comparison with a base load system. At the same time, biogas is now coming into the picture, with some countries showing great ambitions to develop it, like the Scandinavian countries, Germany and more recently France.

In a more distant future, there are also other possibilities for gas to participate in the evolution of the energy sector. The power to gas technologies is under development and should enhance the cooperation between renewables and gas, a duo which is already working to combine gas as a backup to renewables in the production of power but which could also work to transform the excess of renewable electricity into gas, thanks to a methanisation process still to be economically mastered.

16.3 The Internal Gas Market

16.3.1 The Implementation of the Third Energy Package in the Field of Gas

It took a very long time to open the European gas market. In each Member State, there was usually a single player controlling the national market. And each national market was well protected against any incumbent from the neighbouring countries. In addition, long-term contracts concluded between the incumbent and the external supplier of gas, such as Gazprom, Statoil or Sonatrach, included destination clauses, prohibiting any reexport of gas to another country. To ensure the full control, the transmission system was owned and controlled by the main supplier and was not equipped to allow reverse flows. It is well known that the company which is controlling the infrastructure is controlling the market. This was made obvious by the sector enquiry made by DG Competition of the European Commission in 2006, convincing the Commission to propose the third package and the (in) famous ownership unbundling provisions, requiring the separation of infrastructure ownership and management from the production and supply of gas. It was not a surprise that these provisions were the most attacked by the Member States, and particularly France and Germany with many others hiding behind them. The Commission had no other choice than accept a very mild version of unbundling under the name ITO for independent transmission operator, the vertically integrated incumbents keeping the right to own and manage the transmission system provided they ensured some independence of the management and of investment decisions of the subsidiary from the mother company (Chinese walls and so on, compliance officer and of course supervision by the regulatory authority having to certify the compliance of the TSO with the provisions of the third gas directive, subject to scrutiny by the Commission etc.). After the adoption of the third gas directive, we have, however, seen decisions of divestment by several incumbents, particularly the Germans (EON and RWE) and the Italians (ENI) while France maintained GRTGas inside GDF Suez, now ENGIE, and several east European countries are maintaining the ITO model. All in all, EU counts a bit more TSOs that are fully ownership unbundled than TSOs which are still ITO.[6]

After this fierce battle between the Member States, supported by their champions, and the European Commission, things went more smoothly with the creation of ENTSOG, the European Network of Transmission System Operators for Gas, tasked with the elaboration of network codes, under the guidance of ACER, the Agency for the Cooperation of Energy Regulators, to be processed later on by the EU legislators, the Council and the Parliament. During the last 5 years, these bodies have been working hard to elaborate common rules to manage and operate the gas transmission system. At the last GIE Conference held in June 2016 in Sofia, a panel devoted to the implementation of the third package came with the conclusion that the regulatory framework was now largely in place to enable the wellfunctioning of the internal gas market. Without being naive, because a lot remains to be done as witnessed by the numerous infringement procedures engaged by the Commission against the Member States for lack of compliance, this conclusion is fair enough and recognises the huge commitment of all the players to ensure the success of the market.

In this respect, the gas crisis of January 2009 acted also as a wake-up call about the negative consequences of the fragmentation of the European market on the level of security of supply of each Member State and of the European Union as a whole. Indeed, the supply disruption of the Ukrainian route of Russian gas deprived a dozen of east and Central European Member States of part or all of their gas in the middle of the winter. Western Member States companies were full of gas but not able to ensure its transport to the East because of gas pipelines running exclusively from East to West. This situation highlighted the biggest weakness of the European gas market which was the physical impossibility to ensure the free flow of gas throughout the European territory. Hence, the most significant element of the subsequent regulation 994/2010,[7] adopted as a result of the crisis, was the imposition of virtual as well as physical reverse flows at all interconnection points within the EU. Fortunately, the European Commission got a crisis budget in 2010 allowing to finance many of the equipments needed to make possible such reverse flows, in spite of the reluctance of number of incumbents. Today, reverse flows are possible on most pipelines, those exempted should still be equipped, and the paradox is that today Ukraine, as part of the Energy Community Treaty, imports more Russian gas through reverse flows from EU than from Russia itself.

Thus, the reality is that gas, whatever its origin (domestic, piped imports, LNG, storages), may flow East/West, West/East, North/South and South/North. What is preventing this to happen is linked to the lack of interconnections or to some artificial bottlenecks due to regulatory insufficiencies. The 347/2013 infrastructure regulation[8] aimed at solving these physical problems.

16.3.2 The Role of Infrastructures

We explained how the management of infrastructures and their ability to transport gas in all directions played a fundamental role in integrating the European market, leading to convergence of prices never seen before.

It is the combination of a detailed regulatory framework and of a corresponding infrastructure development policy which enabled the working of the internal market.

Trans-European networks (TEN) have been launched at EU level well before the rules of the market were established. The Maastricht Treaty introduced a chapter on TEN for energy, transport and telecommunications as early as 1992 although the first rules governing the energy market came in 1996 for the first gas directive, 2003 for the second and 2009 for the third. Money was already allocated for funding TEN in 1994, but Member States were deciding about the infrastructures to be financed, without any consideration for the market functioning.

It is only with the third package that the link between the market and the need for infrastructures was fully made. Following the regulation 715/2015,[9] the establishment of transmission system operators (TSOs) exclusively dedicated to the management and the development of the gas network, their association at European level within ENTSOG and their obligation to work at a Ten Year Network Development Plan (TYNDP) based on a common analysis of the needs were the decisive elements for the integration of the European network and of the gas market. The first TYNDP was elaborated in 2010, as "a non-binding Community-wide ten-year network development plan, including a European supply adequacy outlook" to be updated every 10 years. National regulatory authorities, acting nationally but having to take the neighbouring countries into account, and ACER, acting to foster the European dimension of the exercise, were able to identify possible investment gaps and to assess the evolution of the supply adequacy. They used various scenarios made more realistic thanks to the gas crisis of January 2009 (e.g. assessing for instance the impact of the disruption of various supply routes, like the Ukrainian one, on the Member States and beyond on the Western Balkans). With the entry into force of the TEN-E regulation in May 2013, the TYNDP acquired a new role, becoming the first step of the identification of the Projects of Common Interest (PCI). This regulation introduced also the need for a detailed cost–benefit analysis (CBA) methodology. An extensive public consultation has to take place. The TYNDP process has gained in robustness and the fourth TYNDP report published in 2015 reflects an in-depth analysis, through the development of models, key indicators, monetisation, alternative gas demand scenarios and gas supply scenarios, of the 259 projects submitted (by transmission, storage and LNG terminal operators), assessing their contribution to the physical integration of the market as well as to the enhancement of security of supply. Last but not least, the 2015 report includes a supply adequacy outlook for the period 2015–2035. In 6 years time, the players of the gas market have been able to create this essential tool to assess the evolution of the gas market and the needs for investments. This is an unprecedented improvement of the past situation where infrastructures were decided nationally on the basis of mere political decisions, leading sometimes to "white elephants". It is, however, clear that, as usual in the energy and infrastructures areas, a "political control" remains, not only based on concerns of security of supply and geopolitical considerations but also based on the need for the approval of all levels of public authorities in order to authorise the building of these infrastructures. Public acceptance of any new infrastructure is today a key element and a key determinant of the speed with which commissioning of new infrastructures may take place.

In this respect, the 2013 TEN-E regulation offered a facilitation of the permit-granting procedure to Projects of Common Interest, mainly cross-border projects like interconnectors, which, as part of the TYNDP developed by ENTSOG, are identified by regional groups of countries to ensure an equal level of effective support on both sides of the border. This is a nice example of regional cooperation and solidarity which aims at facilitating the smooth implementation of any PCI, as well as the financing which may be complemented by EU funds, especially from the Connecting Europe Facility (CEF) established in 2014 for the period up to 2020. Similarly, the list of PCIs is updated every 2 years to take into account new projects but also the progress made with projects adopted previously. The last list has been adopted in 2015 and published as a regulation of the Commission of 18-11-2015. The next list under preparation will be published before end 2017.

As a result of the third package, the regulation on security of gas supply and the TEN-E regulation, a comprehensive and open process, involving all players and stakeholders, has been put in place to identify regularly infrastructure needs to integrate physically the market and to ensure security of supply. This is a major achievement of the European Union, fully in line with the objectives of the Energy Union aiming at optimising the use of the resources and infrastructures of the EU, in a spirit of solidarity.

16.3.2.1 The Liquidity of the Market

The wellfunctioning of the internal market can be measured at the liquidity it is offering to the participants. Coming from national monopolies to a multitude of suppliers and traders is not an overnight operation. With 28 countries coming from different situations and levels of concentration, it is rather normal to see that competition is not working everywhere in favour of the consumers.

Looking at the retail markets essentially governed by national governments, 13 Member States maintain regulated prices for end consumers and 11 for industrial consumers. This is clearly favouring the incumbent as the regulated prices are usually not allowing to cover the costs, preventing any competition to take place. Favouring the incumbent, which is usually state owned or controlled, is still a frequent situation that the European Commission is trying to solve through infringement procedures that are slow and largely inefficient as Member States are buying time. This means also hidden subsidies to fossil fuel and it is running against energy efficiency and a smooth energy transition. It is also favouring all kinds of opaque agreements between external suppliers and national champions, as evidenced by the various agreements concluded between the Russian Federation and Gazprom on the one hand and some Member States and their national champions on the other hand at the time of the South Stream pipeline development between 2009 and 2014. ACER is producing every year a report on the progress of retail markets which is highlighting all the problems still affecting the retail markets, from the regulated prices to the lack of choice of suppliers, the absence of smart meters and the little consideration for consumers' interests. The last report was published in November 2015, covering the year 2014.

Examining the situation of the wholesale market, which is essentially governed by EU rules, things appear to have evolved more favourably with hub pricing beginning to overtake the oil indexed prices of long-term contracts. As a result of an effective diversification of sources of gas, supported by more flexible routes thanks to reverse flows, and as a result of the demand decrease and of an abundant offer, the price of gas has decreased from 12$ per Mbtu in 2012 to 6 in 2016, in comparison with less than 3 in USA. This major decrease in price has, however, not been sufficient to remove coal from the power generation market but has been significant enough to maintain the market share of gas in the energy mix.

Again, the reporting by ACER on market monitoring and by the European Commission, in its last State of the Energy Union, both dated 18 November 2015, showed an improvement of the market functioning.[10] However, their recommendations for a better observance of the rules to ensure competitive and liquid markets and for better physical interconnecting of Member States should not be underestimated. Convergence of wholesale prices between the various regions is encouraging but may be improved.

Regional cooperation may help such improvement, such as BEMIP between the Baltic States, Finland and Poland, the Central East South Europe Gas Connectivity (CESEC) and the High Level Group for gas interconnectivity of the Iberian Peninsula and France.[11]

16.4 Threats and Opportunities

16.4.1 Decarbonisation

With all its intrinsic qualities, including its abundance and its relatively cheap price, gas remains a fossil fuel which has to cope with the main objectives of the energy and climate policies agreed at the December 2015 Paris Conference.[12] If some say that replacing coal by gas would solve the climate change problems, this is a short term view. With half the emissions of coal, gas still emits too much and if it is a nice replacement of coal in the next 10 years, it does not mean that it should not ensure its decarbonisation from 2030 onwards. Today, some banks like the European Investment Bank, followed by some national Development banks, do not borrow anymore to finance projects which are generating more than a certain level of greenhouse gases, de facto eliminating coal-fired power plants. This trend is likely to be followed worldwide but the level of GHG may well be lowered in the coming years to hit equally gas in a foreseeable future.

The ETS now being revised for the period after 2020 will have a growing impact on gas, as it may be expected to see the CO_2 price to reach €25 by 2025. And it is not excluded that some countries are going for more ambitious policies such as a carbon tax to accelerate the transition and stimulate cleaner technologies.

Natural gas producers have now a duty to work hard at decarbonising their fuel. Whether it may be achieved by carbon capture and storage technologies or by any other means is a question to be solved primarily by industry. And it is for the industry to take the initiative and not to wait for the solutions to come from the public authorities as has been seen up to now.

16.4.2 Diversification of Sources and Routes

Security of gas supply is a public good. As such, Member States, like any government in the world, will always consider gas supply as part of their responsibility. For good or less good reasons, the geopolitical dimension of gas will continue to affect the rules governing gas in the market. The increase of import dependency of gas is a reality which requires to be addressed by a genuine diversification policy. This diversification requires to rely on several (some say at least three) geographical sources of gas, several routes such as pipelines and LNG terminals, and from several counterparts, i.e. different supplying companies. In addition, a fully flexible network should be available, connecting the transmission system to the storages and LNG terminals across the European Union where the free flow of gas should be ensured at all times. Domestic resources should also be made available very flexibly at least in case of emergency. Major progress has been made in the last 5 years in this respect, not only in terms of effective cooperation but also in terms of physical infrastructure. This acquis should be well maintained and improved where it is needed.

16.4.3 A Genuine Competition

The efforts made since 2009 to create a liquid gas market, open to all suppliers and enabling consumers to choose their suppliers, have to be pursued particularly by empowering more the consumers. Barriers to entry have to be removed, incumbents have to be challenged, regulated prices have to be removed and if not removed, should not act as a barrier to entry and be linked to hub prices. Hub pricing should be favoured and long-term contracts should be indexed on hub prices. Liquidity of the hubs should be ensured by enough diversification of sources and consolidation of hubs should be sought with common rules for all. Manipulation of the market and particularly of hub prices should be prevented by appropriate monitoring, through a rigorous REMIT regulation implementation and market monitoring,[13] including the use of competition rules governing abuse of dominant position, as some major external suppliers may be tempted by such manipulation. There is a need to create a global gas market to replace the present regional markets dominated by few suppliers and to achieve liquidity of the market everywhere in the European Union. This is the condition for the consumers to trust gas as a commodity and not as a political fuel, subject to various manipulations or abuse as we have seen in the past.

16.5 Conclusions and Recommendations

Gas is a key element of the energy mix today and will remain an important component in the next three decades. Whether it is a bridge fuel, a transition fuel or a destination fuel is not the issue. The energy transition, as enshrined in the Energy Union project of the European Union and as mandated by the Paris agreement of December 2015, is the new rule of the game and requires to reduce the emissions of greenhouse gases as soon as possible. The sense of urgency can only grow in the future.

Gas is still a fossil fuel which is emitting such GHG, meaning that there is a strong incentive to replace it by other fuels which are cleaner.

Renewables are clearly preferred to gas in terms of investments as witnessed recently by the International Energy Agency in its first World Energy Investments Outlook, even if gas appears to be a good complement to renewables in power generation.

In addition, the world of energy is changing fast not only through the new affordability of renewables but also thanks to the progress made in energy efficiency and demand side management which are leading to very significant reductions in consumption, affecting also the other segments of consumption of gas such as industry and buildings.

The gas industry should worry about the ability of gas to cope with the future requirements of the energy sector as dictated by the climate change policies.

If successful in decarbonising the molecule, gas may also expand to new areas such as transport. Biogas may also become an important part of the future gas landscape and infrastructures which have been built to hold between 30 and 60 years might well remain in use. The same may be said of new technologies like power to gas, under development.

The main recommendation is addressed to the gas industry. It is its dynamism and its ability to adapt to change and to climate change which will govern the future of gas. It will be up to the industry to demonstrate to the population and to the policy decision makers that gas is a fuel that deserves to stay in the energy mix because it will also make substantial efforts to reduce drastically its carbon footprint. Again this is going well beyond the various slogans usually heard such as gas being the bridge fuel, the transition fuel or the destination fuel.

Notes

1. See Eurostat pocket book at https://ec.europa.eu/energy/en/statistics/energy-statistical pocketbook. The 2016 edition covers 2014.
2. Article 194 of the Treaty on the Functioning of the European Union is the first explicit legal basis for energy legislation and has been introduced by the Lisbon Treaty which entered into force in 2009.
3. The UK decision made in September 2016 to build Hinkley Point C is the last expression of this willingness but will it mean that it will be

finally built is another question, given the conditions attached to this decision and the high price to be paid over decades.

4. The governance of the EU energy policy is a key issue of subsidiarity: how will the Member States design their energy and climate policies to cope with the 2030 objectives of the EU and with the Paris agreement of December 2015. The non-binding character of the targets for renewable and for energy efficiency, coppled with the binding character of the EU target leaves a great freedom to the Member States to organise their energy policies. See the conclusions of the European Council of 24 October 2014 at http://www.consilium.europa.eu/uedocs/cms_data/docs/pressdata/en/ec/145397.pdf. See also the Communication of the Commission of 25 February 2015 about the Energy Union "A Framework for a Resilient Energy Union with a Forward-Looking Climate Change Policy" COM (2015) 80 final at http://ec.europa.eu/priorities/energy-union/docs/energyunion_en.pdf and for a broader view of the Energy Union, see Andoura Sami and Vinois Jean-Arnold: "From the European Energy Community to the Energy Union: a policy proposal for the short and the long term" preface by Jacques Delors, Institut Jacques Delors, 27 January 2015, Paris at www.delorsinstitute.eu (Andoura and Vinois 2015; European Commission 2015).
5. NordStream2 is "the" dividing pipeline of the Energy Union. Some Member States support it while others are vehemently opposed to it. A lot of articles have been written in 2015–2016 taking position in favour or against. For a European view, see Thomas Pellerin-Carlin and Jean-Arnold Vinois in Natural gas Europe of 16 December 2015.
6. See the third gas directive: directive 2009/73/EC of the European Parliament and of the Council of 13 July 2009 concerning common rules for the internal market in natural gas and repealing Directive 2003/55/EC, OJ L211/94 of 14.08.2009. On the issue of certification of TSOs see Inge Bernaerts, the third internal market package and its implications for electricity and gas infrastructure in the EU and beyond, in The Energy Infrastructure of the European Union, EU Energy Law, volume VIII pages 7–36, Claeys & Casteels.
7. Regulation (EU) n°994/2010 concerning measures to safeguard security of gas supply and repealing Council Directive 2004/67/EC of 20 October 2010.

8. Regulation (EU) n°347/2013 of the European Parliament and of the Council of 17 April 2013 on guidelines for trans-European energy infrastructure and repealing Decision 1364/2006/EC, OJ L115/39.
9. Regulation (EC) n°715/2009 of the European Parliament and of the Council of 13 July 2009 on conditions for access to the natural gas transmission networks and repealing Regulation (EC) 1775/2005, OJ L211/36 of 14.08.2009.
10. These papers are most interesting as they also look at the situation of each Member State and are offering a general assessment and an in-depth analysis. See www.europa.eu on Energy Union. A second State of the Energy Union has been published on 1 February 2017 and it includes a lot of useful information about the implementation of the roadmap for the Energy Union and key indicators.
11. All the papers referring to these regional cooperations are available on www.europa.eu.
12. UN document FCCC/CP/2015/L.9 issued by UNFCCC on 12 December 2015.
13. Regulation (EU) n°1227/2011 of the European Parliament and of the Council of 25 October 2011 on Wholesale Energy Market Integrity and Transparency, OJ L326/1.

Bibliography

Andersen, Svein, Andreas Goldthau, and Nick Sitter (eds.). 2016. *Energy Union: Europe's New Liberal Mercantilism.*, International Political Economy Series London: Palgrave Macmillan.

Andoura, Sami and Jean-Arnold Vinois. 2015. From the European Energy Community to the Energy Union: A policy proposal for the short and the long term, foreword by Jacques Delors. Jacques Delors Institute, January.

Bros, Thierry. 2016. Enlarged TSO's could benefit consumers in Russia and the EU. *Natural Gas World Magazine,* December 14.

European Commission. 2015a. A Framework Strategy for a Resilient Energy Union with a Forward-Looking Climate Change Policy, COM(2015) 80, 25 February 2015. Brussels: European Commission.

European Commission. 2015b. State of the Energy Union 2015, COM(2015) 572 final 18 November 2015. Brussels: European Commission.

European Commission. 2016a. Communication from the Commission to the European Parliament, the Council, the European Economic and Social Committee and the Committee of the Regions on an EU strategy for Liquefied Natural Gas and Gas Storage. Brussels: European Commission.
European Commission. 2016b. Projects of Common Interest. Brussels: European Commission.
Manfred, Hafner and SimoneTagliapietra. 2013. The Globalization of Natural Gas Markets. In *European energy studies*, vol. 6. Netherland: Claeys & Casteels.
International Energy Agency. 2016a. World Energy Outlook 2016 OECD/IEA. Paris.
International Energy Agency. 2016b. Global Gas Security Review 2016 OECD/IEA. Paris.
International Energy Agency. 2016c. Medium-Term Gas Market Report 2016 OECD/IEA. Paris.
Jones, Christopher (ed). 2010. The internal energy market: The third liberalization package. EU Energy Law, vol.1. Claeys & Casteels.
Leal-Arcas, Rafael. 2016. The European Energy Union, European Energy Studies, vol. 8. Claeys & Casteels.
Riley, Alan. 2016. Nord Stream 2: A Legal and Policy Analysis. CEPS Special Report, No. 151, November.
Schmitt von Sydow, Helmut (ed.). 2016. The EU Gas Laws, EU GEO Laws, vol. 2. Claeys & Casteels.
Vinois, Jean-Arnold. (ed.). 2012, The Security of energy supply in the European Union, vol. VI, EU Energy Law, Claeys & Casteels.
Vinois, Jean-Arnold. (ed.). 2014. The energy infrastructure of the European Union, EU Energy Law, vol. 8. Claeys & Casteels.

Author Biography

Jean-Arnold Vinois, a Belgian lawyer, spent most of his professional career in the European Commission. Starting in 1987, he occupied several management posts dealing with internal market, transport, trans-European networks and energy policies and acted lastly as director for the internal market for electricity and gas. At his retirement in 2013, he was appointed Honorary Director. He is now Adviser on the Energy Union at the Jacques Delors Institute in Paris and Berlin.

Index

M. Hafner and S. Tagliapietra (eds.), *The European Gas Markets*,
DOI 10.1007/978-3-319-55801-1

FSC
www.fsc.org
MIX
Papier | Fördert
gute Waldnutzung
FSC® C083411

Kolibri 360 GmbH
Ferdinand-Jühlke-Straße 7
99095 Erfurt, Deutschland
produktsicherheit@kolibri360.de